Image Processing and Intelligent Computing Systems

There is presently a drastic growth in multimedia data. Even during the Covid-19 pandemic, we observed that images helped doctors immensely in the rapid detection of Covid-19 infection in patients. There are many critical applications in which images play a vital role. These applications use raw image data to extract some useful information about the world around us. The quick extraction of valuable information from raw images is one challenge that academicians and professionals face in the present day. This is where image processing comes into action. Image processing's primary purpose is to get an enhanced image or extract some useful information from the raw image data. Therefore, there is a major need for some technique or system that addresses this challenge. Intelligent Systems have emerged as a solution to address quick image information extraction. In simple words, an Intelligent System can be defined as a mathematical model that adapts itself to deal with a problem's dynamicity. These systems learn how to act so an image can reach an objective. An Intelligent System helps accomplish various image-processing functions like enhancement, segmentation, reconstruction, object detection, and morphing. The advent of Intelligent Systems in the image-processing field has leveraged many critical applications for humankind. These critical applications include factory automation, biomedical imaging analysis, decision econometrics, as well as Intelligent Systems and challenges.

Image Processing and Intelligent Computing Systems

Edited by

Prateek Singhal
Sagar Institute of Research Technology-Excellence, India

Abhishek Verma
IIITDM Jabalpur, India

Prabhat Kumar Srivastava
Quantum University, India

Virender Ranga
National Institute of Technology, Kurukshetra

Ram Kumar
Katihar Engineering College, India

CRC Press is an imprint of the
Taylor & Francis Group, an **informa** business

First edition published 2023
by CRC Press
4 Park Square, Milton Park, Abingdon, Oxon, OX14 4RN

and by CRC Press
6000 Broken Sound Parkway NW, Suite 300, Boca Raton, FL 33487-2742

CRC Press is an imprint of Informa UK Limited

British Library Cataloguing-in-Publication Data
A catalogue record for this book is available from the British Library

Library of Congress Cataloging-in-Publication Data
Names: Singhal, Prateek, editor.
Title: Image processing and intelligent computing systems / edited by Prateek Singhal, Sagar Institute of Research Technology, India, Abhishek Verma, IIIT Jabalpur, India, Prabhat Kumar Shrivastava, Quantum University, India Virender Ranga, National Institute of Technology, Kurukshetra Ram Kumar, Katihar Engineering College, India.
Description: First edition. | Boca Raton : CRC Press, [2023] | Includes bibliographical references.
Identifiers: LCCN 2022034487 (print) | LCCN 2022034488 (ebook) | ISBN 9781032213149 (hbk) | ISBN 9781032213156 (pbk) | ISBN 9781003267782 (ebk)
Subjects: LCSH: Imaging systems in medicine. | Image analysis--Data processing. | Artificial intelligence. | Soft computing.
Classification: LCC R857.O6 I453 2023 (print) | LCC R857.O6 (ebook) | DDC 616.07/54--dc23/eng/20220824
LC record available at https://lccn.loc.gov/2022034487
LC ebook record available at https://lccn.loc.gov/2022034488

ISBN: 978-1-032-21314-9 (hbk)
ISBN: 978-1-032-21315-6 (pbk)
ISBN: 978-1-003-26778-2 (ebk)

DOI: 10.1201/9781003267782

Typeset in Times
by SPi Technologies India Pvt Ltd (Straive)

Contents

Editors ix
Contributors xi
Acknowledgement xv

Chapter 1 Digital Image Processing: Theory and Applications 1

Fasel Qadir, Gulnawaz Gani, and Zubair Jeelani

Chapter 2 Content-Based Image Retrieval Using Texture Features 17

Prashant Srivastava, Manish Khare, and Ashish Khare

Chapter 3 Use of Computer Vision Techniques in Healthcare Using MRI Images 35

Sonali D. Patil, Atul B. Kathole, Kapil N. Vhatkar, and Roshani Raut

Chapter 4 Hierarchical Clustering Fuzzy Features Subset Classifier with Ant Colony Optimization for Lung Image Classification 49

Leena Bojaraj and R. Jaikumar

Chapter 5 Health-Mentor: A Personalized Health Monitoring System Using the Internet of Things and Blockchain Technologies 63

M. Sumathi and M. Rajkamal

Chapter 6 Image Analysis Using Artificial Intelligence in Chemical Engineering Processes: Current Trends and Future Directions 79

P. Swapna Reddy and Praveen Kumar Ghodke

Chapter 7 Automatic Vehicle Number Plate Text Detection and Recognition Using MobileNet Architecture for a Single Shot Detection (SSD) Technique 101

Ahmed Mateen Buttar and Muhammad Arslan Anwar

Chapter 8 Medical Image Compression Using a Radial Basic Function Neural Network: Towards Aiding the Teleradiology for Medical Data Storage and Transfer 121

L.R. Jonisha Miriam, A. Lenin Fred, S.N. Kumar, H. Ajay Kumar, Parasuraman Padmanabhan, Balàzs Gulyàs, and I. Christina Jane

Chapter 9 Prospects of Wearable Inertial Sensors for Assessing Performance of Athletes Using Machine Learning Algorithms 137

Ravi Kant Avvari and Priyobroto Basu

Chapter 10 Long Short-Term Memory Neural Network, Bottleneck Distance, and Their Combination for Topological Facial Expression Recognition........ 153

Djamel Bouchaffra, Faycal Ykhlef, and Assia Baouta

Chapter 11 A Comprehensive Assessment of Recent Advances in Cervical Cancer Detection for Automated Screening........ 171

J. Jeyshri and M. Kowsigan

Chapter 12 A Comparative Performance Study of Feature Selection Techniques for the Detection of Parkinson's Disease from Speech........ 185

Faycal Ykhlef and Djamel Bouchaffra

Chapter 13 Enhancing Leaf Disease Identification with GAN for a Limited Training Dataset........ 195

Priyanka Sahu, Anuradha Chug, Amit Prakash Singh, and Dinesh Singh

Chapter 14 A Vision-Based Segmentation Technique Using HSV and YCbCr Color Model........ 207

Shamama Anwar, Subham Kumar Sinha, Snehanshu Vivek, and Vishal Ashank

Chapter 15 Medical Anomaly Detection Using Human Action Recognition........ 215

Mohammad Farukh Hashmi, Praneeth Reddy Kunduru, Sameer Ahmed Mujavar, Sai Shashank Nandigama, and Avinash G. Keskar

Chapter 16 Architecture, Current Challenges, and Research Direction in Designing Optimized, IoT-Based Intelligent Healthcare Systems........ 223

B.S. Rajeshwari, M. Namratha, and A.N. Saritha

Chapter 17 Wireless Body Area Networks (WBANs) – Design Issues and Security Challenges 235

Jyoti Jangir, Khushboo Tripathi, Deepshikha Agarwal, and Abhishek Jain

Chapter 18 Cloud of Things: A Survey on Critical Research Issues 245

Adil Bashir and Saba Hilal

Chapter 19 Evaluating Outdoor Environmental Impacts for Image Understanding and Preparation 267

Roopdeep Kaur, Gour Karmakar, and Feng Xia

Chapter 20 Telemedicine: A New Opportunity for Transforming and Improving Rural India's Healthcare 297

Seema Maitrey, Deepti Seth, Kajal Kansal, and Anil Kumar

Editors

Prateek Singhal is an Assistant Professor in the Department of Computer Science & Engineering at Sagar Institute of Research Technology-Excellence, Bhopal. He is pursuing a PhD degree in Medical Imaging from the Maharishi University of Information Technology, Lucknow, India. He has more than four years of experience in research and teaching, and has published several research articles in SCI/SCIE/Scopus journals and conferences of high repute. He has also authored a book on Cloud Computing. He has various national and international patents and some are granted. He has contributed to IEEE, Elsevier, and other reputed journals. He is part of a team of research advisory members at his present institute. His current areas of interest include Image Processing, Medical Imaging, Human Computation Interface, Neuro-Computing, and the Internet of Things.

Dr. Abhishek Verma is an Assistant Professor in the Department of Computer Science & Engineering at IIITDM Jabalpur, India (an institution of national importance). He obtained his PhD degree (2020) on the Internet of Things Security from the National Institute of Technology Kurukshetra, Haryana, India. He has more than seven years of experience in research and teaching and has published several research articles in international SCI/SCIE/Scopus journals and conferences of high repute. He is an editorial board member of Research Reports on Computer Science (RRCS) and an active review board member of various reputed journals, including IEEE, Springer, Wiley, and Elsevier. His current areas of interest include Information Security, Intrusion Detection, and the Internet of Things.

Dr. Prabhat Kumar Srivastava is a Professor with the Department of Computer Science & Engineering, Quantum University, Roorkee, Uttarakhand, India He earned his doctorate at Sam Higginbottom University of Agriculture and Technology & Science Allahabad U.P. His project was “A Soft Computing Approach for Fuzzy Data Modelling”. He has authored and published a significant number of research articles in peer-reviewed and indexed journals and conferences of high repute. He has also authored a book on Cloud Computing, and holds various national and international patents. He also serves as a review board member for several journals. His research areas are in Networking, Data Structure, and Soft Computing.

Dr. Virender Ranga is currently working as an Assistant Professor with the Department of Computer Science and Engineering, National Institute of Technology, Kurukshetra, India and has more than 15 years of experience in academic, research, and administrative capacities. He is a distinguished researcher, well known in his academic circles for his interdisciplinary research in the areas of wireless and ad hoc & sensor networks. He has contributed to several quality research journals of international repute, published by IEEE, ACM, IET, Elsevier, and others.

Dr. Ram Kumar is presently working as Assistant Professor at the Department of Electrical and Electronics Engineering at Katihar Engineering College, Bihar, India. He has over nine years of experience in academics (teaching/training), research and development, and consulting. He has contributed to several quality research journals of international repute, published by IEEE, ACM, IET, Elsevier, and others. He also possesses various national and international patents and has contributed to a number of publications in reputed international journals, conference proceedings, and book chapters. He has also edited one book.

Contributors

Deepshikha Agarwal
Indian Institute of Information Technology Lucknow
India

Shamama Anwar
Birla Institute of Technology
Mesra, India

Muhammad Arslan Anwar
University of Agriculture Faisalabad
India

Vishal Ashank
Birla Institute of Technology
Mesra, India

Ravi Kant Avvari
NIT Rourkela
India

Adil Bashir
Islamic University of Science and Technology
India

Priyobroto Basu
NIT Rourkela
India

Assia Baouta
Centre for Development of Advanced Technologies
India

Leena Bojaraj
KGiSL Institute of Technology
India

Djamel Bouchaffra
Centre for Development of Advanced Technologies
Algeria

Ahmed Mateen Buttar
University of Agriculture Faisalabad
India

Anuradha Chug
Guru Gobind Singh Indraprastha University
India

A. Lenin Fred
Mar Ephraem College of Engineering and Technology
India

Gulnawaz Gani
University of Kashmir
India

Praveen Kumar Ghodke
National Institute of Technology Calicut, India

Balàzs Gulyàs
Mar Ephraem College of Engineering and Technology
India

Mohammad Farukh Hashmi
NIT Warangal
India

Saba Hilal
Islamic University of Science and Technology
India

R. Jaikumar
KGiSL Institute of Technology
India

Abhishek Jain
Amity University Haryana Gurgaon
India

I. Christina Jane
Amal Jyothi College of Engineering
Kanjirapally, India

Jyoti Jangir
Amity University Haryana Gurgaon
India

Zubair Jeelani
Islamic University of Science and Technology
India

J. Jeyshri
SRM Institute of Science and Technology
India

Kajal Kansal
KIET Group of Institutions
India

Roopdeep Kaur
Federation University Australia
Australia

Gour Karmakar
Federation University Australia
Australia

H. Ajay Kumar
Mar Ephraem College of Engineering and Technology
India

Anil Kumar
Swami Vivekanand Subharti University
India

Atul B. Kathole
Pimpri Chinchwad College of Engineering
India

Avinash G. Keskar
NIT Warangal
India

Ashish Khare
University of Allahabad
India

Manish Khare
Dhirubhai Ambani Institute of Information and Communication Technology
India

S.N. Kumar
Mar Ephraem College of Engineering and Technology
India

M. Kowsigan
SRM Institute of Science and Technology
India

Praneeth Reddy Kunduru
NIT Warangal
India

Seema Maitrey
KIET Group of Institution
India

L.R Jonisha Miriam
Mar Ephraem College of Engineering and Technology
India

Sameer Ahmed Mujavar
NIT Warangal
India

M. Namratha
B.M.S College of Engineering
India

Sai Shashank Nandigama
NIT Warangal
India

Parasuraman Padmanabhan
Mar Ephraem College of Engineering and Technology
India

Sonali D. Patil
Pimpri Chinchwad College of Engineering
India

Amit Prakash Singh
Guru Gobind Singh Indraprastha University
India

Fasel Qadir
University of Kashmir
India

B.S. Rajeshwari
B.M.S College of Engineering
India

M. Rajkamal
IBM Bangalore
India

Roshani Raut
Pimpri Chinchwad College of Engineering
India

P. Swapna Reddy
National Institute of Technology
Calicut, India

Priyanka Sahu
Guru Gobind Singh Indraprastha University
India

Deepti Seth
KIET Group of Institutions
India

A.N. Saritha
B.M.S College of Engineering
India

Dinesh Singh
Indian Agricultural Research Institute
India

Subham Kumar Sinha
Birla Institute of Technology, Mesra
India

Prashant Srivastava
NIIT University
India

M. Sumathi
SASTRA Deemed University
India

Khushboo Tripathi
Amity University Haryana
Gurgaon, India

Kapil N. Vhatkar
Pimpri Chinchwad College of Engineering
India

Snehanshu Vivek
Birla Institute of Technology
Mesra, India

Feng Xia
Federation University Australia
Australia

Faycal Ykhlef
Centre for Development of Advanced Technologies
Algeria

Acknowledgement

We express our heartfelt gratitude to CRC Press (Taylor & Francis Group) and the editorial team for their guidance and support during completion of this book. We are sincerely grateful to our reviewers for their suggestions and illuminating views for each book chapter presented here in *Image Processing and Intelligent Computing Systems*.

1 Digital Image Processing
Theory and Applications

Fasel Qadir and Gulnawaz Gani
University of Kashmir, Srinagar, India
Zubair Jeelani
Islamic University of Science and Technology, Awantipora, India

CONTENTS

1.1 An Introduction to Image Processing....1
1.2 Key Concepts of Image Processing....2
1.2.1 What is Digital Image Processing?....2
1.2.2 Image Matrix Representation....4
1.2.3 Pixel....5
1.2.4 Pixel Neighborhoods....6
1.2.5 How Pixels Are Processed....6
1.2.6 Image Types....7
1.3 Fundamental Steps in Digital Image Processing....8
1.4 Applications of Image Processing....9
1.4.1 Noise....9
1.4.2 Scrambling....11
1.4.3 Forgery....12
1.4.4 Medical....13
1.5 Conclusions and Future Work....14
References....14

1.1 AN INTRODUCTION TO IMAGE PROCESSING

Digital images have become an integral component of our daily lives due to advances in visual media and the wide availability of social platforms for the dissemination of images. Digital image procedures deal with the manipulation of digital images using a digital computer. Broadly speaking, there are three main application areas of image processing. First, image enhancement refers to the manipulation of images so that the output image is clearer than the input. Second, autonomous machines refers to the development of intelligent machines that proceed by means of machine learning. Third, storage and transmission suggests developing efficient computer procedures for the storage and transmission of images over networks [1–4].

DOI: 10.1201/9781003267782-1

In the 1960s, digital images were studied after the invention of digital computers and the related technologies, which included storage, display and transmission. Thus, in the early 1960s, computers and image processing procedures were used to process a wide range of digital images. From the 1960's onwards, however, the field of image processing has expanded rapidly and it has encountered potential applications in different areas of study. Here, we present a brief list of image processing applications in different areas of study [1, 3]. In computer sciences, computer procedures have been successfully applied in the areas of the processing and analysis of digital images such as noise, scrambling, forgeries, segmentation, etc,. In medical sciences, digital images such as X-rays, nuclear magnetic resonance, ultrasonic, etc., are used in patient inspection or, more specifically, in areas such as tumour detection or the detection of any other ailment. In the field of agriculture, satellite digital images of land are used to predict their suitability for the planting of different crops, tracking earth resources, flood and fire control, and other applications related to environmental issues. Similarly, in law enforcement, image processing methods are used to sharpen and analyze fingerprint images. In geography, digital image methods are used to analyze polluted patterns in aerial images. In archeology, restoration image processing methods were successfully applied in the restoration of blurred digital images of damaged artefacts. Image storage, transmission and manipulation applications are used in televisions, teleconferencing, network communications, medical imaging, etc. Radar and sonar imaging and processing applications are used for detection of different targets of aircraft or missile systems.

The main purpose of this chapter is to introduce the concept of digital image processing and some of its more important applications. The remainder of this chapter focuses on the following points. Section 1.2 presents the general and key concepts of image processing, such as the pixel and its neighbors, image representation, image types, image formats, basic image operations, and so on. Section 1.3 presents the fundamentals steps involved in digital image processing, with the aim of introducing the basic concepts of image processing. Section 1.4 presents a few important applications of image processing. Section 1.5 presents the characteristics and limitations image processing. Finally, Section 1.6 presents conclusions.

1.2 KEY CONCEPTS OF IMAGE PROCESSING

This chapter presents a brief introduction to image processing, by outlining its key concepts. The matrix representation of images, which is used for the processing and development of algorithms, is presented. The concepts of the pixel, two common neighborhoods, and how pixels are processed are all provided. In addition, some common image types, such as binary, grayscale, and color, are discussed.

1.2.1 WHAT IS DIGITAL IMAGE PROCESSING?

Digital Image is the most recent buzzword in academic circles, IT industries, business sectors, and social networks. A digital image or still image is defined as a two-dimensional (2D) discrete function, denoted by $I(r, c)$, which represents some data arranged in row–column fashion. In $I(r, c)$, r denotes the image rows ($r = 1, 2, 3, \ldots, R$) and c

denotes the image columns ($c = 1, 2, 3, \ldots, C$). A pair of cartesian coordinate index (r, c) stores binary, or gray, or color, or pseudo intensity value of a particular pixel of a digital image.

Digital image processing refers to the manipulation of digital images through a digital computer. In other words, digital image processing provides platforms/domains (such as Spatial, Fourier, Wavelet, etc) through which digital images can be processed in order to gain more insights than are visible in the original image. A digital image holds real or complex values which are denoted by a finite number of bits, referred to as pixel depth. To process a digital image on a computer, it is first required to convert the image into a matrix form referred as image digitization. To display and maintain image quality of images on computer without flickering, special memory known is used that refreshes the image frames 30 or more times per second, which is known as buffer memory [1, 3]. Among the most famous examples of digital image processing are contrast enhancement, noise reduction, segmentation, compression, histograms, segmentation, and so on.

Depending upon the output of digital images, the field of digital image processing is broadly divided into three types: low-level, mid-level and high-level image processing [1]. Low-level refers to pre-processing image operations, including noise reduction, scaling, and so on. Mid-level refers to basic image processing operations related to edges, segmentation, and so on. Finally, high-level refers to complex image processing operations related to the analysis, recognition and interpretation of image contents for some decision-making. In all these image-processing types, the following three basic steps are performed: (1) Input: a digital image is imported on any image-processing software, such as Matlab, Scilab, Python, and so on. (2) Processing: depending upon the required output, a particular image-processing domain and method/algorithm is used. (3) Output: this produces the output (images or image components) that can also be used as input to the next image-processing tasks. For instance, noise reduction is considered to be the basic step of various image-processing tasks such as segmentation, recognition, and so on.

Currently, image processing, computer vision, computer graphics and artificial intelligence are considered to be overlapping fields. Yet there are no defined boundaries to identify either the beginning and/or the end of a particular field. However, example (1.1) shows a procedure according to which a particular field or a technique can be identified [5]. That is, we can identify whether a particular technique or a procedure belongs to image processing or computer vision or to some other domain. If both input and output are images, then the process is known as image processing. If the inputs is an image but the output is a descriptor (that is, the output provides some information about the image), then the process is known as

Output / Input	*Image*	*Description*
Image	Image Processing	Computer Vision
Description	Computer Graphics	Artificial Intelligence

EXAMPLE 1.1 Procedure for identifying overlapping fields of image processing.

computer vision. If the input is a descriptor and the outputs is an image, then the process is known as computer graphics (that is, generation of images using computers). If both input and output are descriptors, then the process is known as artificial intelligence.

	Output	
Input	**Image**	**Description**
Image	Image Processing	Computer Vision
Description	Computer Graphics	Artificial Intelligence

Digital images can be studied in all dimensions (d's), such as 1d, 2d, 3d, and so on. In 1d, images are considered as collection of pixels distributed either in the x or the y direction. Mostly, such image-processing problems are studied when high-end image processing systems are not available and therefore we need to convert a 2d image into 1d image. Examples of such images are lines and line segments. In 2d, pixels are distributed along 2 dimensions, that is, in both x and y directions. Such images are also known as flat or x,y-images. Examples of such images are photographs, wall paintings, 2d geometrical objects, and so on. Mostly, images are studied in 2ds, as this is considered as the most realistic image view formed using a camera. In 3d, image pixels are distributed along 3 dimensions, such as x, y, and z directions. 3d image processing tasks are usually studied in medical image and computer vision. Due to the non-availability of 3d systems, computer graphics procedures such as projections, are used to give 3d image views. Examples of such images include realistic photographs of clouds, plants, buildings, and so on. Similarly, images are studied in higher dimensions.

1.2.2 Image Matrix Representation

The matrix representation of images refers to the conversion and storage of digital images into a matrix of numbers on a computer system. These matrix numbers are known as image pixels. Matrix representation is required to store, process, and develop algorithms for digital images on a digital computer. The matrix representation of digital images is shown in equation (1.1). $I(r, c)$, on the left-hand side, denotes a digital image. The right-hand side matrix of real numbers denotes the elements of $I(r, c)$, distributed along rows (r) and columns (c). There are no restrictions on either the number of rows or the number of columns. However, limited hardware imposes restrictions on the number of intensity levels present in an image. Typically, the number of intensity levels are denoted by $[1, L]$, where, $L = 2^k$ & k takes an integer value. Furthermore, example (1.2) shows the cartesian coordinate conventions of digital images used for computer processing. As shown in the example, the x-direction moves downwards denotes image rows and y-direction moves from left to right denotes image columns. That is, the top left-hand side of the computer system represents the first pixel of an image, denoted by the coordinates (1,1). The

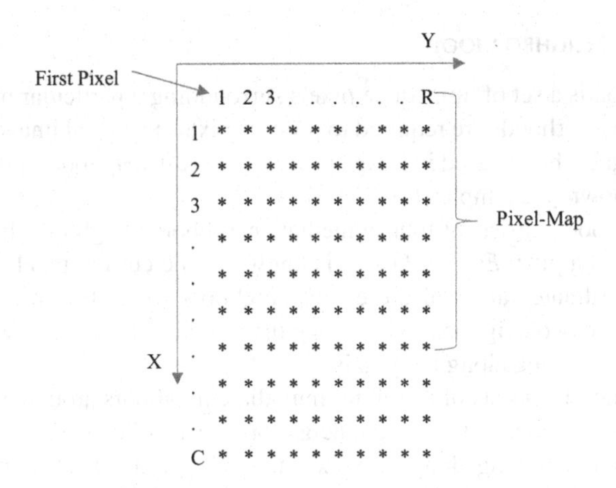

EXAMPLE 1.2 Cartesian conventions of digital images for computer processing.

points represented by dotted notations denotes image pixels known as pixel-map. The last pixel on the pixel map is denoted by (C, R).

$$I(r,c) = \begin{bmatrix} I(1,1) & I(1,2) & \cdots & I(1,R) \\ I(2,1) & I(2,2) & \cdots & I(2,R) \\ \vdots & \vdots & \vdots & \vdots \\ I(C,1) & I(C,2) & \cdots & I(C,R) \end{bmatrix} \quad (1.1)$$

1.2.3 Pixel

A digital image is made up of a finite number of coded cells known as pixels. A pixel, or picture (pix) element (el), is the smallest addressable cell coded to represent a particular brightness of a digital image. Each pixel has a defined location in the digital image that holds a numerical value known as pixel intensity. The four main types of intensity values of a pixel are binary, gray, color, and pseudo. In 1965, Frederic Crockett Billingsley, an image processing engineer, was the first to use the term "pixel" for describing picture elements of video games. To understand the concept of pixel, let's take the example of a Cathode Ray Tube (CRT). The CRT screen is painted with phosphorus material. This is a substance which glows if light is applied on it, like the driving sign boards during night driving when vehicle light is applied to or focused on them. Then, the phosphor-coated screen is divided into tiny/small dots. Each tiny dot is called a pixel. Thus, for a CRT, a pixel is a tiny phosphor dot or pixel on the computer screen. Similarly, for other types of computer screens.

Camera quality, or pixel count or resolution, is measured by counting the number of pixels across the horizontal and vertical directions of a digital image or digital space. For instance, 1000 pixels along horizontal direction and 1000 pixels along vertical direction (denoted by 1000*1000) means a resolution of 1,000,000 pixels or 1 million pixels or 1 megapixel (denoted by 1 Megapixels). Similarly, pixels per inch measure refers to the number of pixels displayed in one inch of a digital image. A higher pixel density per inch represents higher quality of digital images.

1.2.4 Pixel Neighborhoods

A neighborhood is a set of immediate pixels surrounding a particular pixel of a digital image. Neighborhoods are required to process pixels in digital images. Two of the important neighborhoods used in image processing are 4-neighborhood and 8-neighborhood, as shown in example (1.3).

4-neighborhood consists of four immediate neighbors (neighbors that lie at a unit distance) around a pixel $E(x, y)$. $E(x, y)$ is known as the current pixel or pixel under study. The coordinate values of these four neighbors are as follows. $I(x, y - 1)$ & $I(x, y + 1)$ are the two neighbors lying along the x-axis and $(x - 1, y)$ & $(x + 1, y)$ are the two neighbors lying along the y-axis.

8-neighborhood consists of the eight immediate neighbors around a pixel $E(x, y)$. The coordinate locations of these eight neighbors are as follows. $I(x, y - 1)$, $(x, y + 1)$ are the two neighbors lying along the x-axis, $(x - 1, y)$ & $(x + 1, y)$ are the two neighbors lying along the y-axis and $(x - 1, y - 1)$, $(x - 1, y + 1)$, $(x + 1, y - 1)$ & $(x + 1, y + 1)$ are the four neighbors lying along two diagonals (known as diagonal neighbors).

1.2.5 How Pixels Are Processed

Example (1.4) shows a black and white digital image of size 5*5 and a sliding window of size 3*3. A sliding window is used to process the pixels based on the neighborhood by placing it on every pixel of the input image. But the problem arises when processing pixels along image boarders. Suppose we are processing pixels based on an 8-neighborhood in spatial domain. When placing sliding windows $E(x, y)$ pixel on the first pixel on the image, you can see, we do not have values for locations $I(x - 1, y - 1)$, $I(x - 1, y)$, $I(x - 1, y + 1)$, $I(x, y - 1)$, and $I(x + 1, y - 1)$. $E(x, y)$ is known as the current pixel whose next value is to be calculated with the help of its surrounding pixels using sliding window. The same is the case for other corner and border pixels.

(a)

	I(x-1,y)	
I(x,y-1)	E(x,y)	I(x,y+1)
	I(x+1,y)	

(b)

I(x-1,y-1)	I(x-1,y)	I(x-1,y+1)
I(x,y-1)	E(x,y)	I(x,y+1)
I(x+1,y-1)	I(x+1,y)	I(x+1,y+1)

EXAMPLE 1.3 Pixel neighborhoods: (a) 4-neighborhood; (b) 8-neighborhood.

(a)

I(1,1)	I(1,2)	I(1,3)	I(1,4)	I(1,5)
I(2,1)	I(2,2)	I(2,3)	I(2,4)	I(2,5)
I(3,1)	I(3,2)	I(3,3)	I(3,4)	I(3,5)
I(4,1)	I(4,2)	I(4,3)	I(4,4)	I(4,5)
I(5,1)	I(5,2)	I(5,3)	I(5,4)	I(5,5)

(b)

I(x-1,y-1)	I(x-1,y)	I(x-1,y+1)
I(x,y-1)	E(x,y)	I(x,y+1)
I(x+1,y-1)	I(x+1,y)	I(x+1,y+1)

EXAMPLE 1.4 Pixel processing example: (a) image of size 5*5; (b) sliding window of size 3*3.

0	0	0	0	0	0	0
0	I(1,1)	I(1,2)	I(1,3)	I(1,4)	I(1,5)	0
0	I(2,1)	I(2,2)	I(2,3)	I(2,4)	I(2,5)	0
0	I(3,1)	I(3,2)	I(3,3)	I(3,4)	I(3,5)	0
0	I(4,1)	I(4,2)	I(4,3)	I(4,4)	I(4,5)	0
0	I(5,1)	I(5,2)	I(5,3)	I(5,4)	I(5,5)	0
0	0	0	0	0	0	0

(a)

I(5,5)	I(5,1)	I(5,2)	I(5,3)	I(5,4)	I(5,5)	I(5,1)
I(1,5)	I(1,1)	I(1,2)	I(1,3)	I(1,4)	I(1,5)	I(1,1)
I(2,5)	I(2,1)	I(2,2)	I(2,3)	I(2,4)	I(2,5)	I(2,1)
I(3,5)	I(3,1)	I(3,2)	I(3,3)	I(3,4)	I(3,5)	I(3,1)
I(4,5)	I(4,1)	I(4,2)	I(4,3)	I(4,4)	I(4,5)	I(4,1)
I(5,5)	I(5,1)	I(5,2)	I(5,3)	I(5,4)	I(5,5)	I(5,1)
I(1,5)	I(1,1)	I(1,2)	I(1,3)	I(1,4)	I(1,5)	I(1,1)

(b)

EXAMPLE 1.5 Boundary conditions example: (a) Null-padding; (b) Periodic-padding.

The solution for this problem is to add padding. Padding means appending additional rows and columns, so that we can get all values in the sliding window. Here, if we append two rows (one at the top and the second at the bottom) and two columns (one at the extreme left and one at the extreme right) our problem will be solved, as shown in example (1.5). Now, we do not need to process the pixels lying on the borders, because these pixels are not the pixels of the original image. Therefore, in the pixel processing of digital images we exclude the processing of border pixels.

The two standard boundary conditions used for adding values to the padding cells are null- padding and periodic-padding. Null-padding means to assign zero values in the padding locations as shown in example (1.5a). Periodic-padding means the wrapping of the original input image from left to right and from top to bottom, as shown in example (1.5b). That is, assign the extreme left values of the original image to the right padding locations and the extreme right values of the original image to the left padding locations. A similar process is carried out for top and bottom padding locations.

1.2.6 Image Types

Three standard types of digital images are binary, grayscale and color as shown in Figure 1.1 [1–4]. Depending upon the systems' specifications, a particular type of image is used.

A binary image is one in which each pixel got a value from the set {0, 1}. Since there are only two colors, that is 0 (black) and 1 (white), they are also called as

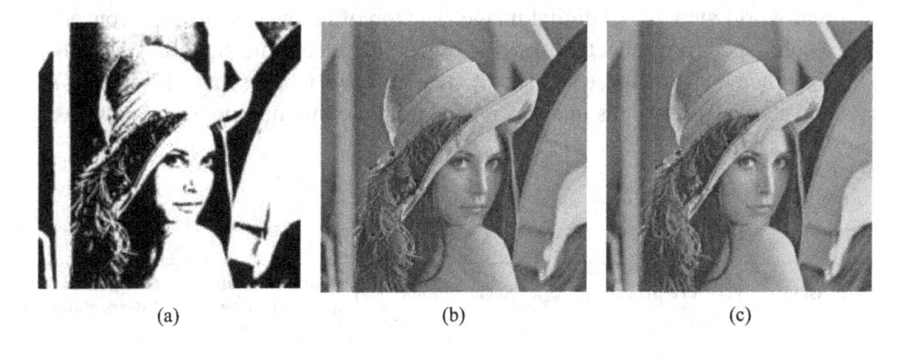

(a) (b) (c)

FIGURE 1.1 Types of digital images: (a) binary; (b) grayscale; (c) shade.

bi-level or black and white images. Pixel depth is used to find the number of bits required to store color values in an image. And the total number of image colors are measured by 2^k, where k can take any positive value and it represents the number of bits required (that is, pixel depth). Since, in binary images there are only two colors to store, so only one bit is required. That is, $2^1 = 2$.

A grayscale image is one in which each pixel got a value from the interval [1, 256]. This interval indicates that grayscale image pixels are stored by eight bits (that is, $2^8 = 256$), where 1 represents black, 256 represents white, 158 represents gray, and 2 to 255 represents shades of gray. 256 gray intensity values or colors are sufficient to recognize most of the natural objects. Generally, grayscale images have intensities of the order of 2^n. Thus, other grayscale images can also be created by changing the value of k between 1 to 8. For instance, if pixels are stored by 4-bits, that is, if $k = 4$, then there are $2^4 = 16$ colors. This image is known as 16-color grayscale image. Such types of images are usually used in medical imaging, such as contrast X-ray images.

A color image is one in which each pixel got a value from three color channels known as Red, Green, and Blue channels. Each channel has a range in the interval [1, 256]. During the computer processing of color images, each channel is represented by three separate matrices, known as the Red, Green, and Blue matrices of an image. This further means that for each color pixel there are three corresponding values. Therefore, the total number of possible colors in a color image are equal to $256^3 = 16{,}777{,}216$. This huge range of intensity values are enough to capture natural object through any lens. Since, there are three channels and each channel is represented by 8 bits, therefore total number of bits required to store a color pixel value are $3 \times 2^3 = 24$.

1.3 FUNDAMENTAL STEPS IN DIGITAL IMAGE PROCESSING

To analyse any real-world view on a digital computer, the following basic or fundamentals steps, as shown in the example (1.6), are performed [1–4].

Sensing: The primary step in image processing is to record real-world views onto a digital computer system. To do this, image sensors such as charged-coupled device used in a camera and a digitizer are used to convert any real-world view into a digital image. The job of image sensors is to convert a color spectrum reflected from the object into electronic signals. The digitizer then converts these signals into real numbers (for instance, 0's and 1's) and stores these values in a specialized part of computer memory known as buffer memory.

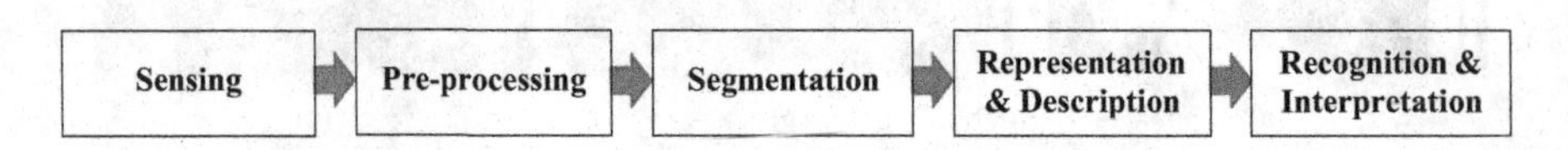

EXAMPLE 1.6 Basic steps of image processing.

Pre-processing: The basic step performed before analysing a digital image is known as image pre-processing. Pre-processing is used to remove basic problems, such as noise, which occurs during image sensing or blurredness or scaling, from digital images so that the input image could be more suitable for processing and provides appropriate results when applying image analysing procedures. Sometimes, post-processing image processing procedures, similar to the procedures used in pre- processing, are applied to the output image for improving the quality of results.

Segmentation: This refers to the partitioning of digital images into multiple segments. Different types of segmentation procedures are applied to digital images, such as thresholding, rugged, autonomous, and so on. Efficient segmentation procedures guarantees the detection of accurate segments in a digital image.

Representation & Description: These steps are required for extracting the regions of interest or particular features from the digital images under study.

Recognition & Interpretation: These steps are used to recognize regions of interest and assign labels to these regions based on its descriptors.

1.4 APPLICATIONS OF IMAGE PROCESSING

This section presents four important image processing applications namely, noise, scrambling, forgery, and medical.

1.4.1 Noise

In image processing, noise is an undesirable data or object that alters or misrepresents the information carried in an image. There are various reasons why images may get corrupted while applying computer procedures of image processing. For instance, during the image acquisition phase glitching of sensors in digital camera, the storage of images in faulty computer memory locations, image transmission over noisy channels, and so on. Depending upon the intensity of the noise signal, images may become corrupted by different types of noises, like salt & pepper, Gaussian, Speckle, and so on [1–4]. A simple illustration of noise filtration is shown in example (1.7).

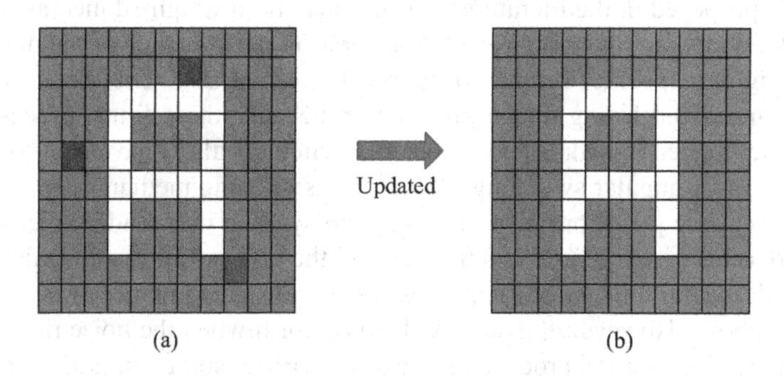

EXAMPLE 1.7 Process of noise filtration; (a) noisy image; (b) filtered image.

Example (1.7a) shows a grayscale image in which gray colors represent background pixels, white colors represent image objects and red colors represent noisy pixels. Example (1.7b) shows the output image after the application of a digital image noise filter. It is shown that the output images does not contain noisy pixels, such as red pixels. Such a process is known as noise filtration.

Digital image processing has a wide range of applications; sometimes images may carry sensitive information as in medical image processing application. If noise is not removed, image processing procedures may yield inappropriate results. Therefore, it is a fundamental step to remove noise from images before processing them in the case of higher-image processing tasks such as edge detection, segmentation, recognition, and so on. This is one of the main motivations why the noise removal field is the most explored and focussed field in image and video processing. Image filtration is a standard process used in all image processing systems. An image filter is a mask or procedure used to remove noisy pixels, by sliding the mask over the pixels, while preserving the details in an image. Different types of image filters are used for the removal of different types of noise. For instance, a 3*3 filter mask is usually used but larger masks (both, even and odd sizes) can also be used in a filtration process. Broadly speaking, image noise filters are divided into two categories: linear filters and non-linear filters.

Average filter, which is a linear filter, removes noise by replacing every image pixel by the average value of pixels in the sliding mask. The restoration performance of average filters is unsatisfactory because this filter removes indiscriminate image details while performing the filtration process. Median filter, which is a non-linear filter, removes noise by replacing every image pixel by the median value of the pixels in the sliding mask. The restoration performance of median filters is satisfactory at low noise rate, below 10 percent. Its performance becomes poorer if the image is corrupted with a high noise rate. However, the performance of the median filter is better than that of the average filter. The standard median filter processes every image pixel irrespective of whether or not the pixel is corrupted with noise. Therefore, the median filter faces the following two major disadvantages. One, the processing of uncorrupted image pixels modifies the original image pixel values, leading to the production of false image information. Second, the process is time-consuming, because it has to process all of the image pixels, and it performs worse if watermarks are present. As such, several variations to the median filter have been proposed in the literature. For instance, both weighted median and center-weighted median filters improve the restoration performance of the median filter by assigning specific weights to the pixels in the mask. The filters discussed above cannot avoid losing image information because these filters process every pixel of the image. To address this problem, switching filters have been proposed. One of the most popular switching filters is the switching median filter. This filter first examines the pixel for noise. If pixel is noisy, then the standard median filter is applied, otherwise the pixel is untreated and the process moves towards the next step. By doing this, it greatly improves restoration performance over the filters discussed above. However, this filter fails to perform when the noise ratio is high. Many efficient restoration procedures, based on various soft computing approaches for noise such as genetic algorithms, neural networks, cellular automata and so on, have been proposed in the literature. However, the image restoration field is still open for the development of efficient restoration procedures, so that stability can be improved further [6–9].

1.4.2 Scrambling

The rapid growth of the internet has greatly improved the development of global communication technologies. At present, information is communicated through various types of multimedia, including digital images, audio and video. In fact, the dissemination of multimedia over the internet is increasing on a daily basis which allows its forgery, an easy task via various multimedia tools such as paint software. Further, due to the availability of powerful computers multimedia content can be easily deciphered by hackers in real time. Sometimes, however, these multimedia types can carry confidential information, such as in defence and medical imaging. In fact, the secure transmission and storage of multimedia, particularly in the cloud, has remained a major challenge. Therefore, they must be protected properly before their transmission over the network. One of the best solutions to protect multimedia content is to use digital scrambling procedures. Based upon the type of input data, scrambling algorithms are categorized into three: image scrambling, audio scrambling and video scrambling. This section focuses solely on the concept of image scrambling, because the concept remains the same for both audio and video scrambling.

Image scrambling refers to the process of changing image pixel locations by reordering them so that it becomes impossible to understand the true contents of the image. By reordering pixel locations, the correlation between adjacent pixels is broken. For instance, example (1.8) shows a simple process of digital image scrambling. The input image is a sub-image of size 4*4. A scrambling matrix is the matrix used for scrambling the locations of the input image. First, the pixels in the input image are sequentially taken and placed at 1's location in the scrambling matrix. When this process is completed, then the remaining pixels of the input matrix are placed at 0's locations in the scrambling matrix. The scrambling sub-image is shown in our image.

A number of image scrambling procedures are available in the literature. For example, Arnold transformation, Fibonacci transformations, linear transformations and so on are used to scramble images. However, most of these procedures provide less security, as they are either linear or affine. Recently, cellular automata has been successfully tested for the development of efficient scrambling procedures [10–12].

230	217	59	177
250	82	212	188
96	123	78	18
99	234	61	198

Input Image

1	0	0	1
0	0	1	0
1	0	1	0
1	0	0	1

Scrambling Matrix

230	188	96	217
123	78	59	18
177	99	250	234
82	61	198	212

Output Image

EXAMPLE 1.8 Digital image scrambling.

1.4.3 Forgery

Multimedia, such as image, video and audio, is used in a wide range of application areas. The widespread availability of advanced multimedia editing tools, such as Adobe Photoshop, makes it easy to edit multimedia content. Sometimes, it becomes extremely difficult to differentiate between original and tempered or forged multimedia, that is, it is difficult to recognize forged multimedia through just our eyes or ears. For example, hundreds and thousands of multimedia are uploaded regularly on social networking websites, but there is no guarantee that all are genuine. Therefore, some mechanism is required to detect forged multimedia. The most common image tempering type is the copy-move forgery. It refers the process of copying some set of pixels from an image and then pasting them in the same image at some target location for content hiding or image manipulation. Figure 1.2 shows a simple example of copy- move image forgery. In this example, the first image is the original image, the second image is the copy-move forgery image and the third image shows the copied and pasted objects in the second image. Since humans are generally very poor in recognizing such forgeries, computer-based procedures are typically used to analyse them.

Two similar regions, the first the original and the second its replica, are used for incorporating the copy-move forgery into an image. Considerable research has been done toward developing efficient methods for detecting copy-move forgery. Almost all of these methods are based on the assumption that similar regions of the copy moved regions produce similar regions that can be identified with the help of matching techniques. The general framework of these methods consists of the following steps. First, is the pre-processing step, which refers the process of manipulating the input image in order to make it suitable for copy-move forgery detection. The second step is the feature extraction step; this refers the process of extracting features from the copy-move forgery image. Third comes the feature matching step, which refers to the process of matching features for identifying a similar region. Finally is the post-processing step, which refers to the process of suppressing false regions in order to ensure the accurate detection of forgeries. Based on the feature extraction procedure, copy-move forgery methods are divided into two types: block-based methods and key-point-based methods. Block-based methods divide

FIGURE 1.2 Copy-move forgery.

the copy-move forgery image into overlapping blocks prior to the extraction of features. Key-point based methods, by contrast, extract features from high-entropy regions. Although key-point based methods are fast, they do also have some limitations, such as poor rates of detection under post-processing attacks. Therefore, the detection of copy-move forgeries remains a challenging issue that needs to be addressed effectively [13–16].

1.4.4 Medical

Medical imaging is currently considered to be the most important application of image processing. Medical imaging refers to the process of acquiring medical images of the body parts by focussing radiations onto tissues and then processing those images on powerful systems in order to analyze diseases and extract clinically relevant information, such as infected cells [17]. All the basic image processing algorithms can be applied to the processing of medical images. Therefore, tremendous development of image processing, in terms of efficient computer procedures and efficient systems, has also developed and augmented the field of medical imaging. Further, powerful digital image processing systems, particularly those based on Graphics Processing Units (GPUs), help radiologists in storing, displaying, processing, transmitting, detecting, and analysing medical images for accurate medical diagnosis [19]. There are different types of medical images, which range from X-rays to magnetic resonance imaging (MRI). The four main imaging radiations used for acquiring medical images are X-rays, gamma rays, ultrasonic, and nuclear magnetic resonance induction.

The analysis and diagnosis of medical images is the backbone of modern medicine. Since it is quite difficult to analyse images manually, computer-based procedures have been developed for the efficient diagnosis of the diseases. The general framework for computer processing of medical images is depicted in example (1.9). Image sensing refers to the acquisition of medical images by exposing body parts to various radiations with the help of sensors such as charged couple devices. Low-level medical image processing represents various steps such as pre-processing, noise reduction, segmentation, and so on. High-level medical image processing also represents various steps, including recognition, classification, representation extraction, and so on. In low-level processing output images are also images; in high-level processing, however, output images are the sub-parts of the input images. Medical image processing has found potential applications in a wide range of medical diseases, such as the evaluation of chronic kidney disease [18]. Although efficient computer-based procedures have been developed, these methods perform poorly in the presence of noise, an issue which needs to be addressed.

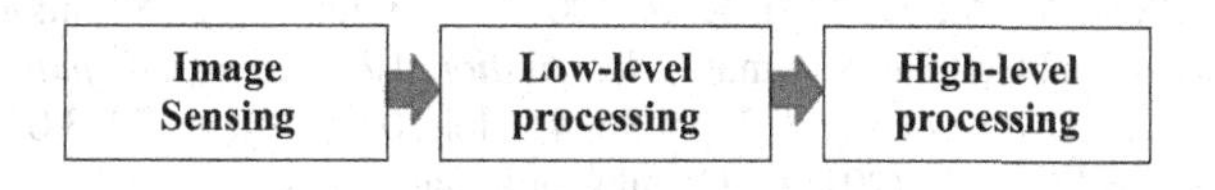

EXAMPLE 1.9 3-step model for medical imaging systems.

1.5 CONCLUSIONS AND FUTURE WORK

The field of image processing is a branch of computer science that refers to the processing of digital images using digital computers. Digital images are used to store, process and analyze visual representations of real objects such as painting, plants, clouds, buildings, etc., on any storage system, such as a computer system. The simplicity of image-processing algorithms has attracted the attention of many researchers in a variety of different fields, from computer science to arts to medical science. It is believed that computer procedures and efficient systems of image processing have the capabilities to eliminate inefficiencies caused by the traditional way of medical imaging analysis.

The aim of this chapter is to introduce the key concept of image processing and its important applications. From this perspective, this chapter first presents the key and general concepts of digital image processing which lays the foundation for presenting the various image-processing applications. This chapter then discusses four important applications: noise, scrambling, forgery and medical. Research gaps in connection to these presented applications are also mentioned in the relevant sections, which will need to be addressed in future studies.

REFERENCES

1. Gonzalez, R. C., & Woods R. E., *Digital image processing*. Fourth Edition, Pearson, NY, 2018.
2. Solomon, C., & Breckon, T., *Fundamentals of digital image processing: A practical approach with examples in Matlab*. A John Wiley & Sons Ltd., Publication, 2011.
3. Jain, A. K., *Fundamentals of digital image processing*. Englewood Cliffs, NJ: Prentice-Hall, Inc., 1989.
4. Tyagi, V. *Understanding digital image processing*. CRC Press, 2018.
5. Srivastava, P., Shukla, A., & Bansal, A., "A comprehensive review on soil classification using deep learning and computer vision techniques," *Multimedia Tools and Applications*, vol. 80, no. 10, pp. 14887–14914, 2021.
6. Qadir, F., & Shoosha, I. Q., "Cellular automata-based efficient method for the removal of high- density impulsive noise from digital images," *International Journal of Information Technology*, vol. 10, no. 4, pp. 529–536, Apr. 2018.
7. Jeelani, Z., & Qadir, F., "Cellular automata-based approach for salt-and-pepper noise filtration," *Journal of King Saud University - Computer and Information Sciences*, Dec. 2018.
8. Qadir, F., Peer, M. A., & Khan, K. A. (2012) "Cellular automata based identification and removal of impulsive noise from corrupted images," *Journal of Global Research in Computer Science*, vol. 3, no. 4, pp. 12–15
9. Qadir, F., Peer, M. A., & Khan, K. A. (2012) An effective image noise filtering algorithm using cellular automata. In: Proceedings of international conference on computer communications and informatics, IEEE explorer, Coimbatore, India, pp. 1–5
10. Qadir, F., Peer, M. A., & Khan, K. A. (2012). "Digital Image Scrambling Based on Two Dimensional Cellular Automata," *International Journal of Computer Network and Information Security*, vol. 5, no. 2, pp. 36–41. doi:10.5815/ijcnis.2013.02.05
11. Jeelani, Z., & Qadir, F. (2018). "Cellular automata-based approach for digital image scrambling," *International Journal of Intelligent Computing and Cybernetics*, vol. 11, no. 3, pp. 353–370. doi:10.1108/ijicc-10-2017-0132

12. Jeelani, Z., & Qadir, F. (2020). "A comparative study of cellular automata-based digital image scrambling techniques," *Evolving Systems*, vol. 12, no. 2, pp. 359–375. doi:10.1007/s12530-020-09326-5.
13. Gani, G., & Qadir, F., "Copy move forgery detection using DCT, PatchMatch and cellular automata," *Multimedia Tools and Applications*, vol. 80, no. 21–23, pp. 32219–32243, Jul. 2021.
14. Gani, G., Jeelani, Z., & Qadir, F., "Cellular automata-based CMF detection under single and multiple post-processing attacks," *Multimedia Systems*, vol. 28, no. 1, pp. 257–266, Jul. 2021.
15. Gani, G., & Qadir, F., "A novel method for digital image copy-move forgery detection and localization using evolving cellular automata and local binary patterns," *Evolving Systems*, vol. 12, no. 2, pp. 503–517, Nov. 2019.
16. Gani, G., & Qadir, F., "A robust copy-move forgery detection technique based on discrete cosine transform and cellular automata," *Journal of Information Security and Applications*, vol. 54, p. 102510, Oct. 2020.
17. Meyer-Baese, A., & Schmid, V., "Introduction" in Pattern Recognition and Signal Analysis in Medical Imaging, Second Edition, Elsevier, ISBN: 978-0-12-409545-8, 2014. https://doi.org/10.1016/B978-0-12-409545-8.00001-7
18. Alnazer, I., Bourdon, P., Urruty, T., Falou, O., Khalil, M., Shahin, A., & Fernandez-Maloigne, C., "Recent advances in medical image processing for the evaluation of chronic kidney disease," *Medical Image Analysis*, vol. 69, p. 101960, Apr. 2021.
19. Eklund, A., Dufort, P., Forsberg, D., & S. M. LaConte, "Medical image processing on the GPU – Past, present and future," *Medical Image Analysis*, vol. 17, no. 8, pp. 1073–1094, Dec. 2013.

2 Content-Based Image Retrieval Using Texture Features

Prashant Srivastava

NIIT University, Neemrana, India

Manish Khare

Dhirubhai Ambani Institute of Information and Communication Technology, Gandhinagar, India

Ashish Khare

Department of Electronics and Communication University of Allahabad, Allahabad, India

CONTENTS

2.1 Introduction ... 17
2.2 The State of the Art ... 19
2.3 Texture Features for CBIR ... 20
2.4 The Proposed Method ... 22
2.5 Experiment and Results ... 24
2.6 Performance Evaluation ... 25
2.7 Retrieval Results ... 26
2.8 Performance Comparison ... 28
2.9 Conclusion ... 30
References ... 31

2.1 INTRODUCTION

In recent decades, the exponential growth of information has led to the need of its proper organization and indexing. A huge amount of information being available in various forms is making the task of accessing relevant information tedious. In order to make the task of access easier, it is imperative, to properly index and organize the data. In this age, where information exists in various forms, multimedia information content is quite popular among people. Multimedia information includes both image and video, which are considered to be quite rich in information content. Image is one of the most popular forms of multimedia information, which is widely accessed and shared among huge numbers of people around the world. Image data

DOI: 10.1201/9781003267782-2

has been one of the most popular sources of information, being shared widely by people through social networking sites across the world. This sharing of multimedia information produces billions of images every day. This led to the new challenge of storage and the retrieval of a large volume of images. The existence of a large number of unorganized images makes it difficult to search and retrieve relevant image data. The field of image retrieval attempts to solve this problem of storage, search and retrieval of data. Image retrieval refers to the searching for and retrieval of images based on either text and keywords or features and attributes of images. Image retrieval systems can be classified into two categories: Text Based Image Retrieval System (TBIR) and Content Based Image Retrieval System (CBIR). Text Based Image Retrieval (TBIR) system performs the searching for and retrieval of relevant images based on keywords, phrases and text. The retrieval accuracy of such systems depends upon how efficiently the users express their query in the form of text, as well as the type of keywords with which the image has been tagged. Such a system requires the manual tagging of a large number of images, which is an onerous and time-consuming task. In addition, such systems are not capable of retrieving visually similar images.

CBIR refers to searching and retrieving images based on features present in the image. Instead of providing a query in the form of text, a query is represented in the form of image or sketch of image. The CBIR system extracts features from the query image and also constructs a feature vector. This feature vector is matched with the feature vector of images in the database to retrieve visually similar images [1]. Such a system has two advantages. First, the CBIR system does not require the manual annotation of images. Secondly, such systems are capable of retrieving visually similar images. The field of CBIR has caught the attention of scientists across the world over the past two decades. With the increase in low-cost image capturing devices, a large number of images being produced on a daily basis. To make the process of searching and retrieval quite easy no manual tagging of images is required. The expression of a query in the form of text differs from user to user. A user searching for a particular image may express a query for it in one way, whereas a second user searching for the same image may express their query in a very different form. This may produce a different result set, which may not satisfy a user looking for a specific image. In the case of CBIR, such problems do not arise as the query is expressed in the form of image; therefore, searching and retrieval takes place based on features present in the image.

The term CBIR came into existence in the late 1980s [2]. When text-based searching was prevalent in all applications of imaging systems, CBIR was the new term at that time. Early CBIR techniques focused on primary features of image such as color feature to retrieve visually similar images. Color, being a visible feature, has been exploited a lot for CBIR. Later on, texture and shape features were also exploited to retrieve visually similar images. For a few years, CBIR systems were mostly dependent on primary features such as color, texture and shape as single feature. However, the use of primary features as a single feature did not produce high retrieval accuracy. To overcome this limitation, the trend of CBIR shifted to a combination of features. The combination of features involved integration of primary features to construct feature vector for retrieval. The combination of multiple features to construct a feature vector not only improved retrieval accuracy but also led to the construction of efficient feature vector extracting more details rather than a single feature [3].

The use of primary features on a single resolution of image for the construction of a feature vector worked efficiently in the case of simple images. However, an image is a complex structure, which contains varying levels of details. For the extraction of such details, single-resolution processing of the image is insufficient. Multiresolution processing of the image aims to overcome this drawback. Multiresolution processing techniques analyze and interpret an image at more than one resolution of the image. This technique exploits multiple resolutions of the image to construct feature vector for image retrieval. Multiresolution technique extracts not only varying level of details in an image, but also features that are left undetected at one scale get detected at another scale. A number of multiresolution techniques, such as wavelet transform [4], curvelet transform [5], contourlet transform [6], and so on, have been proposed which have been frequently exploited for image retrieval. CBIR using multiresolution techniques generally involve either single features being exploited at multiple resolutions of image or combination of features using multiresolution techniques [7–10]. The exploitation of features at the multiple resolution of the image helps in obtaining foreground as well as background details and both coarse and fine details in an image.

Most of the abovementioned method focuses on the processing of primary features of the image. However, human beings recognize an image using semantic features rather than primary features. In order to bridge this semantic gap, numerous CBIR techniques have been proposed which utilize machine learning techniques [11–14]. These techniques construct a knowledge base, which is used to map features present in the image and compare them with those present in the database [15].

2.2 THE STATE OF THE ART

The process of CBIR starts with the process of feature extraction; this is then followed by the construction of a feature vector. Therefore, feature plays a very important role in CBIR as the success of CBIR depends upon the feature vector constructed. When the term CBIR was first proposed, the feature extraction process mostly involved the extraction of primary features, such as color, texture, and shape. The primary features represent visual features efficiently and therefore holds great significance in the field of CBIR. CBIR using color features involves the representation of color features using color histograms [16], color correlograms [17], and color coherence vector [18]. Color is a visible feature of an image and is invariant to certain geometric transformations. Apart from the color feature, the trend of CBIR techniques also witnessed the usage of texture and shape features. There are various methods for representing texture features. The most commonly used feature for representing texture feature is local pattern. A number of local patterns have been proposed, which efficiently represent texture features [19–23]. Local patterns have the property of encoding texture features in the local neighborhood. This results in the construction of an effective feature vector as they efficiently extract local information. Apart from color and texture features, shape is another feature, which has been extensively used for constructing feature vectors [24–29]. The extraction of shape feature generally requires segmenting out objects from the images. Shape features, such as moments and polygonal structure, perform well after segmentation process. However, there are certain shape features which perform efficient shape extraction without requiring segmentation. Histogram of Oriented Gradients (HOG) [30] is one such local shape descriptor, which constructs efficient feature vector for retrieval.

Early CBIR techniques mostly utilized primary features as a single feature to retrieve visually similar images. However, primary features work efficiently when used in combination with each other rather than as individual features. The combination of primary features such as color and texture [31, 32], texture and shape [32, 33], and color, texture and shape [34] prove this. There are a number of advantages of combining features. First, the combinations of features integrate advantages of multiple features. Second, they overcome each other's limitations. Therefore, they produce high retrieval accuracy.

The use of color, texture and shape features construct an efficient feature vector. However, most of the CBIR techniques extract primary features from a single resolution of an image [35–39]. An image may consist of many different types of objects, such as high- as well as low-resolution objects and small as well as large objects. The extraction of features from an image consisting of varying level of details using a single resolution of the image does not prove to be sufficient. Hence, there is a need for a technique which is capable of extracting features at multiple resolutions of image. Multiresolution technique aims to fulfill this objective. Of all of the various multiresolution techniques, wavelet transform is one such technique, which has been extensively used for feature extraction at multiple resolutions of image [3, 15, 24]. Wavelet transform decomposes an image into multiple resolutions by computing coefficients at multiple orientations. However, wavelets suffer from certain drawbacks. Wavelets have limited directionality and fail to represent edges efficiently. These drawbacks are overcome by other multiresolution techniques, such as curvelet transform and contourlet transform. These techniques are highly anisotropic in nature and efficiently represent edges with fewer coefficients. Due to these properties, curvelets and contourlets have been used a lot for extracting features at multiple resolutions of image [25–28].

2.3 TEXTURE FEATURES FOR CBIR

It is a general assumption that the image of an object has uniform intensity over the entire region. However, this is not always the case. Image of certain objects may contain variations in the pattern of intensity values. These patterns are encoded using texture features. Like color, texture is also a visual feature. It is difficult to give an exact definition of texture. Various researchers across the world have proposed a number of definitions for texture [29]. Texture generally refers to the structural arrangement of pixel values. It is difficult to propose a technique which can efficiently encode texture feature for all types of images. Early image retrieval techniques utilized Tamura features [35, 36] and Fourier transform [37] for encoding texture feature for the retrieval of similar images. Tamura features include coarseness, contrast, directionality, regularity and roughness information. These features have been designed on the psychological perception of texture by human beings. Fourier transform, on the other hand, encodes frequency domain information of an image. Fourier transform descriptors are capable of generating different texture patterns, which can efficiently construct discriminative texture features [40–43].

Tamura feature and Fourier transform descriptors work well when applied as a global feature. Texture feature tends to change locally in the image of an object. Hence, these feature descriptors, which are capable of extracting information from the local region, are considered to be more efficient than those which are applied globally. Ojala et al. [38] proposed a local texture descriptor, named the Local Binary Pattern (LBP). This encodes information in the local neighborhood by obtaining difference between center pixel and neighboring pixels. LBP proved to be a breakthrough in the area of texture-based CBIR. Being an efficient local feature descriptor, it is able to capture local information from the entire image. It is also invariant to grayscale transformations. LBP proved to be an efficient feature descriptor not only for CBIR, but also for other applications of computer vision. Inspired by the concept and success of LBP, a number of local patterns have been proposed to extract texture feature in an image. Tan and Triggs [39] proposed the concept of Local Ternary Pattern (LTP). While LBP encodes local information into two values (0,1), LTP encodes into three values (0, 1, –1) according to a certain threshold value. It is then divided into two patterns based on its positive and negative components. LTP extracts more information than LBP and, therefore, considered as a better feature descriptor. Due to these features, LTP has been extensively used for constructing a feature vector for CBIR [40, 41]. Extending the concept of LBP and LTP, Zhang et al. [42] proposed the concept of Local Derivative Pattern (LDP). LDP extracts more information than either LBP or LTP as it attempts to extract a high-order derivative pattern. Motivated by the useful properties and success of LBP, LDP and LTP, Murala et al. [44] proposed Local Tetra Pattern (LTrP) which encodes the relationship between the center pixel and neighborhood pixels on the basis of directions computed using first-order derivatives in vertical and horizontal directions. The method extracts more details than LBP, LTP, and LTP, and produces high retrieval accuracy. Local patterns obtain a change in intensity values in the local neighborhood and efficiently represent the texture features of an image.

Texture provides information about how intensity values are structurally arranged. It generally gives information about roughness, coarseness, smoothness etc. In order to extract such kinds of information from an image, feature descriptors that provide details about the spatial arrangement of intensity values are required. A number of feature descriptors with this property have been proposed. Liu et al. [44] proposed a descriptor named a Multi-Texton Histogram, which combined the properties of a co-occurrence matrix and a histogram. It efficiently represents the spatial correlation of color and texture orientation. Liu et al. [45] proposed the concept of Microstructure Descriptor, which attempted to construct a feature vector by identifying color in a microstructure with similar edge orientation. Wang et al. [46] proposed the concept of a Structure Element's Descriptor, which combines the advantages of statistical and structural texture description methods. Zhao et al. [47] proposed a new descriptor, named Multi-Trend Structure, which analyzes correlation among intensity values in a local region. The advantage of such a kind of descriptor is that they exploit the local neighborhood to extract a relationship between intensity values, thereby determining changes in intensity values among different regions of the same image.

The use of a single resolution of an image to determine the structural arrangement of intensity values has been extensively used. However, such a descriptor, which exploits the single resolution of an image, fails to extract varying level of details. To overcome this limitation, a number of multiresolution texture descriptors, which combine multiresolution processing technique with texture feature descriptors, have been proposed. There have been a number or recent proposals in this regard: Srivastava and Khare [15] proposed a multiresolution feature descriptor named a Local Binary Curvelet Co-occurrence Pattern. This descriptor combined LBP, curvelet transform and GLCM to construct a feature vector for retrieval. Similarly, Moghaddam et al. [24] proposed a descriptor named Wavelet Correlogram which integrated wavelet transform with a color correlogram descriptor. Khare et al. [48] proposed a technique for CBIR which computes the DWT of an LBP image. The advantage of such feature descriptors is that they extract texture features at multiple orientation and scales. Thus, the features that are left undetected at one scale become detected at another scale of image.

2.4 THE PROPOSED METHOD

Texture feature is considered to be one of the most important primary features of an image which has been extensively used to construct features vector for CBIR. Texture feature has been used both as single feature as well as in combination with other features. An efficient texture descriptor, when combined with other primary feature descriptors, produces high retrieval accuracy. This section demonstrates the effectiveness of the texture feature in the field of CBIR. The feature vector is constructed by combining the texture feature and the shape feature of the image. The texture feature is extracted using Local Ternary Pattern while the shape feature is extracted using Geometric moments. The combination of texture and shape feature has been exploited at multiple resolutions of image in order to extract varying level of details in an image. The multiresolution decomposition of image has been performed using DWT. The proposed method attempts to extract the shape feature from the texture feature at multiple resolutions of image. The method consists of the following steps:

1. Computation of the DWT coefficients of the grayscale image
2. Computation of the LTP descriptor of the resulting DWT coefficients
3. Computation of Geometric moments of the LTP descriptor
4. Similarity measurement

The schematic diagram of the proposed method is shown in Figure 2.1.

The first step of the proposed method is the computation of DWT coefficients of the grayscale image. The application of DWT on the grayscale image produces one approximate coefficient matrix and three detail coefficient matrices. The detail coefficient matrix consists of coefficients at three different orientations: horizontal, vertical and diagonal. Each of these detail coefficient matrices are used to construct the

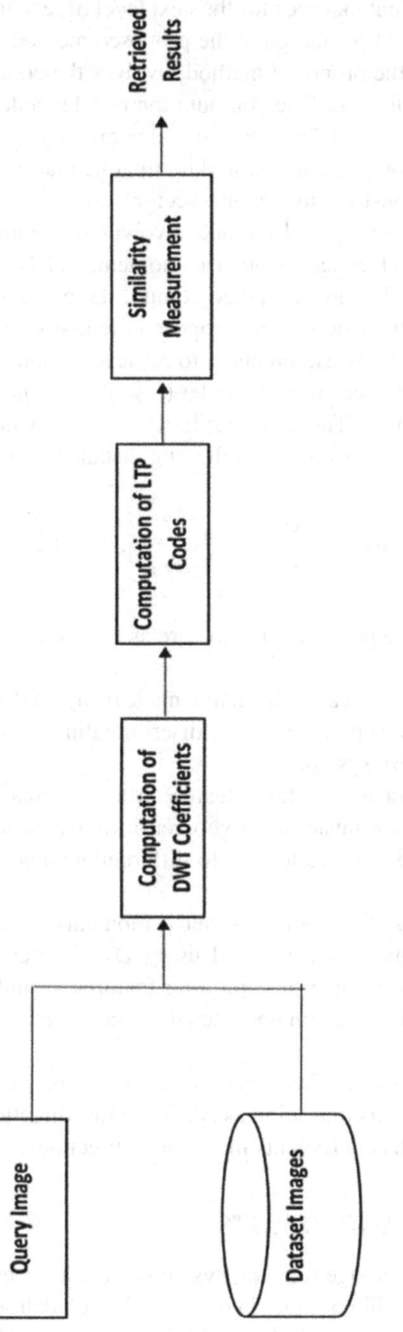

FIGURE 2.1 Schematic diagram of the proposed method.

feature vector. In the next phase, the approximate coefficient matrix is again subjected to DWT decomposition, which again results in one approximate coefficient and three detail coefficient matrices for the next level of resolution. This procedure is repeated for four levels of resolution in the proposed method.

The second step of the proposed method involves the computation of LTP codes of resulting DWT coefficients. The computation of LTP codes results in two matrices: Upper LTP and Lower LTP. For each detail coefficient matrices, Upper LTP and Lower LTP codes are computed and stored in separate matrices. Each of these matrices are considered to construct the feature vector.

The third step of the proposed method involves the computation of Geometric moments of resulting LTP codes. Geometric moments of LTP codes extract the shape feature from the texture feature computed at multiple resolutions of the image.

The fourth and the final step of the proposed method is similarity measurement. The purpose of similarity measurement is to retrieve visually similar images. Let f_Q be the feature vector of query image and let $(f_{DB1}, f_{DB2}, \ldots f_{DBn})$ be the set of feature vectors of database images. Then, the similarity measurement between query image and database image is done using the following distance formula:

$$Similarity(S) = \sum_{i=1}^{n} \left| \frac{f_{DBi} - f_Q}{1 + f_{DBi} + f_Q} \right|, \quad i = 1, 2, \ldots, n \tag{2.1}$$

The advantages of the proposed method are as follows:

1. It extracts the texture feature from the image using LTP, which is an efficient texture descriptor as it gathers more discriminating features from the image than under the LBP system.
2. It efficiently extracts the shape feature from the image using Geometric moments which are invariant to geometric transformations and provide a sufficiently discriminative feature to differentiate among objects of different shapes.
3. The combinations of LTP and Geometric moments are exploited at multiple resolutions of image, decomposed using DWT, to extract varying levels of details. The advantage of exploiting features at multiple resolutions is that the features that remain undetected at one scale are detected at another scale.
4. LTP does not provide directional information, but its combination with DWT overcomes this limitation as DWT obtains directional information by computing wavelet coefficients in multiple directions.

2.5 EXPERIMENT AND RESULTS

To test the efficiency of image retrieval systems, there are a number of image datasets such as Corel image library, Olivia dataset, Caltech dataset, GHIM-10K dataset etc. which are used as benchmark datasets for this purpose. These datasets contain a

FIGURE 2.2 Sample images from Corel-1K dataset.

wide variety of natural images classified into different categories. In this chapter, the performance of the proposed method has been tested on Corel-1K dataset [49]. Corel dataset contains a wide variety of natural images, which prove to be sufficient to evaluate the performance of image retrieval systems. Corel-1K dataset consist of one thousand images divided into ten categories, each category consisting of one hundred images. The size of each image of this dataset is either 256 × 384 or 384 × 256 pixels. The sample images of Corel-1K dataset is shown in Figure 2.2. While performing the experiment each image of the dataset is taken as query image. If the retrieved images belong to the same category as that of the query image, the retrieval is considered to be successful; otherwise, the retrieval fails.

2.6 PERFORMANCE EVALUATION

Performance of the proposed method has been evaluated in terms of precision and recall. Precision is defined as the ratio of the total number of relevant images retrieved to the total number of images retrieved. Mathematically, precision can be formulated as

$$P = \frac{I_R}{T_R} \tag{2.2}$$

where I_R denotes the total number of relevant images retrieved and T_R denotes the total number of images retrieved.

Recall is defined as the ratio of total number of relevant images retrieved to the total number of relevant images in the database. Mathematically, recall can be formulated as

$$R = \frac{I_R}{C_R} \tag{2.3}$$

where I_R denotes the total number of relevant images retrieved and C_R denotes the total number of relevant images in the database. In this experiment, $TR = 10$ and $CR = 100$.

2.7 RETRIEVAL RESULTS

The application of DWT on a grayscale image produces one approximation coefficient matrix and three detail coefficient matrices: horizontal detail, which consists of coefficients in the horizontal direction; vertical detail, consisting of coefficients computed in the vertical direction; and diagonal detail, consisting of coefficients in the diagonal direction. LTP codes of each of these three detail coefficient matrices are computed and stored in three separate matrices. Computation of LTP codes of each detail coefficient matrix results in two LTP matrices: Upper LTP matrix and Lower LTP matrix. Therefore, for three detail coefficient matrices, six LTP matrices are generated. Geometric moments of each of these detail coefficient matrices are computed and stored separately. These Geometric moment values are used as a feature vector to retrieve visually similar images. The retrieval process produces six sets of similar images. Union of all these sets is taken to produce the final image set. Recall is computed by counting the total number of relevant images in the final image set. Similarly, for precision, top *n* matches for each set is counted and then union operation is applied on all sets to produce final image set. Mathematically, this can be stated as follows: Let f_H be the set of similar images obtained from the horizontal detail feature vector, f_V be the set of similar images obtained from the vertical detail feature vector, and f_D be the set of similar images obtained from the diagonal detail feature vector. Then, the final set of similar images denoted by f_{RS} is given as

$$f_{RS} = f_H \cup f_V \cup f_D \quad (2.4)$$

Similarly, let f_H^n be the set of top *n* images obtained from the horizontal detail feature vector, f_V^n be the set of top *n* images obtained from the vertical detail feature vector, and f_D^n be the set of top *n* images obtained from the diagonal detail feature vector. Then, the final set of top *n* images denoted by f_{PS}^n is given as

$$f_{PS}^n = f_H^n \cup f_V^n \cup f_D^n \quad (2.5)$$

The above procedure is repeated for four levels of resolution. In each level, the relevant image set of the previous level is also considered to produce the final image set for current level. Retrieval is considered to be successful if the values of precision and recall are high.

Table 2.1 shows the values of precision and recall for four levels of resolution on Corel-1K dataset. Figure 2.3 shows the plot between recall vs. level of resolution and precision vs. level of resolution respectively.

From the above experimental observations, it is clearly observed that the average values of precision and recall increase with the level of resolution. This is due to multiresolution processing that each level attempts to gather details which were undetected at previous levels. The proposed method constructs the feature vector at each level of resolution of image. The features that are left undetected at the previous level of resolution become detected at another level. Due to this phenomenon, precision and recall values increase at different levels of resolution.

TABLE 2.1
Average Precision and Recall Values for Four Levels of Resolutions on Corel-1K Dataset

	Recall (%)	Precision (%)
Level 1	39.13	70.34
Level 2	58.16	93.32
Level 3	74.58	98.18
Level 4	85.04	99.53

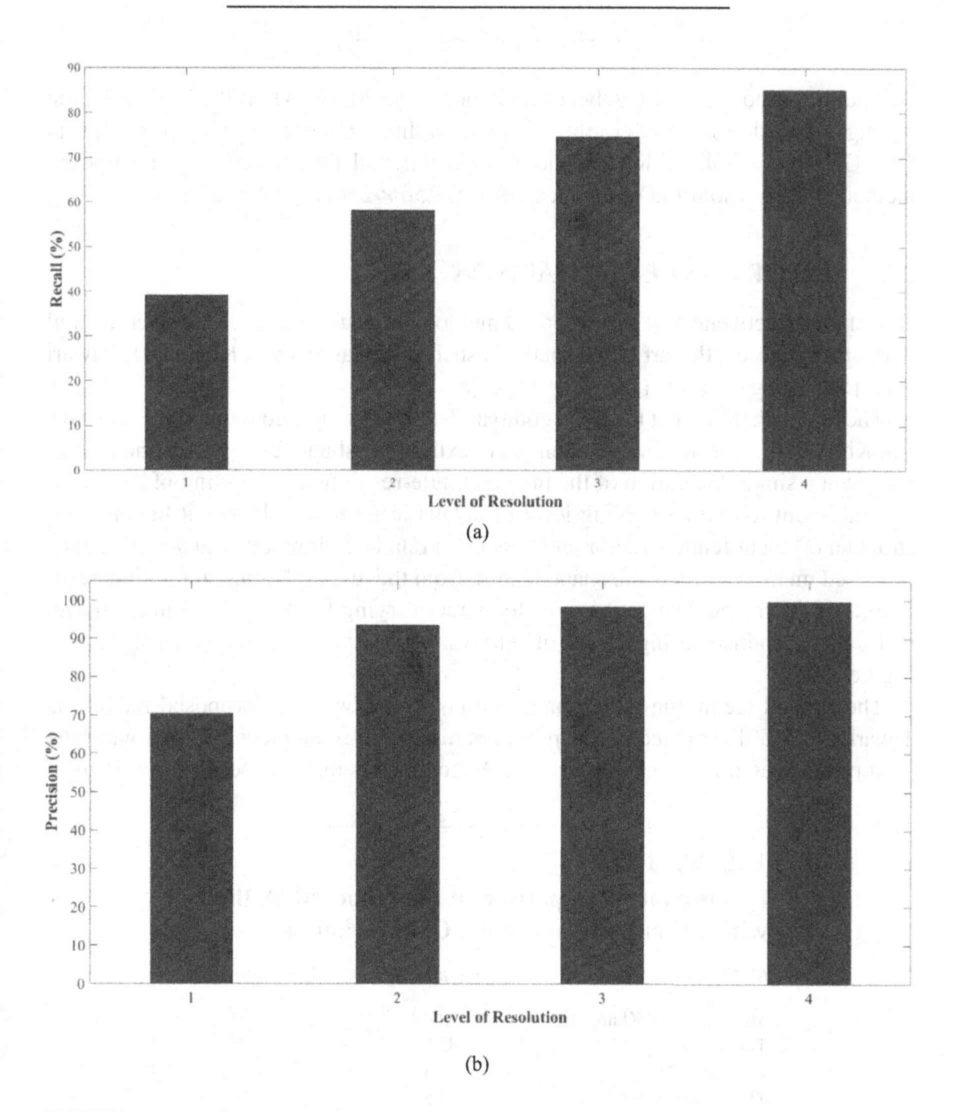

FIGURE 2.3 (a) Average recall vs. Level of resolution; (b) Average precision vs. Level of resolution for Corel-1K.

TABLE 2.2
Average Retrieval Time for Four Levels of Resolutions on Corel-1K Dataset

Level	Time (sec)
Level 1	9.48
Level 2	9.95
Level 3	12.04
Level 4	12.74

The proposed method has been implemented in MATLAB R2017a on a laptop having a Windows 10 operating system, an Intel Core i7-8550U processor at 1.80 GHz and 8 GB of RAM. The average retrieval time taken by the proposed method at each resolution of an image of size 256 × 256 is shown in Table 2.2.

2.8 PERFORMANCE COMPARISON

To test the effectiveness of the proposed method, its performance has been compared with other state-of- the-art CBIR methods such as Srivastava and Khare [10], Tiwari et al. [50], Zeng et al. [51], Zhao et al. [47].

The first technique that has been compared with the proposed method is Srivastava and Khare [10]. This technique attempts to extract the shape feature from the texture feature at a single resolution of the image. Single-resolution processing of the image is insufficient to extract varying level of details in an image. Hence it fails to construct an efficient feature vector and thus has a relatively low retrieval accuracy. The proposed method extracts the shape feature from the texture feature at multiple resolutions of the image. Hence it is able to extract varying levels of detail in an image, and thus it produces a high level of retrieval accuracy, as shown in Table 2.3 and Figure 2.4.

The second technique which has been compared with the proposed method is Tiwari et al. [50]. This technique proposes histogram refinement for improving the performance of the texture descriptor. Although the method performs well as it

TABLE 2.3
Performance Comparison of the Proposed Method with Other State-of-the-Art CBIR Techniques

Methods	Recall (%)	Precision (%)
Srivastava and Khare [10]	72.09	53.70
Tiwari et al. [50]	45.79	79.78
Zeng et al. [51]	16.11	80.57
Zhou et al. [47]	11.89	79.28
Proposed Method	**85.04**	**99.53**

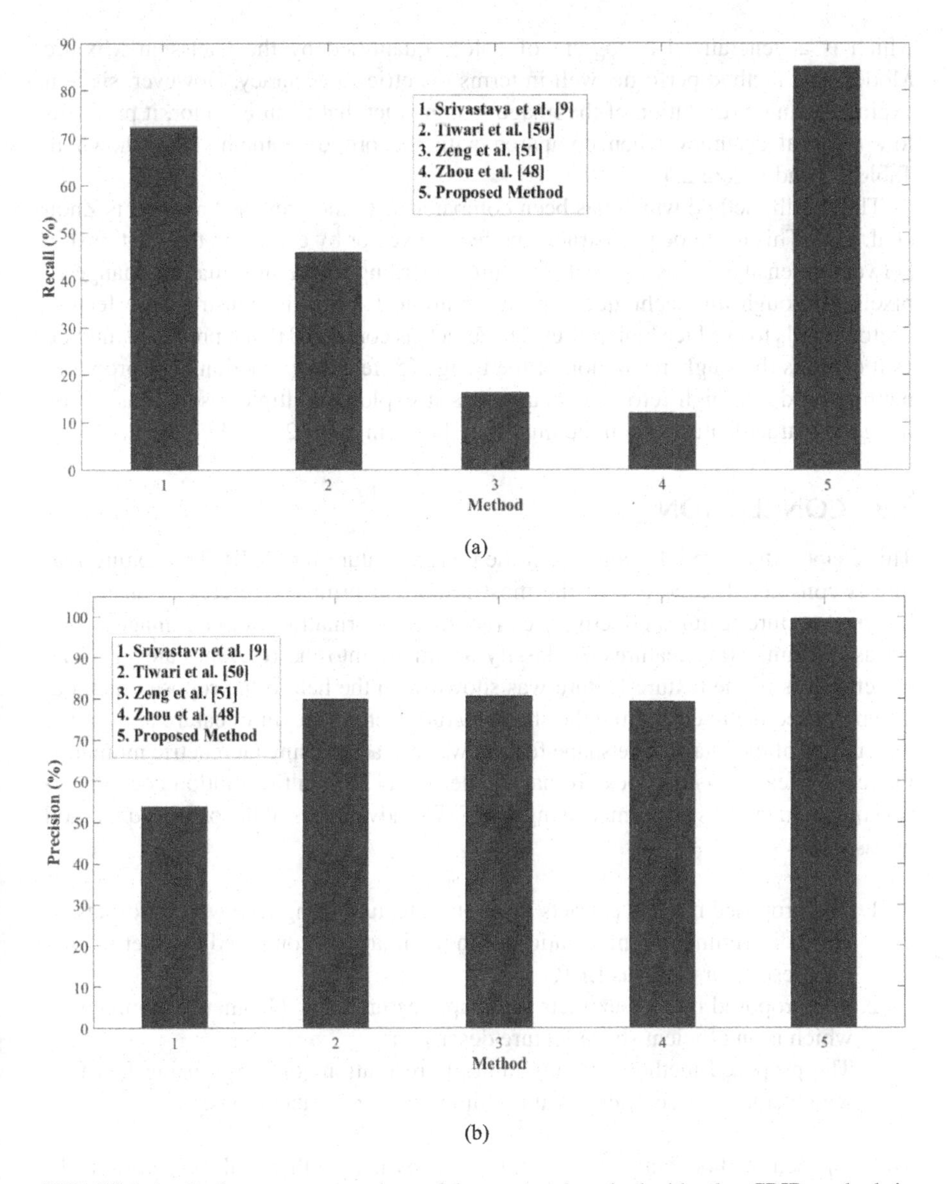

FIGURE 2.4 Performance comparison of the proposed method with other CBIR methods in terms of (a) Recall and (b) Precision.

constructs the feature vector based on analysis of pixels in the local neighborhood, it fails to extract varying level of details since it exploits a single resolution of the image. The proposed method, on the other hand, exploits multiple resolutions of the image to construct a feature vector and hence produces high retrieval accuracy, as shown in Table 2.3 and Figure 2.4.

The third method which has been compared with the proposed method is that of Zeng et al. [51]. This technique proposes a novel feature descriptor Spatiogram, which is a generalized histogram of colors quantified by the Gaussian Mixture Model. The method performs well in terms of retrieval accuracy. However, since it exploits a single resolution of the image to construct that feature vector, it produces low retrieval accuracy when compared with the proposed method, as shown in Table 2.3 and Figure 2.4.

The fourth method which has been compared with the proposed method is Zhou et al. [47]. This technique constructs the feature vector by exploring the relationship between intensity values in local structure according to the information change of pixels. Although this technique exploits multiple features to construct the feature vector, it fails to produce high retrieval accuracy as compared to the proposed method as it exploits the single resolution of the image for feature extraction. The proposed method produces high retrieval accuracy as it exploits multiple resolutions of the image to extract features from the image as shown in Table 2.3 and Figure 2.4.

2.9 CONCLUSION

This chapter discussed the concept of the texture feature for CBIR. The texture feature is considered to be one of the most important primary features of an image. Efficient texture features effectively extract local information from an image which act as discriminating features to classify an image into the relevant category. The effectiveness of the texture feature was shown with the help of a proposed method. The proposed method captured the shape feature from the texture feature at multiple resolutions of the image. The shape feature was extracted using Geometric moments, the texture feature using Local Ternary Pattern, and the multiresolution decomposition of the image was performed using DWT. The advantages of the proposed method are as follows:

1. The proposed method extracts the texture feature using LTP which extracts more discriminating information from the image as compared to other texture descriptors such as LBP.
2. The proposed method extracts the shape feature using Geometric moments which is an efficient shape feature descriptor.
3. The proposed method exploits multiple resolutions of image using DWT which computes coefficients at multiple scale and orientations.

The proposed method can be further improved by using other multiresolution techniques, which are better than DWT, such as curvelet transform or contourlet transform. Also, incorporating intelligent techniques such as deep learning can help in the extraction of semantic features, along with primary features which can further improve retrieval accuracy.

Like other primary features, texture is a visible property of an object, which is recognized when visualized by human beings. Hence, the application of the texture feature is not only limited to the field of image classification and retrieval. Various applications of computer vision, such as Human Activity Recognition, Object Recognition, Pose Estimation and so on, extensively use the texture feature to construct the feature vector. The importance of the texture feature can be understood by the fact that numerous texture feature descriptors have been proposed in the recent past. These descriptors have shown promising results for the extraction of low-level features in an image. The combination of these descriptors with computational intelligence techniques can not only help in improving retrieval accuracy, but also help in bridging the semantic gap.

REFERENCES

1. Rui, Y., Huang, T. S., & Chang, S. F. (1999). Image retrieval: Current techniques, promising directions, and open issues. *Journal of Visual Communication and Image Representation*, 10(1), 39–62.
2. Smeulders, A. W., Worring, M., Santini, S., Gupta, A., & Jain, R. (2000). Content-based image retrieval at the end of the early years. *IEEE Transactions on Pattern Analysis and Machine Intelligence*, 22(12), 1349–1380.
3. Yil̇dizer, E., Balci, A. M., Jarada, T. N., & Alhajj, R. (2012). Integrating wavelets with clustering and indexing for effective content-based image retrieval. *Knowledge-Based Systems*, 31, 55–66.
4. Mallat, S. G. (1989). A theory for multiresolution signal decomposition: the wavelet representation. *IEEE Transactions on Pattern Analysis and Machine Intelligence*, 11(7), 674–693.
5. Starck, J. L., Candès, E. J., & Donoho, D. L. (2002). The curvelet transform for image denoising. *IEEE Transactions on Image Processing*, 11(6), 670–684.
6. Do, M. N., & Vetterli, M. (2005). The contourlet transform: an efficient directional multiresolution image representation. *IEEE Transactions on Image Processing*, 14(12), 2091–2106.
7. Srivastava, P., & Khare, A. (2019). Content-based image retrieval using local ternary wavelet gradient pattern. *Multimedia Tools and Applications*, 78(24), 34297–34322.
8. Agarwal, S., Verma, A. K., & Singh, P. (2013, March). Content based image retrieval using discrete wavelet transform and edge histogram descriptor. In *2013 International Conference on Information Systems and Computer Networks* (pp. 19–23). IEEE.
9. Ashraf, R., Ahmed, M., Jabbar, S., Khalid, S., Ahmad, A., Din, S., & Jeon, G. (2018). Content based image retrieval by using color descriptor and discrete wavelet transform. *Journal of Medical Systems*, 42(3), 44.
10. Srivastava, P., Binh, N. T., & Khare, A. (2014). Content-based image retrieval using moments of local ternary pattern. *Mobile Networks and Applications*, 19(5), 618–625.
11. Park, S. B., Lee, J. W., & Kim, S. K. (2004). Content-based image classification using a neural network. *Pattern Recognition Letters*, 25(3), 287–300.
12. Xiaoling, W., & Kanglin, X. (2005). Application of the fuzzy logic in content-based image retrieval. *Journal of Computer Science & Technology*, 5(1), 19–24.
13. Torres, R. D. S., Falcão, A. X., Gonçalves, M. A., Papa, J. P., Zhang, B., Fan, W., & Fox, E. A. (2009). A genetic programming framework for content-based image retrieval. *Pattern Recognition*, 42(2), 283–292.
14. Wan, J., Wang, D., Hoi, S. C. H., Wu, P., Zhu, J., Zhang, Y., & Li, J. (2014, November). Deep learning for content-based image retrieval: A comprehensive study. In *Proceedings of the 22nd ACM international conference on Multimedia* (pp. 157–166).
15. Srivastava, P., & Khare, A. (2017). Integration of wavelet transform, local binary patterns and moments for content-based image retrieval. *Journal of Visual Communication and Image Representation*, 42, 78–103.
16. Pass, G., & Zabih, R. (1996, December). Histogram refinement for content-based image retrieval. In *Proceedings Third IEEE Workshop on Applications of Computer Vision. WACV'96* (pp. 96–102). IEEE.

17. Huang, J., Kumar, S. R., Mitra, M., Zhu, W. J., & Zabih, R. (1997, June). Image indexing using color correlograms. In *Proceedings of IEEE computer society conference on Computer Vision and Pattern Recognition* (pp. 762–768). IEEE.
18. Pass, G., Zabih, R., & Miller, J. (1997, February). Comparing images using color coherence vectors. In *Proceedings of the fourth ACM international conference on Multimedia* (pp. 65–73).
19. Murala, S., & Wu, Q. J. (2014). Expert content-based image retrieval system using robust local patterns. *Journal of Visual Communication and Image Representation*, 25(6), 1324–1334.
20. Murala, S., Maheshwari, R. P., & Balasubramanian, R. (2012). Directional local extrema patterns: a new descriptor for content based image retrieval. *International Journal of Multimedia Information Retrieval*, 1(3), 191–203.
21. Dubey, S. R., Singh, S. K., & Singh, R. K. (2016). Multichannel decoded local binary patterns for content-based image retrieval. *IEEE Transactions on Image Processing*, 25(9), 4018–4032.
22. Fadaei, S., Amirfattahi, R., & Ahmadzadeh, M. R. (2017). Local derivative radial patterns: a new texture descriptor for content-based image retrieval. *Signal Processing*, 137, 274–286.
23. Verma, M., & Raman, B. (2015). Center symmetric local binary co-occurrence pattern for texture, face and bio-medical image retrieval. *Journal of Visual Communication and Image Representation*, 32, 224–236.
24. Moghaddam, H. A., Khajoie, T. T., & Rouhi, A. H. (2003, September). A new algorithm for image indexing and retrieval using wavelet correlogram. In *Proceedings 2003 International Conference on Image Processing (Cat. No. 03CH37429)* (Vol. 3, pp. III–497). IEEE.
25. Sumana, I. J., Islam, M. M., Zhang, D., & Lu, G. (2008, October). Content based image retrieval using curvelet transform. In *2008 IEEE 10th workshop on multimedia signal processing* (pp. 11–16). IEEE.
26. Srivastava, P., & Khare, A. (2018). Content-based image retrieval using local binary curvelet co- occurrence pattern—a multiresolution technique. *The Computer Journal*, 61(3), 369–385.
27. Arun, K. S., & Menon, H. P. (2009). Content based medical image retrieval by combining rotation invariant contourlet features and fourier descriptors. *International Journal of Recent Trends in Engineering*, 2(2), 35.
28. Romdhane, R., Mahersia, H., & Hamrouni, K. (2008, April). A novel content image retrieval method based on contourlet. In *2008 3rd International Conference on Information and Communication Technologies: From Theory to Applications* (pp. 1–5). IEEE.
29. Tuceryan, M., & Jain, A. K. (1993). Texture analysis. In *Handbook of pattern recognition and computer vision* (pp. 235–276).
30. Dalal, N., & Triggs, B. (2005, June). Histograms of oriented gradients for human detection. In *2005 IEEE computer society conference on computer vision and pattern recognition (CVPR'05)* (Vol. 1, pp. 886–893). IEEE.
31. Yue, J., Li, Z., Liu, L., & Fu, Z. (2011). Content-based image retrieval using color and texture fused features. *Mathematical and Computer Modelling*, 54(3–4), 1121–1127.
32. Chun, Y. D., Kim, N. C., & Jang, I. H. (2008). Content-based image retrieval using multiresolution color and texture features. *IEEE Transactions on Multimedia*, 10(6), 1073–1084.
33. Fu, X., Li, Y., Harrison, R., & Belkasim, S. (2006, August). Content-based image retrieval using gabor-zernike features. In *18th International Conference on Pattern Recognition (ICPR'06)* (Vol. 2, pp. 417–420). IEEE.

34. Srivastava, P., Binh, N. T., & Khare, A. (2014). Content-based image retrieval using moments of local ternary pattern. *Mobile Networks and Applications*, 19(5), 618–625.
35. Tamura, H., Mori, S., & Yamawaki, T. (1978). Textural features corresponding to visual perception. *IEEE Transactions on Systems, Man, and Cybernetics*, 8(6), 460–473.
36. Howarth, P., & Rüger, S. (2004, July). Evaluation of texture features for content-based image retrieval. In *International conference on image and video retrieval* (pp. 326–334). Springer, Berlin, Heidelberg.
37. Zhou, F., Feng, J. F., & Shi, Q. Y. (2001, October). Texture feature based on local Fourier transform. In *Proceedings 2001 International Conference on Image Processing (Cat. No. 01CH37205)* (Vol. 2, pp. 610–613). IEEE.
38. Ojala, T., Pietikäinen, M., & Mäenpää, T. (2000, June). Gray scale and rotation invariant texture classification with local binary patterns. In *European Conference on Computer Vision* (pp. 404–420). Springer, Berlin, Heidelberg.
39. Tan, X., & Triggs, B. (2010). Enhanced local texture feature sets for face recognition under difficult lighting conditions. *IEEE Transactions on Image Processing*, 19(6), 1635–1650.
40. Vipparthi, S. K., & Nagar, S. K. (2015). Directional local ternary patterns for multimedia image indexing and retrieval. *International Journal of Signal and Imaging Systems Engineering*, 8(3), 137–145.
41. Agarwal, M., Singhal, A., & Lall, B. (2019). Multi-channel local ternary pattern for content-based image retrieval. *Pattern Analysis and Applications*, 22(4), 1585–1596.
42. Zhang, B., Gao, Y., Zhao, S., & Liu, J. (2009). Local derivative pattern versus local binary pattern: face recognition with high-order local pattern descriptor. *IEEE Transactions on Image Processing*, 19(2), 533–544.
43. Murala, S., Maheshwari, R. P., & Balasubramanian, R. (2012). Local tetra patterns: a new feature descriptor for content-based image retrieval. *IEEE Transactions on Image Processing*, 21(5), 2874–2886.
44. Liu, G. H., Zhang, L., Hou, Y. K., Li, Z. Y., & Yang, J. Y. (2010). Image retrieval based on multi-texton histogram. *Pattern Recognition*, 43(7), 2380–2389.
45. Liu, G. H., Li, Z. Y., Zhang, L., & Xu, Y. (2011). Image retrieval based on micro-structure descriptor. *Pattern Recognition*, 44(9), 2123–2133.
46. Wang, X., & Wang, Z. (2013). A novel method for image retrieval based on structure elements' descriptor. *Journal of Visual Communication and Image Representation*, 24(1), 63–74.
47. Zhao, M., Zhang, H., & Sun, J. (2016). A novel image retrieval method based on multi-trend structure descriptor. *Journal of Visual Communication and Image Representation*, 38, 73–81.
48. Khare, M., Srivastava, P., Gwak, J., & Khare, A. (2018, March). A multiresolution approach for content-based image retrieval using wavelet transform of local binary pattern. In *Asian Conference on Intelligent Information and Database Systems* (pp. 529–538). Springer, Cham.
49. http://wang.ist.psu.edu/docs/related/ Accessed April 2014.
50. Tiwari, A. K., Kanhangad, V., & Pachori, R. B. (2017). Histogram refinement for texture descriptor based image retrieval. *Signal Processing: Image Communication*, 53, 73–85.
51. Zeng, S., Huang, R., Wang, H., & Kang, Z. (2016). Image retrieval using spatiograms of colors quantized by Gaussian Mixture Models. *Neurocomputing*, 171, 673–684.
52. Alsmadi, M. K. (2020). Content-based image retrieval using color, shape and texture descriptors and features. *Arabian Journal for Science and Engineering*, 1–14.

3 Use of Computer Vision Techniques in Healthcare Using MRI Images

Sonali D. Patil, Atul B. Kathole, Kapil N. Vhatkar, and Roshani Raut

Pimpri Chinchwad College of Engineering, Pune, India

CONTENTS

3.1 Introduction 35
3.1.1 Difficulties and Opportunities 36
3.1.2 Obstacles in the Realm of Medical Imaging 36
3.2 Analysis of Medical Images 37
3.2.1 Typical Applications of AI in Medical Imaging Include the Following 38
3.3 Computer in Healthcare, Computer Vision 38
3.3.1 CV and AI in Health Imaging 38
3.4 Applications of Computer Vision in Healthcare 40
3.5 Critical Achievement Factor 45
3.6 Discussion and Conclusions 45
References 46

3.1 INTRODUCTION

It is fascinating to consider the technical advancements made in medicine over the past few decades. They have not only advanced our understanding of the architecture and functioning of the many tissues that comprise the anthropological body, but they have also aided in the early detection and action of a variety of disorders in a variety of fields of remedy. This has been achieved significantly through advances in both computer vision (CV) and artificial intelligence (AI). In a nutshell, these technologies enable us to capture, develop, analyze, and comprehend an endless number of stationary and dynamic pictures in real time, resulting in a complete understanding of each illness and a more accurate patient choice for early intervention.

Since many analytic methods that exist to date are hostile, luxurious, and too composite for standardization in the majority of the world, supported diagnosis via CV and AI represents a possible solution in enabling the early detection of a wide variety of diseases, improved behavior and follow-up, and a reduction in the well-being care costs associated with each patient.

DOI: 10.1201/9781003267782-3

The combination of high-performance computers and machine learning (ML) enables the accurate and efficient diagnosis of a sizeable amount of medical imaging data. AI enables the automated quantification of detailed medical images, resulting in improved diagnostic accuracy [1]. Additionally, AI and CV can eliminate considerable intra- and inter-observer inconsistency, undermining the clinical value of the results.

Over the past few decades, research into computer vision, image processing, and design recognition has achieved significant advances. In addition, medical imaging has gained increased devotion due to its critical role in well-being care applications. Researchers have released many fundamental research and data demonstrating the growth in medical imaging and its use in healthcare. Since these study fields have positioned clinicians to develop from the seat to the bedside, this volume will feature progressive computer vision approaches for health care manufacturing and appraisal articles that will inspire ongoing energies to comprehend the difficulties that frequently arise in this arena [2].

3.1.1 Difficulties and Opportunities

- Increasing demand (a growing number of senior people, new healthcare marketplaces).
- An unmistakable trend in the direction of reduced radiation exposure.
- Expensive equipment and facilities – critical for the community sector; private sector usage is also increasing.
- Inadequate radiologists – a strong case for automating analysis that is now performed manually.
- Significant growth in the amount, quality, and intricacy of imaging data – a critical requirement for information compression, packing, and lookup/access efficiency.
- Sharing information, skills, and consequences is challenging. At the moment, most information is locked away in PACS (picture archiving and communiqué scheme for DICOM-standard information (Digital Imaging and Transportations in Medicine)).

3.1.2 Obstacles in the Realm of Medical Imaging

- Investing in meaningful research requires a significant investment in development expenditures.
- Data collection for testing/training purposes may be complex/exclusive (hardware tends to be very exclusive).
- Clinical authentication of proposed procedures is required, time-consuming, and costly.
- It is necessary to adhere to a variety of local and global policies.
- Possibly relatively stagnant market – clients may find it difficult and expensive to integrate new services with current systems.

The primary objective of medical image analysis is to improve the efficiency of clinical examination and medical intervention, in other words, to see through the skin and bone into the internal organs and detect abnormalities:

- On the one hand, medical imaging delves into the inner workings of anatomy and physiology.
- On the other hand, medical image analysis enables the detection of anomalies and understanding their origins and consequences.
- Additionally, we highlight some significant areas addressed by the study. The first section discusses medical image investigation for healthcare. The second section discusses computer vision for predictive analytics and treatment. The third section discusses vital issues in medicinal pictures. The last section discusses machine learning methods for medical images [3].

3.2 ANALYSIS OF MEDICAL IMAGES

This aspect of the subject is concerned with advancing and developing novel approaches for medical image analysis. First, it is critical to integrate multimodal data obtained from several diagnostic imaging methods to complete the area under inspection. As a result, picture co-registration has become crucial in qualitative visual evaluation, and multiparametric investigation examines applications. Researchers have noted and have evaluated the presentation between the outdated co-registration approaches functional in terms of PET and MR developed as sole modalities and the revealed consequences with the indirect co-registration of a hybrid PET/MR, in composite anatomical areas such as the head and neck (HN).

The investigational findings indicate that hybrid PET/MR scans register more precisely than retrospectively coregistered pictures [4].

One of the most critical aspects of medical image analysis is feature extraction. In Turkey, a novel and compelling feature cooperative with a multistage cataloging scheme for breast cancer detection is developed. This ensemble is employed in a computer-aided diagnostic (CAD) arrangement for breast cancer analysis. Four characteristics were concatenated as feature vectors in this novel method: local outline pattern-based, arithmetic, and frequency area features. Eight well-known classifiers were employed in a multistage cataloging system. The high cataloging accuracy demonstrated that the suggested multistage cataloging scheme is more successful for breast cancer diagnosis than the single-stage classification.

Currently, the conventional method to diminish colorectal cancer-related death is to conduct routine transmission for polyps, resulting in a high number of polyp misses and an inability to evaluate polyp malignancy visually. In Spain and Canada, D. Vazquez et al. offer an expanded standard for colonoscopy picture separation and build a new robust standard for colonoscopy image examination in “A Benchmark for Endoluminal Scene Segmentation of Colonoscopy Images.” They demonstrate that training a conventional fully convolutional network (FCN) can outperform previous research results in endoluminal scene segmentation [5].

3.2.1 Typical Applications of AI in Medical Imaging Include the Following

Diagnostic Support: On a typical day, radiologists analyze hundreds of pictures. By emphasizing or quantifying the questionable region of the picture, an AI program may aid radiologists in saving time spent on diagnosis.

Screening and Triaging: When a radiologist has a long queue of photos to review, an AI program may assess and triage the images in PACS, ensuring that the essential cases get immediate attention.

Monitoring: To study therapy response (for example, in Oncology), pictures of sick tissues collected at several time points are aligned and compared. The size change of unhealthy tissue may help in determining the therapy response.

Charting: Following the medical practitioner's examination of the AI tool's results, the typical AI product often provides the essential inputs for clinical charting that would have to be manually entered otherwise.

3.3 COMPUTER IN HEALTHCARE, COMPUTER VISION

Computer vision is concerned with the comprehension of images and videos, and it entails tasks such as object identification, categorization, and segmentation of images. Recent developments in picture categorization and object identification significantly impact medical imaging. Numerous research has revealed encouraging outcomes in the challenging diagnostic tasks associated with dermatology, radiology, and pathology. Deep-learning procedures may assist clinicians by giving second opinions and detecting potentially problematic areas in photos [6].

Convolutional Neural Networks (CNN) have shown human-level performance in object classification challenges, in which a neural network learns to categorize the objects included inside an image. These convolutional neural networks (CNN) have exhibited excellent performance in transfer learning, a process in which a CNN is first trained on a large dataset (e.g., ImageNet) unrelated to the job at hand and then fine-tuned on a much smaller dataset relating to the task at hand (e.g., medical images) [7].

3.3.1 CV and AI in Health Imaging

According to Patel et al. [8], among the first efforts on artificial intelligence in medication originate from the 1970s, when AI was previously a well-established science and the name was coined at the illustrious 1956 Dartmouth College meeting [9]. While several investigators developed model-based image processing algorithms based on AI, they did not refer to them as such, even when their techniques were in this domain [8]. Since then, other claims have emerged, culminating in establishing the phrase artificial intelligence in medicine. Initially, the majority of approaches were based on proficient schemes. More freshly, Kononenko [10] emphasizes that the purpose of AI in medication is to increase the intelligence of computers, and one of the fundamental needs for intelligence is the capacity to learn. Machine learning is motivated by this principle, and some of the first practical procedures in medication were based on naïve Bayesian techniques [10].

AI is a vast field of mainframe science that focuses on developing automated ways for resolving issues that need human intellect in the traditional sense. CV can address a wide variety of topics, including picture segmentation, object identification, detection, and reconstruction. It tries to represent and comprehend the graphical environment by removing relevant data from digital pictures, often stimulated by complicated visual works performed by humans. While it has existed since the 1960s, it remains an unresolved and exciting issue, to the point that processers have just lately been able to deliver meaningful explanations in a variety of request domains. It is an interdisciplinary field of study that is inextricably linked to artificial intelligence.

On the other hand, machine learning is a subset of artificial intelligence that develops systems that can robotically learn from information and explanations. The utmost effective CV systems have been created using machine learning methods. Among the greatest extensively used machine learning techniques are Support Vector Machines (SVM), Arbitrary Forests, Regression (linear and logistic), K- Means, k-nearest neighbors (k-NN), Linear Discriminant Examination, Naive Bayes (NB), etc. [11].

Deep learning (DL) is a branch of machine learning that has developed an exciting and widespread method in CV due to its remarkable achievements in complicated challenges involving the processing and interpretation of visual input. The word "deep" refers to neural network models with numerous layers. There has been a surge in curiosity in applying deep learning representations to health concerns [10–12]. For instance, deep neural networks have demonstrated remarkable categorization tasks involving skin lesions. Annual competitions have many significant examples [13].

DL representations are instance: convolutional neuronal networks (CNNs), recurrent neuronal networks, long short-term remembrance, generative argumentative systems, etc. Current advancements in this discipline have shown astounding precision and quantifiable outcomes. Figure 3.1 depicts a graphic relationship between CV and AI in medicinal imaging [14].

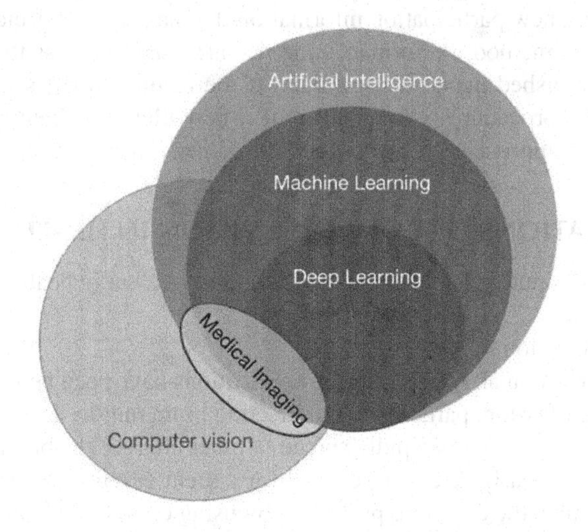

FIGURE 3.1 Relation between computer vision and artificial intelligence.

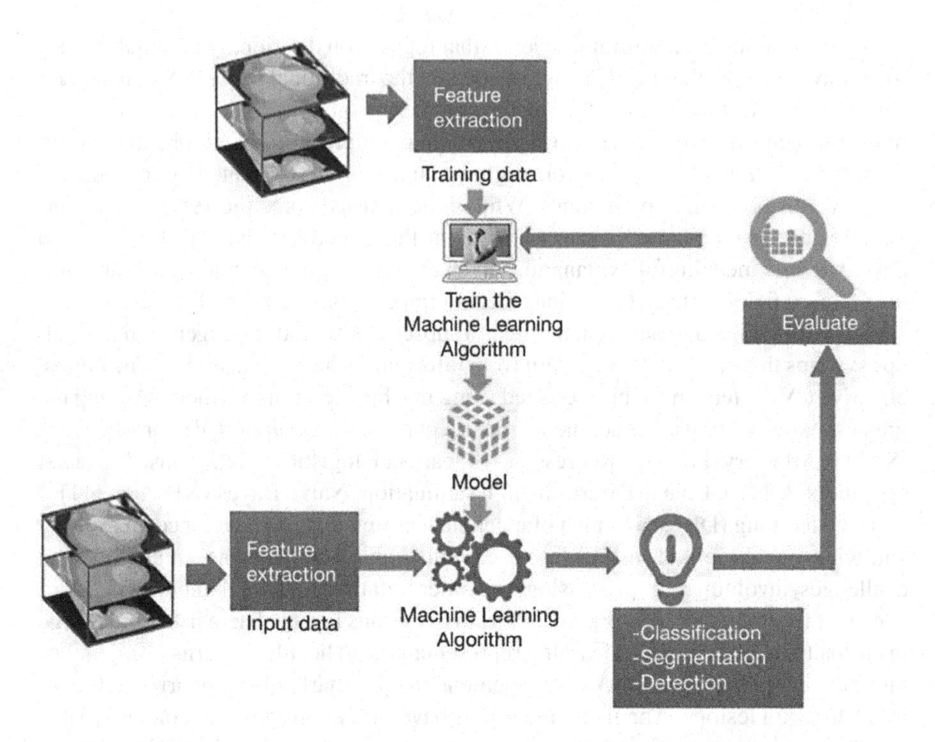

FIGURE 3.2 Working of image processing with CV machine learning approach.

In medical imaging, Figure 3.2 illustrates the interplay between CV and AI. Extracting significant features from medical picture databases is a vital initial step in training a new machine learning model. The training process is designed to produce a prototype that has mastered a given task such as separation, cataloging, recognition, and appreciation using the training information. Following that, the prototypical is validated using new participation information that has been subjected to the same feature extraction method. Performance metrics are used to assess the outcomes of the work accomplished using this test data. If the results do not satisfy the user's expectations, the procedure is frequent until a new blend of feature removal and machine learning approaches achieves the desired performance level.

3.4 APPLICATIONS OF COMPUTER VISION IN HEALTHCARE

The following computer vision applications are now present in healthcare:

1. Tumor Detection
   Computer vision and deep learning applications have been beneficial in the medical profession, particularly in detecting brain tumors accurately. If left untreated, brain tumors rapidly spread to other areas of the brain and spinal cord, making early discovery critical for patient survival. Medical experts may automate the detecting procedure by using computer vision tools.

In healthcare, computer vision methods such as Mask-R Convolutional Neural Networks (Mask R-CNN) may significantly reduce the probability of human error in the identification of brain cancers.

2. Medical Imaging

Computer vision has been employed in various healthcare applications to aid physicians in making more informed treatment choices. Medical imaging, or medical image analysis, is one technique that visualizes specific organs and tissues to facilitate a more precise diagnosis.

With medical image analysis, clinicians and surgeons may better view the patient's interior organs to detect any problems or anomalies. Medical imaging encompasses a variety of specialties, including X-ray radiography, ultrasound, MRI, and endoscopy.

3. Cancer Detection

Surprisingly, deep-learning computer vision models have attained physician-level accuracy in diagnosing moles from melanomas. Skin cancer, for example, is notoriously difficult to identify early because the symptoms often mimic those of common skin conditions. As a result, scientists have developed computer vision programs that successfully distinguish between malignant and non-cancerous skin lesions [15].

Additionally, research has shown various benefits to diagnosing breast cancer utilizing computer vision and deep learning systems. It may assist automate the detection process and lessen the likelihood of human mistakes by being trained on an extensive collection of photos, including healthy and malignant tissue.

With fast technological advancements, healthcare computer vision systems may soon be utilized to diagnose more forms of cancer, such as bone and lung cancer.

4. Medical Training

Computer vision is frequently employed, not just for medical diagnosis but also for skill training in medicine. Currently, surgeons do not acquire abilities just via real experience in the operating room. Rather than that, simulation-based surgical platforms have established themselves as a viable medium for teaching and evaluating surgical skills.

Trainees have the chance to practice their surgical skills before entering the operating room via surgical simulation. It enables them to get extensive feedback and evaluation of their performance, helping them better understand patient care and safety before performing surgery on them.

Computer vision may also be used to assess the surgical quality by caring activity stages, detecting desperate measures, and analyzing time consumed by individuals in certain places.

5. Combating Covid-19

The epidemic of Covid-19 has created a significant challenge to the worldwide healthcare system. With nations all over the globe battling illness, computer vision has the potential to make a substantial contribution to overcoming this obstacle.

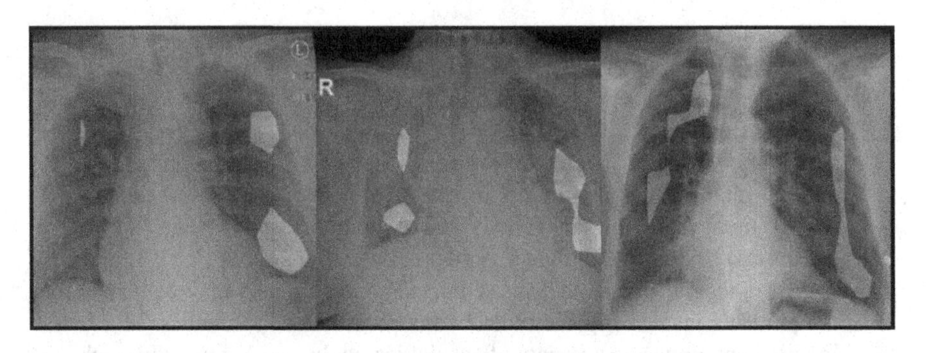

FIGURE 3.3 Computer vision to detect specific features of Covid-19.

Computer vision applications may assist in diagnosing, managing, treating, and preventing Covid-19 due to fast technology breakthroughs. When combined with computer vision technologies such as COVID-Net, digital chest x-ray radiography pictures may readily diagnose illness in patients. The prototype application, built by Darwin AI in Canada, achieved a covid diagnostic accuracy of about 92.4 percent.

Masked face identification, which is frequently used to enforce and monitor policies to limit the transmission of pandemic illnesses, is performed using computer vision Figure 3.3.

For additional information, see our post on Eight Computer Vision Applications for Coronavirus Control in 2021.

6. Health Monitoring

Medical practitioners rely on computer vision and artificial intelligence to track their patients' health and fitness. Doctors and surgeons may make more informed judgments in less time, especially during crises, with the assistance of these analyses.

Computer vision models may be used to quantify the volume of blood lost during procedures to detect whether the patient has progressed to a critical stage. Gauss Surgical's Triton program is one such application that efficiently monitors and calculates blood loss during surgery. It assists surgeons in determining how much blood the patient will need during or after surgery.

7. Machine-assisted Diagnosis

In recent years, computer vision advancements in healthcare have resulted in more precise disease diagnoses. Computer vision improvements have shown that they are superior to human specialists in recognizing patterns that accurately identify illnesses.

These technologies are advantageous because they enable clinicians to diagnose malignancy by identifying minute changes in tumors. Such instruments may assist in discovering, preventing, and treating a variety of illnesses by scanning medical photographs.

8. Timely Detection of Disease
For specific disorders, such as cancer and tumors, quick identification and treatment are critical to the patient's survival. Early detection increases a patient's chances of survival.
Computer vision programs are trained on massive volumes of data comprised of hundreds of photos, enabling them to detect even the tiniest change accurately. Consequently, medical practitioners can spot such minute alterations that would have gone unnoticed by their eyes otherwise.
9. Home-based Patient Rehabilitation and Monitoring
Numerous individuals choose to recover at home rather than in a hospital after an illness. Medical practitioners may digitally provide physical therapy to patients and monitor their progress using computer vision apps. Home training is more convenient, but it is also more cost-effective.
A well-researched subject is computer vision-based fall detection, intending to reduce reliance and care costs in the senior population using deep learning-based fall detection systems. Additionally, computer vision technology may assist in the non-intrusive monitoring of patients and the elderly. The author of this article suggests reading our in-depth post on Fall Detection: A Vision Deep Learning Application for further information on this subject.
Another approach of monitoring patients using computer vision is to analyze standardized medical exams such as the TUG test using video-assisted analysis (Timed Up and test). The computer vision system determines the time required to complete a basic evaluation test to determine functional mobility. The TUG test may be used to determine a person's risk of falling and their ability to walk with balance [16].
10. Machine Learning Procedures for Medicinal Images
There is a rapid global increase in the number of elderly adults, and it will have a significant influence on the healthcare structure. Elders could not historically care for themselves, which is why healthcare and nursing robots have garnered considerable interest in current years. While somatosensory knowledge has been integrated into identifying geriatric action and engagement with healthcare professionals, conventional detection methods are usually monomodal. To advance an effective and suitable communication subordinate scheme for nurses and affected persons with dementia. It suggest two innovative multimodal scant autoencoder outlines based on wave and intellectual structures. After preprocessing the depth picture, motion is removed, and then EEG signals are captured to determine the mental characteristic. The suggested innovative approach is based on multimodal deep neuronal networks and is intended for patients with specific requirements who have dementia. The input characteristics of the networks include the following: (1) Using the depth image sensor, we retrieved motion characteristics; and (2) EEG structures. The yield layer is responsible for determining the assistance required by the affected person. The experimental findings demonstrate that the suggested method streamlines the appreciation process and achieves 96.5 percent and 96.4 percent

(precision and recall rate) for the scrambled information set, respectively, and 90.9 percent and 92.6 percent, respectively, for the unceasing information set. Additionally, the suggested algorithms streamline data collecting and dispensation while maintaining a higher action recognition ratio than the conventional technique.

N. D. Kamarudin et al. in Malaysia and Japan "A Fast SVM-Based Tongue's Color Arrangement Aided by k-Means Clustering Identifiers and Color Attributes as Computer-Assisted Tool for Tongue Diagnosis" suggest a two-stage organization scheme for tongue color analysis assisted with the planned gathering identifiers. The diagnostic technique is very beneficial for detecting imbalances inside the body in its initial stages [17]. It can identify three tongue colors: red, light red, and deep red.

11. Analytical analytics and treatment with computer vision
Computer vision techniques have shown significant use in surgery and illness treatment. Recently, fast prototyping and three-dimensional (3D) modeling skills have fueled the expansion of therapeutic imaging modalities such as CT and MRI. P. Gargiulo et al. in Iceland, "New Directions in 3D Medical Modeling: 3D-Printing Anatomy and Functions in Neurosurgical Planning". The authors deliver an excellent treatment strategy for advanced neurosurgery research.

The old are prone to reduce, which may cause physical hurt and therefore have severe negative psychological consequences. In Taiwan, T. H. Lin et al.'s "Fall Prevention Shoes Using Camera-Based Line-Laser Obstacle Detection System" introduces an intriguing line-laser obstacle recognition system for preventing senior falls. A laser line travels through a flat plane at a specific elevation above the pounded in the scheme. The optical axis of a photographic camera is inclined at a predetermined angle to the flat, allowing the camera to examine the laser design and detect possible impediments. Regrettably, this system is intended for indoor usage only and is unsuitable for outdoor use.

Identifying human activity (HAR) is a well-studied subject in computer vision. In China, S. Zhang et al.'s "A Review on Human Activity Recognition Using Vision-Based Method" summarises different HAR methodologies and their evolutions compared to typical classical literature. The authors discuss the advancements in picture representation and classification techniques for vision-based activity identification. Global, local, and depth-based representations are the most frequently used methodologies. They classify human actions into primitives, actions/activities, and communications.

Additionally, they outline the cataloging strategies used in HAR applications, which comprise seven different classification algorithms, ranging from the traditional DTW to the most recent deep learning. Finally, they examine the difficulty of implementing existing HAR techniques in real-world systems or applications, despite the recent success of HAR approaches. Additionally, their study recommends three following directions.

3.5 CRITICAL ACHIEVEMENT FACTOR

Accuracy: Numerous datasets have previously been used to train AI products. However, all Deep Learning-based solutions exhibit 'bias' due to the training data employed. For each new installation, it is critical to verify that the AI/ML product is accurately forecasted.

Seamless Integration: A self-contained artificial intelligence product is impossible, and it must work in unison with the rest of the application environment. To maximize efficiency, the integration should be well-designed. People do not want the doctor to remain glued to the screen for an extended time while the photographs are uploaded.

Training: When physicians use the device, they are also taught how to use it, and they must be educated about proper product use.

Productivity Metrics: Typically, the business case is built on productivity gains. Metrics should be included in the integration or inside the AI solution to confirm productivity growth.

Data Security: The technology of artificial intelligence and computer vision has been widely democratized. Numerous nations have successful AI product firms. While the AI vendor may be registered in their respective country, it is critical to verify the product's hosting location and the location of developers who will have access to their data.

Medical imaging may benefit from computer vision-based AI/ML algorithms to increase physician efficiency. These apps may assist with diagnostics to a significant degree. At the moment, these artificial intelligence solutions cannot take the position of a radiologist or pathologist. Additionally, regulatory authorities such as the Food and Drug Administration permit these goods in diagnostic procedures.

3.6 DISCUSSION AND CONCLUSIONS

This chapter discussed various CV and machine learning instances effectively employed in medical applications across multiple organs, diseases, and imaging modalities. This knowledge can improve community health by expanding the capacities of general practitioners and standardizing decision-making processes, particularly in places with a shortage of medical experts in community clinics. While these benefits have pushed for continuous updating of performance data across several apps, specific common difficulties need addressing. While several databases have arisen over the past decade, there is still a necessity for more data in the medicinal profession; databases with skilled annotated ground truth to develop and execute these approaches remain critical. Benchmark publicly available datasets may be very helpful for comparing current methodologies, identifying good tactics, and assisting researchers in developing new ways. The absence of existing benchmark databases is due to various variables, including clinical environment limits and a shortage of medical specialists prepared to annotate vast volumes of data. The latter is complex since manual segmentation is time-consuming and disposed to mistakes due to inter and intraobserver inconsistency.

Although an substantial amount of energy has been put into this area of research, with many years of effort and significant financial properties, there is still only a limited availability of adequate and balanced information to assess the performance of various methods applied in the medicinal field, in comparison with the abundance of widely available information sets in other areas, such as ImageNet, COCO, and Google's Open Images. While it is evident that new medical procedures are required to address the issues mentioned above, it is also critical to stimulate study on new AI approaches that are less reliant on large amounts of information and are less computationally intensive. While transmission learning and information augmentation are often employed to solve limited datasets, studying how to enhance cross-domain and crossmodal learning and expansion in the therapeutic area. Meta-learning is a potential ML paradigm, and this encompasses several techniques to apply previously acquired information to specific activities. This creates new opportunities for medical imaging to overcome the limitation of information sets. For example, a prototypical trained to categorize anatomical assemblies in a particular modality, such as CT, may use that information to organize the identical assemblies in other modalities, such as MRI or ultrasound. Similarly, a prototypical trained to section a specific assembly, such as a heart void, may utilize that information to study other openings to an area without building a new prototype from the start.

Additionally, advances in computer power have increased the popularity of these approaches, posing new difficulties to the systematic community in creating and applying completely mechanized real-time experimental duties to assist with complicated analysis and treatments.

REFERENCES

1. Sunnybrook Cardiac Data 2009. Cardiac MR Left Ventricle Segmentation Challenge. Available online: http://www.cardiacatlas.org/studies/sunnybrook-cardiacdata/. [Online; accessed: January 26 2020].
2. CAMUS Database. Available online: https://www.creatis.insa-lyon.fr/Challenge/camus. 2019. [Online; accessed: December 1 2019].
3. Carbajal-Degante E, Avendaño S, Ledesma L, Olveres J, Escalante-Ramírez B. Active contours for multiregion segmentation with a convolutional neural network initialization. *SPIE Photonics Europe Conference*, 2020:36–44.
4. Avendaño S, Olveres J, Escalante-Ramírez B. Segmentación de Imágenes Médicas mediante UNet. In: Reunión Internacional de Inteligencia Artificial y sus Aplicaciones RIIAA 2.0, Aug 2019.
5. Zhang J, Gajjala S, Agrawal P, Tison GH, Hallock LA, Beussink-Nelson L, Lassen MH, Fan E, Aras MA, Jordan C, Fleischmann KE, Melisko M, Qasim A, Shah SJ, Bajcsy R, Deo RC. Fully automated echocardiogram interpretation in clinical practice. *Circulation* 2018;138:1623–1635.
6. Mira C, Moya-Albor E, Escalante-Ramírez B, Olveres J, Brieva J, Venegas E. 3D hermite transform optical flow estimation in left ventricle CT sequences. *Sensors (Basel)* 2020;20:595.
7. World Health Organization. Cardiovascular diseases (CVD) 2019. Available online: https://www.who.int/health-topics/cardiovascular-diseases/. [Online; accessed: May 11 2020].

8. Patel VL, Shortliffe EH, Stefanelli M, Szolovits P, Berthold MR, Bellazzi R, Abu-Hanna A. The coming of age of artificial intelligence in medicine. *Artif Intell Med* 2009;46:5–17.
9. Moor J. The Dartmouth College Artificial Intelligence Conference: The next fifty years. *AI Magazine* 2006;27:87.
10. Kononenko I. Machine learning for medical diagnosis: history, state of the art and perspective. *Artif Intell Med* 2001;23:89–109.
11. Krizhevsky A, Sutskever I, Hinton GE. ImageNet Classification with Deep Convolutional Neural Networks. In: Pereira F, Burges CJC, Bottou L, Weinberger KQ, editors. *NeurIPS Proceedings (2018).* Curran Associates, Inc., 2012:1097–1105.
12. Shen D, Wu G, Suk HI. Deep learning in medical image analysis. *Annu Rev Biomed Eng* 2017;19:221–248.
13. The ISIC 2020 Challenge Dataset. 2020 Jun. [accessed 29 Jun, 2020]. Available online: https://challenge2020.isicarchive.com/
14. Cao Z, Duan L, Yang G, Yue T, Chen Q. An experimental study on breast lesion detection and classification from ultrasound images using deep learning architectures. *BMC Med Imaging* 2019;19:51.
15. Chiao JY, Chen KY, Liao KY, Hsieh PH, Zhang G, Huang TC. Detection and classification the breast tumors using mask R-CNN on sonograms. *Medicine (Baltimore)* 2019;98:e15200.
16. Mahbod A, Ecker R, Ellinger I. Skin lesion classification using hybrid deep neural networks. arXiv preprint arXiv:170208434. 2017.
17. Harangi B. Skin lesion classification with ensembles of deep convolutional neural networks. *J Biomed Inform* 2018;86:25–32.

4 Hierarchical Clustering Fuzzy Features Subset Classifier with Ant Colony Optimization for Lung Image Classification

Leena Bojaraj and R. Jaikumar
KGiSL Institute of Technology

CONTENTS

4.1 Introduction 49
4.2 Literature Review 51
4.3 System Design 52
4.4 Result and Discussion 56
4.5 Conclusion 60
References 60

4.1 INTRODUCTION

Data mining reveals an excessive volume of the dataset through a thorough analysis resulting from the unmanageable growth of global data. The hidden patterns and concealed relationships between the variables have been revealed thanks to data analytics. Data are accumulated in every facet of life in the digital era, thanks to the growth of computerised database systems. To investigate and extract hidden knowledge from the lung image dataset, knowledge extraction and representation approaches are routinely used. In the medical field, the accuracy of the disease diagnosis plays a vital role as it leads to further treatment of the patient (Uppaluri et al., 1997). So the prime objective dissertation is to improve the diagnostic accuracy of the medical expert system by lung image datasets (Figure 4.1),

- Employing feature optimization techniques to select most significant feature subset in the medical data (Kalimuthu, 2021).

DOI: 10.1201/9781003267782-4

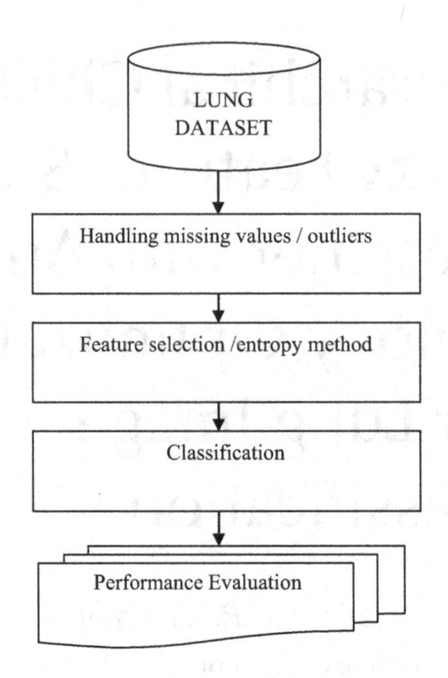

FIGURE 4.1 Lung dataset classification system framework.

- Constructing various classifier models (two-class) to train and test the clinical data (Amin et al., 2013).
- Optimizing classifier parameters and fuzzy rules by using single and hybrid optimization techniques (Anifah et al., 2017).

Machine learning and data mining techniques have to be customised in an efficient manner in order to ensure complete capability of gathered data. The lungs are protected by the pleura and the thin fluid act as a lung smoothly which assists expansion while breathing. For example, in areas such as medical science and astronomy there is widespread use of the predictive ability of big data and data mining. The third party resources do almost all computations on private data, leading to concerns with regard to user data privacy (Alam et al., 2018). Predictive analytics answers the query "What could happen?" by understanding and estimating the forthcoming issues using statistical methods and various forecasting methods. It uses techniques like machine learning, statistics and data mining for predicting the future (Potghan et al., 2018; Kuncheva, 2004). Prescriptive analytics answers the query "What should we do?" by complicated data received from the descriptive and predictive analyses. This model detects the best alternative to minimize or capitalize marketing, finance and other sectors through the use of optimization methods. For example, one selects prescriptive analytics if one wants to make the choice of a perfect way to the carriage of stuff from one industry to another location at a minimal cost (Tafti et al., 2018).

There is a rapid increase in the death rate of patients dying from different types of cancer and lung infection. Early diagnosis can reduce the death count. Effective diagnostic methods and tools such as medical expert systems and improvements in treatment methodologies have saved a considerable number of lives (Jayaraj and Sathiamoorthy, 2019). The survival duration of patients after proper diagnosis and treatment has increased. The researcher has proposed the integrated method HCFFSCACO for overcoming the challenge and providing the effective classification of microarray gene datasets inclusive of efficiency maintenance (Polat & Güneş, 2007) and (Detterbeck et al., 2017). The layout of this chapter is as follows. Section 4.2: Predictions and related research based on big data medical datasets The Hybrid Hierarchical Clustering Feature Subsets Classifier with Ant Colony Optimization Algorithm is discussed in Section 4.3, and the experimental results of the proposed and present systems are compared in Section 4.4. Finally, the work's conclusion and future scope are presented in the Section 4.5.

4.2 LITERATURE REVIEW

Many recent types of research are focused on secure classification due to the rapidly increasing development of internet usage throughout the information world. The pattern recognition and data mining methods employment in risk prediction systems in the domain of cardiovascular medicine was introduced by Peter & Somasundaram (2012). There were a few restrictions in the usual medical scoring systems; there was an occurrence of intrinsic linear variable input set combinations and therefore these were not adapted to model nonlinear difficult medical domains interactions. These restrictions had been tackled here by means of a classification pattern that indirectly identifies difficult nonlinear affiliations among dependent and independent variables and the capability to identify each probable interactions among predictor variables.

Anthimopoulos et al. (2016) discussed the realistic issue of Chinese hospital handling with cardiovascular patients' data to create an early detection and prediction risk. To consider entire multi-techniques benefits and minimized bias, top 6 subclassifiers had been chosen to structure an ensemble system; a regulated voting system had been employed to create final consequences that composed of risk prediction and poise. The system revealed a high degree of accuracy of 79.3 percent for 2628 instances experiments of authentic patients. The risk prediction confidence and algorithm precision had revealed greater importance in practical usage for doctors' diagnosing.

Shen et al. (2015) suggest an innovative image annotation method that scales a huge amount of keywords and it is a speedy and efficient scheme. The performance analysis explicit that for a large amount of keywords the proposed methods scales up in annotation output accuracy with minimum run time. This method can be improved through the inclusion of additional training data and increased conditional probabilities for annotating images that have mutually exclusive hints. Further, Saxena and Sharma (2015) wanted to create a system that could successfully discover the rules for predicting a patient's risk level based on a health parameter in lung diseases. The rules were ranked according to the requirements of the user. The system's performance was evaluated using precision classification, and the results suggested that this system had a higher capability for exact levelling. Radhimeenakshi (2016), using different machine learning algorithms such as SVM and ANN, was able to combine disease dataset classifications. On the basis of accuracy and training duration, an investigation was done between two methods. The Cleveland Database and the Statlog Database were used, both of which were retrieved from the UCI Machine Learning dataset vault. SVM and ANN were used to split the data into two classes. The study also looked at the results from both datasets (Nadkarni and Borkar 2019).

Wijaya and Prihatmanto (2013) employed machine learning to forecast cancer disease development. Data were collected using devices such as smartphones and smart chairs. Data on cancer rates were gathered on a server using the Internet. To gather enough data for forecasts, system approaches were used for a year. Over the course of a year, potential heart disease forecasts boosted a person's knowledge of heart disease. This approach was also meant to reduce the number of patients who died from lung disease. Sabab et al. (2016) and Rajathi and Radhamani (2016) used different data mining methods to optimise the study of lung cancer and pneumonia disease prognosis. The authors propose a feature selection strategy for improving the projected classifier pattern. SMO, Nave Bayes, and C4.5 Decision Tree algorithms achieved precision rates of 87.8 percent, 86.80 percent, and 79.9 percent, respectively, thanks to feature selection approaches that helped to improve precision by eliminating a few low-ranked features. Kalimuthu et al. (2021) have suggested a schema to attain an image density dataset main impact in the classification of machine learning in which the training data are scattered and every piece of distributed data is of a huge volume. The proposed schema bypasses security process in mappers and in Reducer minimum amount of cryptographic process is used to attain the preservation of privacy with reasonable computation cost. The limitation here is that the distributed feature selection process is not to be able to be achieved by this schema. Rodrigues et al. (2018) is made up of two of the most powerful data mining tools: neural networks and genetic algorithms. To initialize neural network weights, the hybrid system used the global optimization merits of a genetic algorithm. When measured across backpropagation, the learning was quick, consistent, and exact. The Matlab-based algorithm accurately predicted cardiac illness with an accuracy of 89 percent. Disease prediction data mining approaches, feature selection techniques, classifiers techniques, and optimization strategies were all summarized in this literature. Big data mostly deal with unlabelled data. In such cases, proper feature selection and feature extraction have compelling roles in which researchers fail. The consideration of heterogeneous data instead of homogeneous data for the classification of the lung dataset is needed. The focus of researchers is on improving the classification accuracy rather than giving prime importance to the privacy of the data. Hence this necessitates the investigation of the privacy-preserving algorithm with the suitable feature selection and feature extraction algorithms to classify heterogeneous data in the data mining and machine learning era.

4.3 SYSTEM DESIGN

In this research we presents a study to design and evaluate approaches to handle missing values, attribute noise and imbalanced class distribution in datasets to Predict. In this section, a brief description in HCFFSCACO in knowledge discovery is presented. The goal of this step is to choose the best classification approach for a given lung image dataset. Because no generalization can be made about the optimal classification approach, including this step has mandated the necessity to test each and every prediction and analysis for a given dataset empirically. Classification is the finding of a model for describing as well as distinguishing the classes or the concepts of data for being able to utilize the model for predicting of the class of the object class labels that are unknown. This model is further based on data object analysis that has known class labels. There are various techniques of classification in data mining, including HCFFSC (Abadeh et al. (2008) and Meenachi and Ramakrishnan 2020).

The HCFFSC architecture, the actual number of codes to be chosen and how the weights have to be set between the features in datasets at the time of training and evaluation of results are all completely covered. The function of activation is mentioned, together with the rate of learning, the momentum and the pruning Kumar et al. (2015) and Cai et al. (2015). The HCFFSC can work on errors better than that of the traditional computer programs (as in a scenario of a faulty statement in the program which can halt everything when the HCFFSC will handle errors better using features such as subset selections). Here in this work the optimized ACO, along with the HCFFSC and the ACO, are proposed. The blooming prominence and advancements seen in machine learning in the latest generation have inspired researchers to have a comprehensive investigation. There are various data mining issues for data classification, which need to be focused on. Thus, the handling of such a huge volume of lung patient reports is considered to be a substantial undertaking and demonstrates that the existing method, Hierarchical Clustering Fuzzy Features Subset Classification (HCFFSC), is perfectly appropriate for the rapid handling and classifications of large volume of lung patient records. The feature selection types we come up with a filter-based feature selection method HCFFSC that detects worthy feature subsets to the hierarchical learning classification model for enhancing the classification performance of medical dataset. After the feature subset selection further data feature selection is an optimization problem, which is based on the principle of picking a subset of attributes which are most significant in deciding the class label, It reduces the dimension of the data. During the training process, the presence of instances with missing values can lead to the degradation of accuracy and the performance of the classification model. By dealing with these missing values appropriately the performance of the model can be improved. Case Deletion is a simple and commonly used missing value handling techniques which is used to delete the instances with missing values. Ant Colony Optimization (ACO), a non-greedy local heuristic approach, is used to solve optimization issues. Because of its generous nature, this algorithm can achieve the global maxima without getting struck into local ones. It derives its name from the metallurgical annealing process, a technique that involves heating followed by controlled cooling of a material like steel so as to increase the size of the crystals. The proposed HCFFSC is a filter-dependent feature selection model for classification of lung data in patient reports. It consists of two key steps, namely, Highly Correlated Fuzzy Classification (HCFS) utilized for identifying appropriate feature subset to the classifier and the hierarchical learning for efficient classification of the feature subset driven from HCFFSC. The HCFFSC method flow diagram is exhibited in Figure 4.2. The entropy and conditional entropy are estimated for every feature set detected from the lung dataset for the measurement of symmetric uncertainty (SU). The HCFFSC algorithm efficiency was incorporated

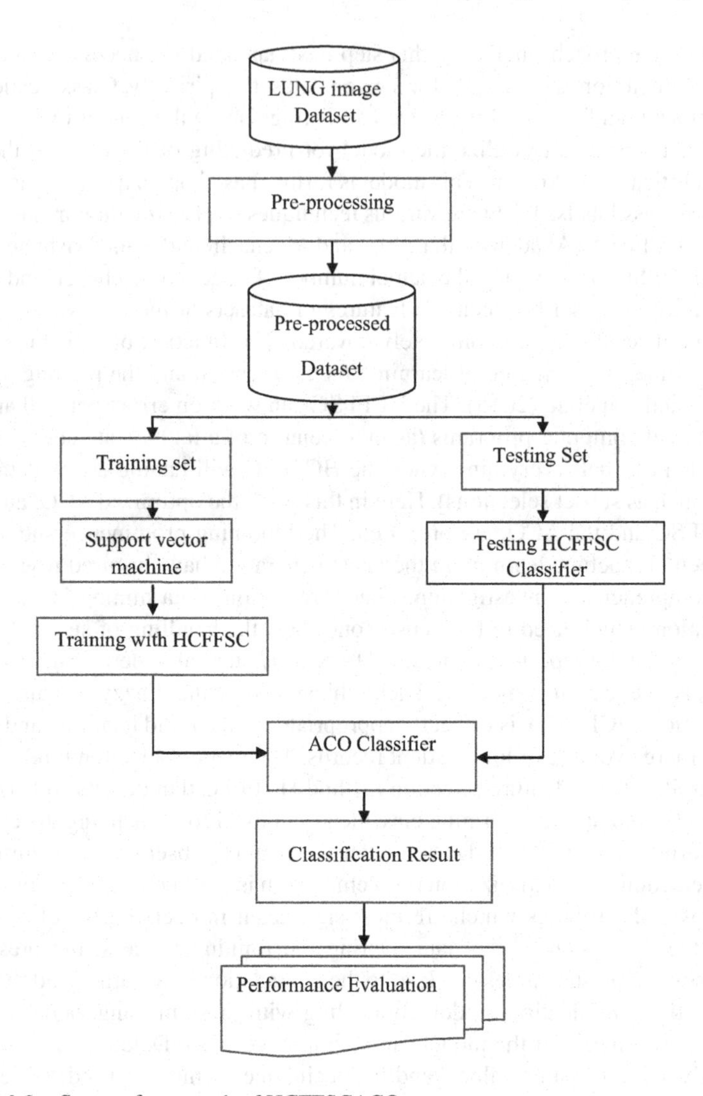

FIGURE 4.2 System framework of HCFFSCACO.

with optimization to predict lung disease. The investigation had been carried out in two phases. The dataset employed here was Streptococcus Pyogenes bacteria that cause pneumonia fever, as Acute Rheumatic Fever (ARF). A novel algorithm Hierarchical Clustering Fuzzy Classification had been incorporated in the present approach and the same was examined on the basis of precision and error rate Dhasal et al. (2012) and Alwan and Ku-Mahamud (2013).

Figure 4.3 explains the hyper-cube framework for ant colony optimization (HCFSACO) algorithm pseudocode for implementation, the purpose of eliminating redundant features lies in reducing the running time of clustering activity and it is a complex process. The purpose of eliminating irrelevant features has no association with the target category. Hence, in the HCFFSC method, these two difficulties are primarily focussed for their elimination. For the data points, such as patient records,

Input: Medical Data set (*Ds*) with the features and the class label

Output: Feature Sub-Set For HL Algorithm

Primary Process: //Predetermine the relevance threshold ( )

(Ds)=Threshold; Detect association among feature and class, Check Symmetric Uncertainty with threshold, Expel redundant features.

Tree Construction:

Find association (Correlation) between features and Characterize features as nodes

Characterize correlation values as edge weight

Construct MST using algorithm

Expel irrelevant features

Expel irrelevant features

Tree Partitioning:

Correlate F – correlation value by predetermined threshold value

If F – correlation < t Then eliminate the relevant edge

Feature sub-set selection:

Design cluster based on the feature relevancy

Cluster head is formed by Feature with highest cluster prominence

Deliver resulting feature sub-sets

FIGURE 4.3 Pseudocode for Hierarchical Clustering Feature Subset with ant colony optimization algorithm.

that are linearly separable, it produces classified output with maximum margin. The hierarchical learning technique is chosen in this work over other choices of conventional techniques due to its prominent suitability towards ‘5V’ characteristics of big data. It attempts to reduce the tradeoff between various performance metrics and related big data characteristics, even in higher dimensions Nóbrega et al. (2018) and Verleden et al. (2014).

If a searching procedure in the GA is executed, the ranger or the scrounger will have chances of discovering a location that is better and the current producer of other members will fail to discover a better location (Akanskha et al., 2021). The factor of constriction is that one other variable that ensures convergence and overfitting being the problem acquires more specifications while training. The ranger or the scrounger having a better location in the next session and the producer and the other members in the previous search session carries out the activity of scrounging (Vieira et al., 2007). This fitness function is designated to ith individual is a least-squared error function as per Equation 4.1:

$$F_i = \frac{1}{2}\sum_{p=1}^{P}\sum_{k=1}^{K}(d_{kp} - y_{kp}^{i})^2 \tag{4.1}$$

The error in the training set can be driven to a small value by means of minimizing the error function but as its side effect, the problems of overfitting may sometimes occur and result in a generalization error which may be large. So for improving the performance of the performance of the ACO the previous stopping strategies are suggested. The rate of error validation has been observed during the training period. If the error of validation takes place for a particular set iterations process the training to final.

4.4 RESULT AND DISCUSSION

The proposed methodology is applied by making use of PYTHON3.6IDE on Intel(R) Core (TP) i3-2410M CPU @ 3.20GHz and 8GB RAM. If no prior knowledge is available, result validation is performed in the dataset before and after feature selection using learning algorithms. All of the comparison is done by evaluating the classification performance of the reducts. Accuracy is measured where is perfectly identified count of true positive records and is the absolute count of positive records for infected person category. The performance analysis in Table 4.1 exhibits the enhanced False positive, True positive, and Fscore of HCFFSCACO-Hierarchical clustering method over other existing methods. The pictorial representation of performance analysis is shown in Figure 4.4. The cancer imaging archive (tcia) lung dataset are used to train and test the results. Figure 4.4 illustrates the input image as enhanced by a suitable median filter and segmented results, meaning that finally the classification obtained results are displayed.

Figure 4.4(a) shows the present pre-processed output results using a mean median filter for lung input image, and Figure 4.4(b) shows the k-means segmentation results for lung pre-processed results; finally, our proposed HCFFSCACO results are presented sequentially. The corresponding performance measures are listed and explained further in Tables 4.1 to 4.3.

Table 4.1 explains the model based on the proposed HCFFSCACO yields the maximum false positive rate fscore, and the true positive rate for lung cancer and lung-infected datasets (81.32, 91.25, and 84.12), at the nearest Support vector machine provides the only 54.85, 87.54, and 71.78 values comparatively proposed approach provides the better result in the difference terms between of 27.12, 4.25, and 12.87 values.

TABLE 4.1
Performance Comparison Lung Dataset

Methods	False Positive	True Positive	F-score
Hierarchical clustering Fuzzy features subset with ACO Classification	31.01	78.45	54.15
Support vector machine	43.65	81.87	61.78
Neural network	54.85	87.54	71.78
Decision tree	81.32	91.25	84.12

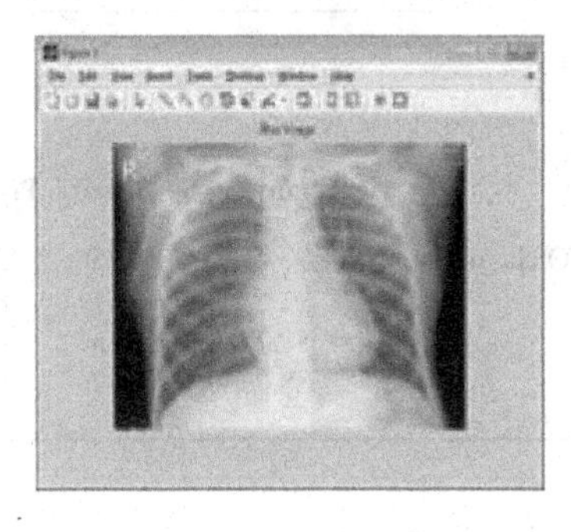
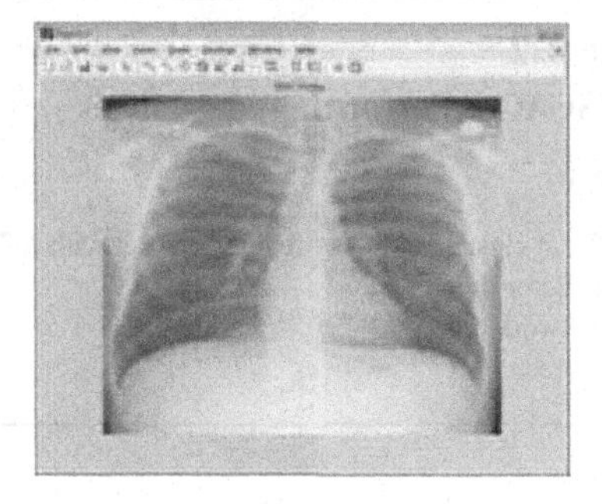

4a. Input image s for (a) cancer and (b) lung infection lung dataset

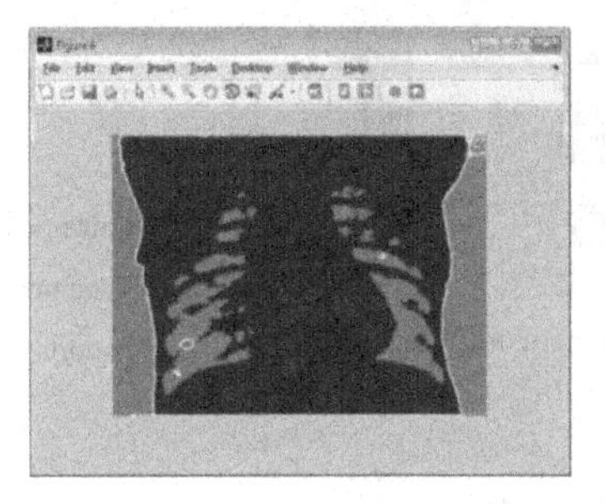
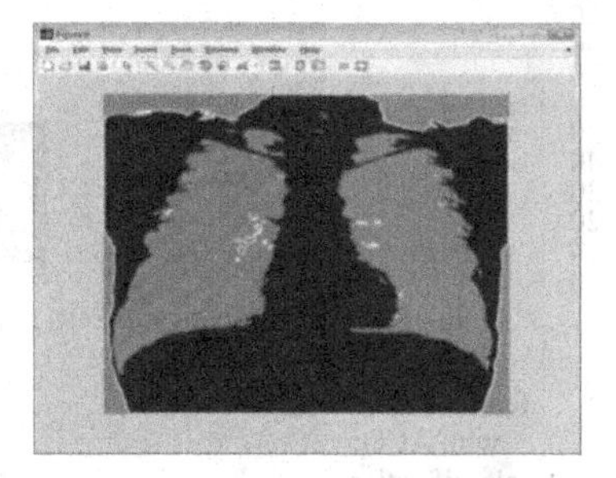

4b. Segmentation results for (a) cancer and (b) lung infection lung dataset

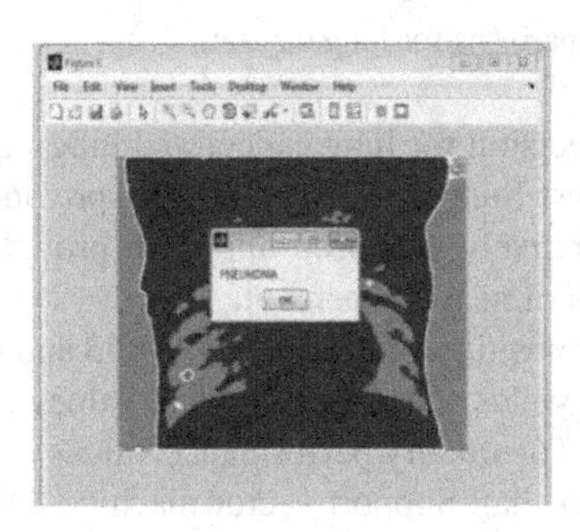
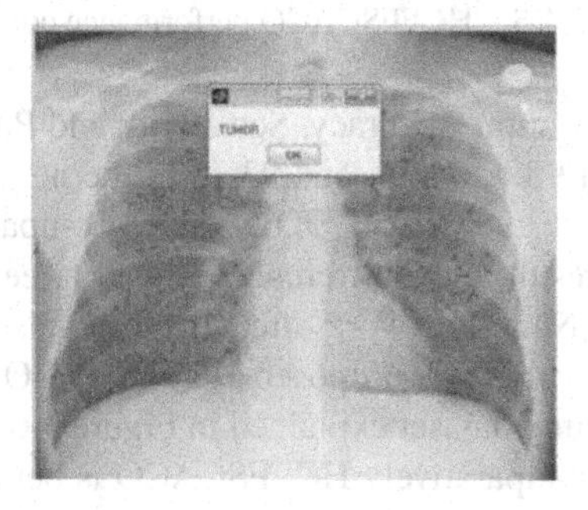

4c. Classification results for (a) cancer and (b) lung infection lung dataset

FIGURE 4.4 Lung dataset results for HCFFSCACO.

The Neural network offers the least false positive values of 31.02 percent, true positive values for 78.45 and a fscore value of 54.12 (Table 4.2). While the proposed HCFFSCACO yields the quality matrix values for the lung dataset explained in Figure 4.5. The quality performance values exposed to the comparatively HCFFSCACO is better than Support vector machine, DT, and Neural network.

Table 4.2 shows performance parameters such as accuracy, sensitivity, and precision. Table 4.1 explains that the model based on tha Proposed HCFFSCACO yields

TABLE 4.2
HCFFSCACO Performance Analysis

Approach	Accuracy	Sensitivity	Precision
Hierarchical clustering Fuzzy features subset with ACO classification	93.4	94.25	93.25
Support vector machine	91.6	90.81	88.65
Neural network	89.7	87.53	89.85
Decision tree	88.6	89.62	90.91

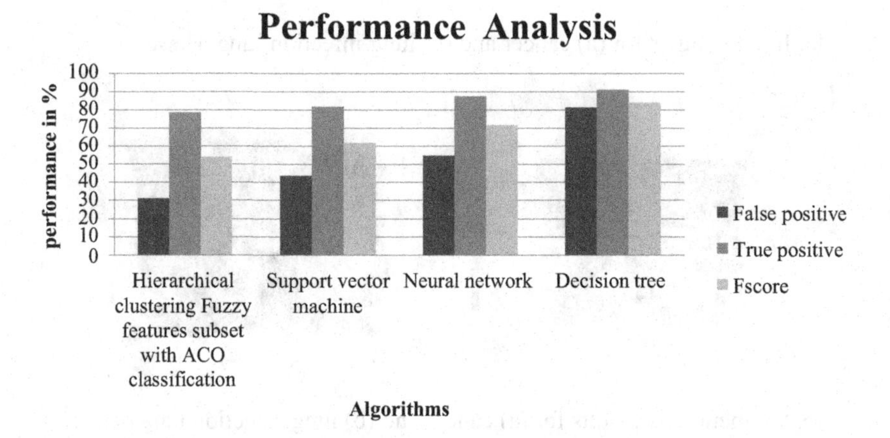

FIGURE 4.5 HCFFSCACO performance comparisons for lung dataset.

the maximum Accuracy, Sensitivity and Precision for lung cancer and lung-infected Dataset 93.4, 94.25, and 93.25 at the nearest Support vector machine provides the only 91.6, 90.81 and 88.65 values comparatively proposed approach provides the better result in the difference terms between of nearly 2 percent.

The Neural network offers the least false positive values of 89.7, 87.53 and 89.85, respectively. The proposed HCFFSCACO yields the quality matrix values for the micro lung dataset explained in Figure 4.6. The quality performance values exposed to the comparatively HCFFSCACO is better than Support vector machine, DT, and Neural network.

The decision tree offers the least false positive values of 88.6, 89.62, and 90.91, respectively, while the proposed HCFFSCACO yields the quality matrix values for the micro lung dataset explained in Figure 4.6. The quality performance values exposed to the comparatively HCFFSCACO is better than Support vector machine, DT, and Neural network. Finally, comparatively all the dataset proved the proposed system performance achieve better efficiency and sensitivity and other parameters.

Table 4.3 shows the performance parameters such as accuracy, sensitivity, and precision yield time duration for proposed and conventional approaches explained. Table 4.3 explains the model based on Proposed HCFFSCACO yields the minimum time duration to achieve better Accuracy, Sensitivity and Precision for the lung dataset. The time duration for all approaches are 8.20, 16.10, 19.22 and 21.50,

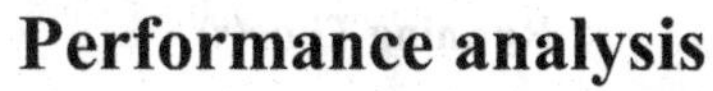

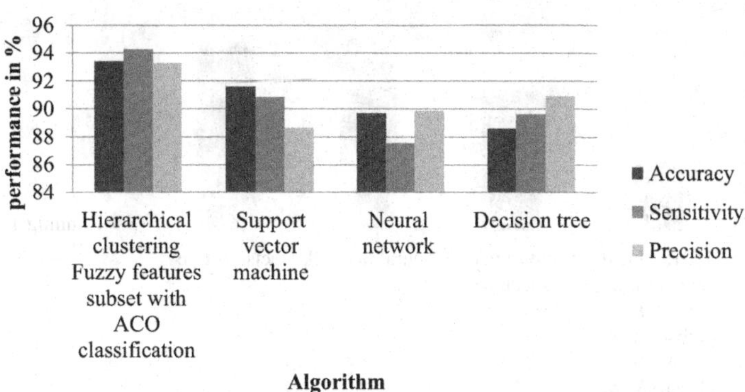

FIGURE 4.6 HCFFSCACO performance analyses.

TABLE 4.3
HCFFSCACO Time Duration Analysis

Approach	Running Time(s)
Hierarchical clustering Fuzzy features subset with ACO classification	8.20
Support vector machine	16.10
Neural network	19.22
Decision tree	21.50

respectively. The HCFFSCACO achieves 8.20 seconds in 93.4 percent efficiency. Other methods not attained time duration and accuracy, proposed HCFFSCACO proved better performances.

The overall lung image data classification accuracy of Hierarchical clustering is depicted in Figures 4.5 and 4.6. The fuzzy features subset classification approach achieves a 1.8 percent better result than the existing support vector machine, which only achieves 91.6 percent. Other methods include a 89.7 percent for the neural network and 88.6 percent for the decision tree. The running time parameter of the Hierarchical Clustering technique is 8.20 seconds, whereas the SVM method is 16.10 seconds, the neural network method is 19.22 seconds, and the decision tree method is 21.50 seconds. The Hierarchical Clustering method has a four times faster running time than the SVM method. The Hierarchical Clustering Fuzzy Features Subset Classification approach attained the best accuracy result for the shortest time period. The performance result indicates that the developed classifier can be used for classifying the selected dataset to assist the clinician in decision-making. After verifying the results, it is proved that designed research can be used for real-time data classification in various environments without any error whatsoever. The system is able to provide the solutions for the problems faced in real time and without delay perfect achievement is succeeded (Figure 4.7).

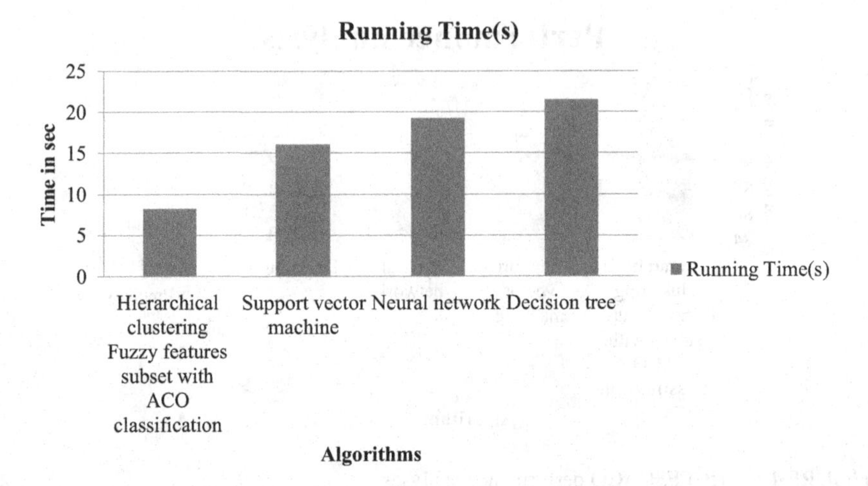

FIGURE 4.7 Hierarchical clustering fuzzy features subset classification running time analyses.

4.5 CONCLUSION

A big data patient record classification has been performed in this work for a medical dataset using the HCFFSC method, the hierarchical clustering method and the max-margin classifier. This work necessitates the use of the feature selection process prior to the classification process while analyzing a huge dataset. Since feeding the classifier model with entire features may cause barriers to the classification performance, a hybrid method named Hierarchical Clustering Fuzzy Features Subset Classification (HCFFSCACO) is initiated to enhance the classification performance lung datasets in medical diagnosis systems. The use of the decision support system has revolutionized patient care, industry analysis and treatment in the healthcare industry. This work is done to ascertain the imperative need for feature selection to be done prior to big data classification process and it is seen that the feature selection cannot be neglected during the classification process. Furthermore, our approach outperforms traditional methods by removing the bottleneck generated by the classification algorithm without compromising classification performance. Furthermore, hybrid classification approaches using two or more classifiers may enable knowledge engineers to design efficient decision support systems in real-world scenarios. In future, the application of hybrid optimization techniques and bio-inspired artificial intelligence approaches would yield better classifier models that can be used for the design and development of decision support systems to improve the efficiency.

REFERENCES

Abadeh, M. S., Habibi, J., & Soroush, E. (2008). Induction of fuzzy classification systems via evolutionary ACO-based algorithms. *Computer*, *35*, 37.

Akanskha, E., Sahoo, A., Gulati, K., & Sharma, N. (2021, June). Hybrid Classifier Based on Binary Neural Network and Fuzzy Ant Colony Optimization Algorithm. In *2021 5th*

International Conference on Trends in Electronics and Informatics (ICOEI) (pp. 1613–1619). IEEE.

Alam, J., Alam, S. & Hossan, A. (2018). "Multi-Stage Lung Cancer Detection and Prediction Using Multi-class SVM Classifier," 2018, IC4ME2, Rajshahi, pp. 1–4.

Alwan, H. B. & Ku-Mahamud, K. R. (2013). Formulating new enhanced pattern classification algorithms based on ACO-SVM. *International Journal of Mathematical Models and Methods in Applied Sciences*, *7*(7), 700–707.

Amin, S.U., Agarwal, K. & Beg, R. (2013). 'Genetic neural network based data mining in prediction of heart disease using risk factors', In *Information & Communication Technologies (ICT), 2013 IEEE Conference on IEEE*, pp. 1227–1231.

Anifah, L., Haryanto, R. Harimurti, Z. Permatasari, P. W. Rusimamto and A. R. Muhamad (2017). "Cancer lungs detection on CT scan image using artificial neural network back-propagation based gray level co-occurrence matrices feature," 2017 ICACSIS, Bali, pp. 327–332.

Anthimopoulos, M., Christodoulidis, S., Ebner, L., Christe, A., & Mougiakakou, S. (2016). Lung pattern classification for interstitial lung diseases using a deep convolutional neural network. *IEEE Transactions on Medical Imaging*, *35*(5), 1207–1216.

Cai, Z., Xu, D., Zhang, Q., Zhang, J., Ngai, S. M., & Shao, J. (2015). Classification of lung cancer using ensemble-based feature selection and machine learning methods. *Molecular BioSystems*, *11*(3), 791–800.

Detterbeck, F. C., Boffa, D. J., Kim, A. W., & Tanoue, L. T. (2017). The eighth edition lung cancer stage classification. *Chest*, *151*(1), 193–203.

Dhasal, P., Shrivastava, S. S., Gupta, H., & Kumar, P. (2012). An optimized feature selection for image classification based on SVM-ACO. *International Journal of Advanced Computer Research*, *2*(3), 123.

Jayaraj, D. & Sathiamoorthy, S. (2019). "Random Forest based Classification Model for Lung Cancer Prediction on Computer Tomography Images," 2019, ICSSIT, Tirunelveli, India, pp. 100–104.

Kalimuthu, S. (2021). Sentiment analysis on social media for emotional prediction during COVID-19 pandemic using efficient machine learning approach. *Computational Intelligence and Healthcare Informatics*, 215, 215–230.

Kalimuthu, S., Naït-Abdesselam, F., & Jaishankar, B. (2021). Multimedia Data Protection Using Hybridized Crystal Payload Algorithm With Chicken Swarm Optimization. In *Multidisciplinary Approach to Modern Digital Steganography* (pp. 235–257). IGI Global.

Kumar, D., Wong, A., & Clausi, D. A. (2015, June). Lung nodule classification using deep features in CT images. In *2015 12th Conference on Computer and Robot Vision* (pp. 133–138). IEEE.

Kuncheva, L.I. (2004). *Combining Pattern Classifiers Methods and Algorithms*. John Wiley & Sons, New Jersey.

Meenachi, L. & Ramakrishnan, S. (2020). Differential evolution and ACO based global optimal feature selection with fuzzy rough set for cancer data classification. *Soft Computing*, *24*, 18463–18475.

Nadkarni, N.S. & Borkar, S. (2019). "Detection of Lung Cancer in CT Images using Image Processing," *2019 3rd, ICOEI, Tirunelveli, India*, pp. 863–866,

Nóbrega, R.V.M.d., Peixoto, S.A., da Silva, S.P.P., & Filho, P.P.R. (2018). "Lung Nodule Classification via Deep Transfer Learning in CT Lung Images," *2018 IEEE (CBMS), Karlstad*, pp. 244–249.

Peter, T.J. & Somasundaram, K. (2012). 'An empirical study on prediction of heart disease using classification data mining techniques', In *Advances in Engineering, Science and Management (ICAESM), 2012 International Conference on IEEE*, pp. 514–518.

Polat, K. & Güneş, S. (2007). "An expert system approach based on principal component analysis and adaptive neuro-fuzzy inference system to diagnosis of diabetes disease", *Digital Signal Processing*, 17(4), 702–710.

Potghan, S., Rajamenakshi, R., & Bhise, A. (2018, March). Multi-layer perceptron based lung tumor classification. In *2018 Second International Conference on Electronics, Communication and Aerospace Technology (ICECA)* (pp. 499–502). IEEE.

Radhimeenakshi, S. (2016). 'Classification and prediction of heart disease risk using data mining techniques of Support Vector Machine and Artificial Neural Network', In *Computing for Sustainable Global Development (INDIACom), 2016 3rd International Conference on IEEE*, pp. 3107–3111.

Rajathi, S., & Radhamani, G. (2016, March). Prediction and analysis of Rheumatic heart disease using kNN classification with ACO. In *2016 International Conference on Data Mining and Advanced Computing (SAPIENCE)* (pp. 68–73). IEEE.

Rodrigues, M.B., Da Nobrega, R.V.M., Alves, S.S.A., Reboucas Filho, P.R., Duarte, J.B.F., Sangaiah, A.K., & De Albuquerque, V.H.C. (2018). Health of things algorithms for malignancy level classification of lung nodules. *IEEE Access*, *6*, 18592–18601.

Sabab, S.A., Munshi, M.A.R., & Pritom, A.I. (2016). 'Cardiovascular disease prognosis using effective classification and feature selection technique', In *Medical Engineering, Health Informatics and Technology (MediTec), 2016 International Conference on IEEE*, pp. 1–6.

Saxena, K. & Sharma, R. (2015). 'Efficient heart disease prediction system using decision tree', In *Computing, Communication & Automation (ICCCA), 2015 International Conference on IEEE*, pp. 72–77.

Shen, W., Zhou, M., Yang, F., Yang, C., & Tian, J. (2015, June). Multi-scale convolutional neural networks for lung nodule classification. In *International conference on information processing in medical imaging* (pp. 588–599). Springer, Cham.

Tafti, A.P., Bashiri, F.S., LaRose, E., & Peissig, P. (2018). "Diagnostic Classification of Lung CT Images Using Deep 3D Multi-Scale Convolutional Neural Network," 2018 IEEE (ICHI), New York, NY, pp. 412–414.

Uppaluri, R., Mitsa, T., Sonka, M., Hoffman, E.A., & McLennan, G. (1997). Quantification of pulmonary emphysema from lung computed tomography images. *American Journal of Respiratory and Critical Care Medicine* 156(1), 248–254.

Verleden, G.M., Raghu, G., Meyer, K.C., Glanville, A.R., & Corris, P. (2014). A new classification system for chronic lung allograft dysfunction. *Journal of Heart and Lung Transplant*, 33, 127–133.

Vieira, S.M., Sousa, J.M., & Runkler, T.A. (2007, June). Ant colony optimization applied to feature selection in fuzzy classifiers. In *International Fuzzy Systems Association World Congress* (pp. 778–788). Springer, Berlin, Heidelberg.

Wijaya, R. & Prihatmanto, A.S. (2013). 'Preliminary design of estimation heart disease by using machine learning ANN within one year', In *Rural Information & Communication Technology and Electric-Vehicle Technology (rICT & ICeV-T), 2013 Joint International Conference on IEEE*, pp. 1–4.

5 Health-Mentor
A Personalized Health Monitoring System Using the Internet of Things and Blockchain Technologies

M. Sumathi
SASTRA Deemed University, Thanjavur, India

M. Rajkamal
IBM, Bangalore, India

CONTENTS

5.1 Introduction 64
5.2 Related Works 65
5.3 IoT-Based Health Monitoring 65
5.4 Machine Learning-Based Health Data Classification 66
5.5 Blockchain-Based Health Data Transfer and Storage 67
5.6 Summary of Existing Techniques 68
5.7 Research Gap in the Existing Technique 68
5.8 Objective of the Proposed Work 68
5.8.1 Proposed Health-Mentor System 68
5.9 IoT Data Collection 69
5.10 Normal and Abnormal Data Classification 70
5.11 Block Generation and Transfer 70
5.12 Block Analysis and Recommendation System 72
5.13 Experimental Results 72
5.14 Machine Learning Algorithm-Based Normal and Abnormal Data Classification 72
5.15 Block Construction and Transfer Analysis 74
5.16 Block Analysis and Recommender System Analysis 75
5.17 Conclusion and Future Work 76
References 76

DOI: 10.1201/9781003267782-5

5.1 INTRODUCTION

At present, the sharing and storage of patient health information in a secured way is a prominent requirement in the healthcare sector. In conventional storage, before sharing to care providers or stored in a secure storage location the patient's entire information is converted into ciphertext form leads to higher computational complexities and the paper-based documentation leads to transfer and maintenance difficulties. Hence, the Electronic Health Record (EHR) is introduced into healthcare sectors. Before 2010, only around 10 percent of healthcare records had been stored in an electronic format. Today, more than 90 percent of healthcare records are maintained in an electronic format. The advantages of EHR are easy maintenance, sharing and storage. The other side of the EHR is to maintain or sharing a record in a secure way is a challenging task [1]. Initially, the EHR had been maintained in a centralized storage in an encrypted form by the third party service providers and the EHR's are accessed/altered by them. Furthermore, the centralized storage leads to data loss when a storage system fails. To avoid these issues a new storage method is required [2]. The EHR is accessible by the different care providers (doctors, nurse, medical students and pharmacy members) for providing different services to patients. The EHR contains confidential sensitive data (CSD) about the patient. Hence, to protect the CSD and assign access control (AC) to care providers is an essential task. Conventionally, the patient should stay in the hospital and health complaints would be monitored by the care providers. Due to the present-day development of automation, the patient's health condition is monitored from the remote locations and patients are not required to stay in the hospitals. This automation process is achieved through the IoT devices [3].

Generally, IoT sensors measure patient health condition at frequent and regular time intervals and transfer enormous amounts of data to health monitoring centers for the purposes of analysis. When a large amount of data is transferred through the network increases transmission time and require the use of a high bandwidth network. The EHR contains different diversity of data like numerical data, scan images and hand written medical prescription. Hence, the EHR record is large in size and the transfer of these large size data is a complicated task [4]. Typically, the observed IoT EHR contains both normal and abnormal data, the abnormal data are considered as a CSD instead of entire data (ED). To classify the normal and abnormal data, a classification technique is applied to EHR. Afterwards, the CSD is transferred to analysis. The CSD transmission takes lesser transmission time than the ED transmission. Likewise in data storage, the CSD is stored securely instead of ED. Thus, the storage size requirement is also reduced [5]. In the proposed work, the IoT based personalized wearable health monitoring system and blockchain based CSD sharing between the patients and care takers are going to be discussed.

In a proposed technique, patient health information such as patient blood pressure, walking time, sleep cycle, heartbeat, temperature, breathing, stress and oxygen levels are captured by a smart wearable IoT device. Usually, IoT devices produce an enormous amount of data and not all of this is critical. Similarly, the afore-mentioned values differ from person to person. Hence, a personalized monitoring device is required to monitor the individual's patient health. The wearable device is customized for the individual and fixes the threshold for each parameter based on the

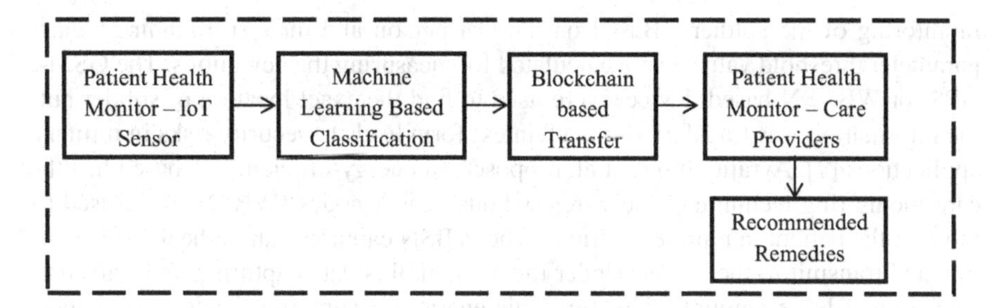

FIGURE 5.1 Workflow of the proposed method.

patient's needs. If abnormalities are occurring in these parameters, those abnormal values are stored in the block and immediately sent to a nearby hospital. The information stored in the block is considered as CSD about the patient. Hence, the CSD is maintained securely in a private blockchain and the normal information is stored in cloud storage for the purpose of future analysis. Figure 5.1 shows the work flow of the proposed method.

The remainder of this chapter is organized as follows: In Section 5.2, the chapter analyses the existing works related to the IoT, machine learning and blockchain-based healthcare monitoring techniques are analyzed in depth in order to identify the features and limitations. In Section 5.3, we introduce the proposed health-mentor: a personalized health monitoring system using the Internet of Things (IoT) and blockchain technologies are discussed with the necessary architecture and algorithms. In Section 5.4, the experimental results of the proposed technique are analyzed with different parameters and in Section 5.5 the security analysis of the proposed technique is discussed. Finally, in Section 5.6 the proposed technique is concluded with the future enhancements.

5.2 RELATED WORKS

In this section, the existing works related to EHR is analyzed in different aspects such as IoT-based health monitoring, machine learning-based health data classification and blockchain-based health data transfer and storage with its features and limitations.

5.3 IoT-BASED HEALTH MONITORING

Dahlia Sam et al. proposed an IoT health monitoring technique. The powered mobile hearing and electronic wristbands were used to monitor the blood pressure, temperature and heartbeat of the patient. The atmega328 is used to measure the pulse and heartbeat, and a LM35 sensor is used to measure the temperature. The patient health information is monitored by different sensors and transfer to remote centers in a plaintext form leads to high cost and security risks [6]. Vijay Anand et al. designed an IoT-based health monitoring system for military applications. The ATmega328 device was programmed to monitor the temperature, heart rate and continuous health monitoring of the soldiers. Based on the test person and their surroundings, each parameter threshold values were calculated for measuring the deviations. The GSM-, GPS- or WBASN-based devices were used to find the exact location of solider but, the information was transferred in a plaintext form leads to security risks in military applications [7]. Avrajit Ghosh et al. proposed an energy-efficient IoT-based healthcare monitoring technique. The wireless body sensor nodes (WBSN) were used to monitor the patient in remote locations. The WBSN captures patient health information and transmit to the server. Under this system, this data capturing and transmission takes a large amount of energy. This energy consumption leads to data loss; hence the iterative threshold and sparse encoding was used for reducing the energy and bandwidth consumption. The encoding scheme compresses the data and transfer to receiver needs an efficient decompression technique otherwise, this transmission is also leads to data loss [8].

Samira Akhbarifar et al. proposed the IoT-based health monitoring technique. The patient's general information was collected for identifying the patient uniquely along with the medical IoT sensor information. Afterwards, the collected information was encrypted by block encryption and transferred to healthcare centers for analysis. The disease prediction algorithm was used to predict both the type of disease and its severity. The EHR confidentiality is maintained by block encryption and the security level depends upon the key size [9]. Vedanarayanan et al. utilized the IoT for secured clinical information transmission and the remote monitoring of patients' health conditions. By using an arduino device, the patient health information is observed and the encoded the observed information for transferring to doctors' personal devices. Through encoding, the secure transmission is achieved in the IoT data transfer [10].

5.4 MACHINE LEARNING-BASED HEALTH DATA CLASSIFICATION

Li Hong tan et al. discussed the IoT-based student health monitoring system. The student health information was collected by IoT devices and stored in a cloud. To identify the abnormal data, the support vector machine (SVM) classification technique was applied to health data. The two-dimensional hyper-plane was used for classifying the data as normal (value 1) and abnormal (value –1). Likewise, student's mental, social and emotional feelings were analyzed by non-linear SVM classifier. The external parameters are not analyzed in deeply [11]. Wang Huifeng et al. analyzed the sportsperson health monitoring system using the ensemble Bayesian deep classifier. The Bayesian classifier maximizes the prediction accuracy and works well in dealing with a large volume of data. Large numbers of hidden layers were used for the processing of large volumes of IoT data. The boosting learning algorithm was used to reduce the levels of weak information participation in the decision. The weak classifiers were removed by a feature ensemble value to improve the monitoring process. The prediction accuracy depends on the number of hidden layers and weight values [12].

Nonita Sharma et al. proposed a system of ontology-based remote patient monitoring. The ECG, PPG, temperature and accelerometer information was given to kernel multi-view canonical correlation analysis-based feature extraction model. The minimal-redundancy and maximal relevance technique is used to select the relevant features. Afterwards, the SVM and KNN classification technique performances were compared to find the suitable technique. The performance comparison was done by tenfold cross validation to predict Covid-19 [13]. Mahesh Ashok Mahant et al. proposed the supervised machine learning-based clinical data classification. The k-means and K-nearest algorithms were used to predict the health risk of children under five years old. The K-Nearest algorithms produced better prediction results than the k-means clustering [14]. Trong Thanh Han et al. proposed the machine learning algorithm for the classification of infected patients from others using vital signs. Through the use of the medical radar, the patient health information was acquired. This acquired data passed to filters for the elimination of interferences other than heart parameters. The filtered data were given as an input to machine learning algorithms such as naïve Bayes, support vector machine, decision tree and logistic regression. Among these algorithm decision trees classification technique provides higher prediction accuracy [15].

5.5 BLOCKCHAIN-BASED HEALTH DATA TRANSFER AND STORAGE

Aruna Sri et al. proposed the blockchain-based secure medical data sharing technique using consensus. This consensus was used to validate the interoperability and proof of word for accessing and sharing of medical records. The centralized trust-based verification was offered the trusted access and communication between parties was involved in the network. The blockchain technique was used for efficient data management, claims /bill management, medication adherence etc. The EHR data security was done by AES and Euler totient functions [16]. Rubal Jeet et al. investigated the need for blockchain in the healthcare applications. The key benefits of blockchain in the field of healthcare are decentralized data management (patients, doctors, hospitals and payers are works independently without any intermediary), immutable audit trail (the her cannot be changed in future), data provenance (blockchain provides efficient data management and maintenance), robustness and availability (blockchain maintain history of data in a ledger). Hence, blockchain is preferable in EHR maintenance [17].

Asad Abbas et al. discussed the blockchain-based secured data management in healthcare applications. Data security was established between personal servers and medical devices. The higher trust value, accuracy and precision ratio was achieved by less response time and latency ratio [18]. Dheeraj Mohan et al. used private Ethereum blockchain for secure data transfer in medical applications. The elliptic curve integrated encryption scheme was used for the double encryption of medical data. The encrypted medical records are transferred to a decentralized node and the proof of authority was used to minimize the computational time. This technique provides better security to medical data with minimal transaction latency, but double encryption leads to high computational complexity [19]. Rajakumar Arul et al. have conducted a study of blockchain-based healthcare systems. The distributed ledger-based data transmission was discussed to remove the central authority-based data transmission. Similarly, through the distributed ledger the EHR integrity was maintained. The backpropagation learning technique was used to verify the data integrity, improve the honesty, and avoid healthcare service failures and postponements in services [20].

5.6 SUMMARY OF EXISTING TECHNIQUES

- The IoT sensors are helpful in measuring patient health conditions on a 24/7 basis.
- Machine learning techniques are used to predict diseases in an accurate way.
- Blockchain technique is used to transfer the data in a secure way between different parties involved in the healthcare sectors.

5.7 RESEARCH GAP IN THE EXISTING TECHNIQUE

- The cost of remote health monitoring technique via the IoT is high.
- Machine learning techniques are used for disease prediction, but not applied to security aspects and data reduction. Similarly, the accuracy of machine learning technique depends on training and test data ratio.
- The major constraint on the blockchain is block size. The block size-based data transmission is not focused.
- The healthcare decision taken is done by the corresponding pre-history based hospitals not by the quick response with nearby hospital.
- Recommendation system is not yet discussed.

These research gaps lead the following objective of the proposed technique.

5.8 OBJECTIVE OF THE PROPOSED WORK

1. Remote health monitoring via IoT medical sensors enables the use of wearable devices with lesser cost.
2. Classify and predict the normal and abnormal health data using machine learning algorithms.
3. Using SHA256 algorithm to generate the block with abnormal health data for providing confidentiality, integrity and availability to EHR.
4. Transfer the block to nearby healthcare provider's analysis and find the remedies by a recommendation system and forward the recommendation to patient in a block.

5.8.1 Proposed Health-Mentor System

The proposed health-mentor: a personalized health monitoring system using IoT and blockchain technologies collect health information from human body through wearable IoT devices such as a wristwatch. The information is then transferred to their mobile devices like mobile phones or personalized computers for normal and abnormal data classification. The abnormal data is identified as sensitive and stored in a block before transfer to nearby health center for quick decision-making. The nearby health center is identified by a shortest path algorithm. The nearby health center feed the block information to recommendation system for identifying the remedies and the send a reply to patient. Figure 5.2 shows the overall architecture of the proposed technique.

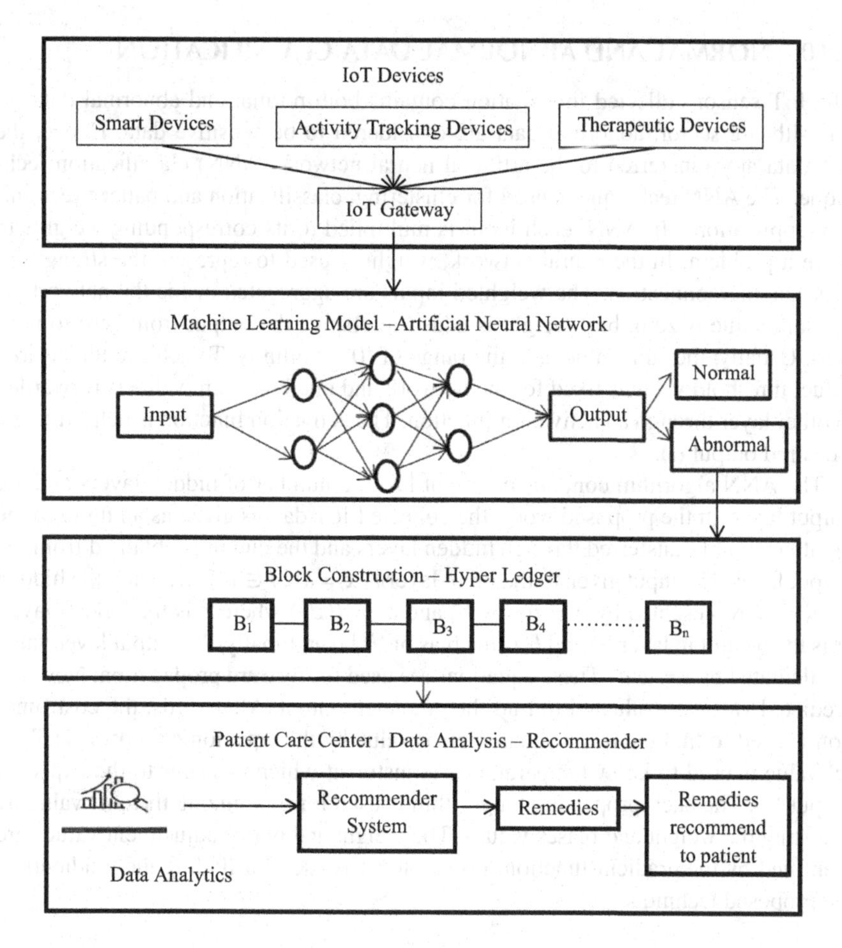

FIGURE 5.2 System architecture.

5.9 IoT DATA COLLECTION

In IoT data collection, the patient health data are collected from the IoT sensors. The wearable IoT device is an integrated device containing a smart device, an activity tracking device and therapeutic devices. A smart device is an example of a smart skin patch which collects information about the heart rate, body temperature and ECG monitor. The activity tracker is used to collect fitness information such as pressure, movement, oxygen level, heartbeat rate and blood pressure. This information is helpful to the recommendation of sleep schedule, eating habits and plan for the general activities. The therapeutic device is used to collect the information about the insulin level and pain information.

Based on these three different devices the patient health information is collected at frequent intervals. This collected information is transferred to patient personal device to classification.

5.10 NORMAL AND ABNORMAL DATA CLASSIFICATION

The IoT sensor collected information contains both normal and abnormal data. In a healthcare sector, abnormal data are considered to be sensitive data. Hence, the IoT data are transferred to the artificial neural network (ANN) classification technique. The ANN technique is used for clustering, classification and pattern recognition applications. In ANN, each input is multiplied to its corresponding weights to solve a problem. In the neural network, weight is used to represent the strength of neurons interconnection. The weighted inputs are aggregated inside the network. If the sum value is zero, bias weight is added to change the output from zero to non-zero. Usually, the sum value is in the range of '0' to infinity. To achieve the desired value, threshold value is fixed for the network and one layer sum value is forwarded to other layer through an activation function. The activation function is helpful to get a desired output (i).

The ANN algorithm contains the input layer, a number of hidden layers and the output layer. In the proposed work, the collected IoT data is given as an input to the input layer and transferred through hidden layers and the output is obtained from the output layer. The input given to the input layer is $x_i = a^1$, $i \in 1, 2, \ldots n$ and the hidden values are represented by $z^i = w^i x + b^i$, and $a^2 = f(z^{(i)})$ where z^i is the hidden layer, w^i is the weight in layer 'i' and b^i is the bias of 'i' layer. Finally, the output layer value is calculated by $s = w^i a^i$. These equations are used for forward propagation. Now, the predicted value is evaluated to find the predicted output. Afterwards, the cost function is used to find the mean squared error value by the equation $c = \cos(s, y)$. This 'c' value is used to know the parameter adjustment which is nearer to the expected output 'y'. The back-propagation algorithm is used for minimizing the cost value by adjusting the weight and biases values. The weight and biases adjustment values are identified by the gradient function. Equation 5.1 is used for finding the gradients in the proposed technique.

$$\frac{\delta c}{\delta x} = \left[\frac{\delta c}{\delta x_1}, \frac{\delta c}{\delta x_2}, \ldots \frac{\delta c}{\delta x_n} \right] \tag{5.1}$$

Through the use of the gradient function, the level of change is required in input value is measured. Generally, the initial value of 'w' and 'b' are chosen randomly. Based on this ANN technique the abnormal data are classified from the normal data.

5.11 BLOCK GENERATION AND TRANSFER

The classified abnormal data are stored in a block before they are transferred for the purposes of analysis. The key benefits of the blockchain technique are secured, consistent data transfer between the members are involved in the network in a decentralized manner. This decentralized data storage avoids the dangers of data loss. Similarly, the blocks which are stored in the blockchain ledgers are unalterable. Thus, unauthorized alterations are impossible. Additionally, the output hash value is fixed for variable input size and provides efficient data transfer with lesser bandwidth size and lower transmission time. Hence, the blockchain-based sensitive data transfer is the preferred approach in the proposed work. Basically, blocks are constructed by the hash values. The hash values are generated by the hash code generation algorithm. When compared to MD5 hash code generation, SHA256 hash code generation algorithm avoids collisions. Hence, SHA256 is used for block generation. Equation 5.2 is used for block construction.

$$Bloc(ASD) = SHA256(PB_{HV}, ASD, Nonce, PID, T_S) \tag{5.2}$$

The block based on Abnormal Sensitive Data (ASD) consists of previous block hash value (PB_{HV}), ADS, Nonce, patient ID (PID) and the timestamp (T_S). Depends on PB_{HV}, the data consistency is maintained, Nonce and T_S are used for representing the time of block generation and PID contains the patient identification information. The constructed block is transferred to a nearby care provider center for rapid decision-making. The nearby center is identified by the single source to multiple destinations shortest path (SSMDSP) algorithm. The SSMDSP algorithm compares the distance between the source locations to each healthcare center $H = \{h_1, h_2, \ldots, h_n$, where $h_i \in SSMDSP\}$. Among these distances, the minimum distance healthcare center is chosen and block is transferred.

ALGORITHM: EHR DATA TRANSMISSION AND ANALYSIS

Input: IoT Sensor Data, ANN Algorithm, Hyper-ledger, Recommender System

Output: Sensitive Data, Block Generation, Remedies Recommendation

1. Collect the Patient health data by IoT sensors
2. Transfer the Collected data to ANN Classification to classify normal and abnormal data

$$x_i = a_i^1, i \in 1, 2, \ldots n$$
$$z^i = w^i x + b^i, \text{ and } a^2 = f(z^{(i)})$$
$$s = w^i a^i$$
$$c = cost(s, y)$$
$$\frac{\delta c}{\delta x} = \left[\frac{\delta c}{\delta x_1}, \frac{\delta c}{\delta x_2}, \ldots, \frac{\delta c}{\delta x_n}\right]$$

3. Transfer sensitive abnormal data to block generation.
4. Transfer generated block to nearest care provider center.

$$Block(ASD) = SHA256(PB_{HV}, ASD, Nonce, PID, T_S)$$

5. Analysis the block and send to recommendation system
6. Identify the remedies and transfer to patient

5.12 BLOCK ANALYSIS AND RECOMMENDATION SYSTEM

The transferred block is analyzed by experts in the healthcare center and fed into the recommender system. Then, the recommender system finds the suitable remedy to the patient based on historical data analysis. The recommendation system works behind the gender and age groups. Weight is assigned to each abnormal attributes and finds the final score for the corresponding patient. If the final score is greater than the average weighted score, the recommender system, recommends admitting the patient to the hospital; otherwise, it suggests a range of remedies to the patient. Equation 5.3 is used to find the average score of the weighted '*n*' attributes.

$$Average\,Score = \frac{\sum_{n=1}^{n} Abnormal_{Weight}}{n} \tag{5.3}$$

Based on abnormalities, the remedies are suggested by the recommender system by the historical data analysis.

5.13 EXPERIMENTAL RESULTS

In the proposed technique, the data are collected from the human body by an IoT wearable device from 100 members. An Anaconda Juypter notebook-based Python tool is used for machine learning algorithm analysis. The threshold value is fixed for each parameter as temperature 98.6 degrees F, blood pressure 80 to 120 mmHg, heartbeat 60 to 100 per minute, sleeping time 7 to 9 hours per day, breathing rate 12 to 16 per minute, stress level 5.1 to 10 point scale and oxygen level 95 percent or above, Gender 1 indicates male and 0 indicates female. If any deviation occurs in these parameters, the corresponding user information is classified as abnormal and considered for block construction. The hyper-ledger is used for block generation based on abnormal data and the SSMDSP algorithm used to find the nearest care center. The data analysis and recommendation system is also implemented by the Python language. In this section, the experimental results of the proposed technique is discussed with the existing technique results.

5.14 MACHINE LEARNING ALGORITHM-BASED NORMAL AND ABNORMAL DATA CLASSIFICATION

The IoT data are classified into normal and abnormal data by an ANN algorithm. The proposed technique is compared with other machine learning algorithms such as the Support Vector Machine (SVM), K-Nearest Neighbor (KNN), Naïve Bayes (NB), Random Forest (RF) and Logistic Regression (LR). When compared to SVM, KNN, NB, RF and LR, the proposed ANN technique provides a higher accuracy rate. When the training and test data ratio is fixed as 80:20, the accuracy rate is less for

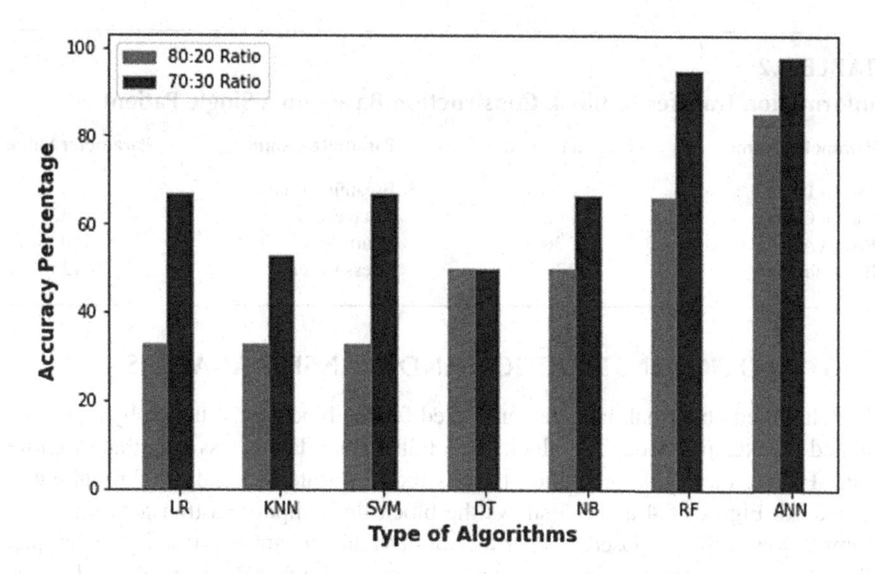

FIGURE 5.3 Classification accuracy comparison – existing and proposed.

TABLE 5.1
Normal and Abnormal Data Prediction Using Random Forest

PID	Age	Sex	Temp	BP	Breathing Rate	Sleep Cycle	Heart Beat	Stress Level	Oxygen Level	Normal	Entire Data Size	Abnormal Data Size
1	20	1	98.6	120	15	7	100	5.8	98	0		
2	35	1	98.6	110	14	8	86	6	97	0	No need to Transfer and Store	
3	40	0	98.6	85	12	8	100	10	95	0		
4	30	0	98.6	130	10	5	110	12	98	1	100%	50%
5	65	1	100	105	10	10	80	4	85	1	100%	50%
6	70	1	99.2	110	12	5	100	15	80	1	100%	40%
7	10	0	105	125	10	10	120	12	92	1	100%	80%
8	15	1	96.5	80	14	8	100	8	97	1	100%	10%
9	25	0	107	110	15	5	100	12	98	1	100%	30%

all algorithms. Hence, the training and test data ration is fixed as 70:30 for the proposed technique. Figure 5.3 shows the classification accuracy rate of the existing and proposed technique. The proposed ANN algorithm provides a higher accuracy rate (98 percent) than the other algorithms. Hence, ANN is used for normal and abnormal data classification.

After classification, the abnormal data prediction is done by random forest algorithm. Due to majority voting, the random forest technique produces better prediction accuracy than other machine learning algorithms. Table 5.1 shows the normal and abnormal data prediction based on random forest. Table 5.2 shows the information which is transferred to block construction.

TABLE 5.2
Information Transfer to Block Construction Based on a Single Patient

Parameter Name	Parameter Value	Parameter Name	Parameter Value
Patient ID	4	Breathing Rate	10
Patient Gender	0	Sleep cycle	5
Patient Age	30	Heart Beat	110
Blood Pressure	130	Stress Level	12

5.15 BLOCK CONSTRUCTION AND TRANSFER ANALYSIS

The classified abnormal data are considered for the block construction by a permissioned blockchain mode. The block size of abnormal data is less than that of entire data. Hence, the block generation time is reduced (latency) and the throughput is increased. Figures 5.4 and 5.5 shows the block throughput and transaction latency comparison of the proposed and permission-less blockchain construction technique. Transaction throughput is measured by the successful transactions per second is varied from 200 to 1400. Network latency is measured by the total time taken by a block to be executed in a network.

When compared to the permission-less network, the permissioned network throughput and transaction latency is lower. In the permission-less network, the increased number of users need to verify the transaction and update the ledger. Hence, the transaction time is higher than under the proposed system. Similarly, in a proposed technique, the block size is lesser than entire data size. Thus, the processing complexity is less than the existing technique. Hence, the proposed technique takes lesser latency time and higher throughput is proven.

The major benefit of the blockchain technique is to provide data integrity, confidentiality and availability to authorized users. The hash value-based blocks provide data

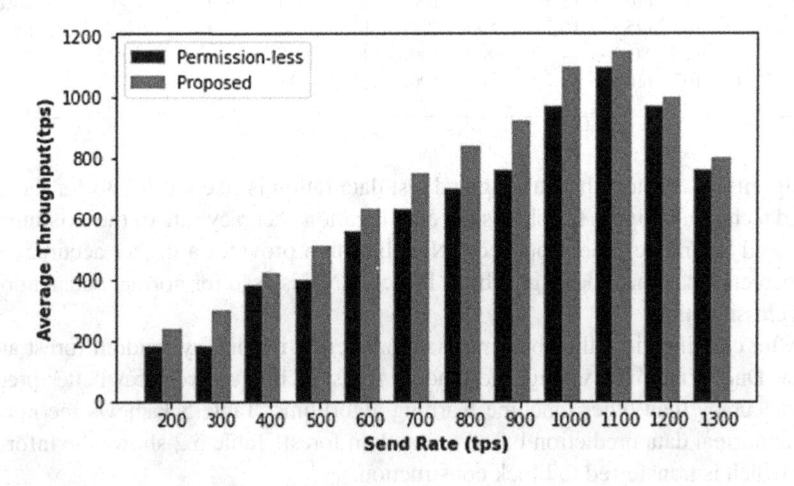

FIGURE 5.4 Average throughput comparison.

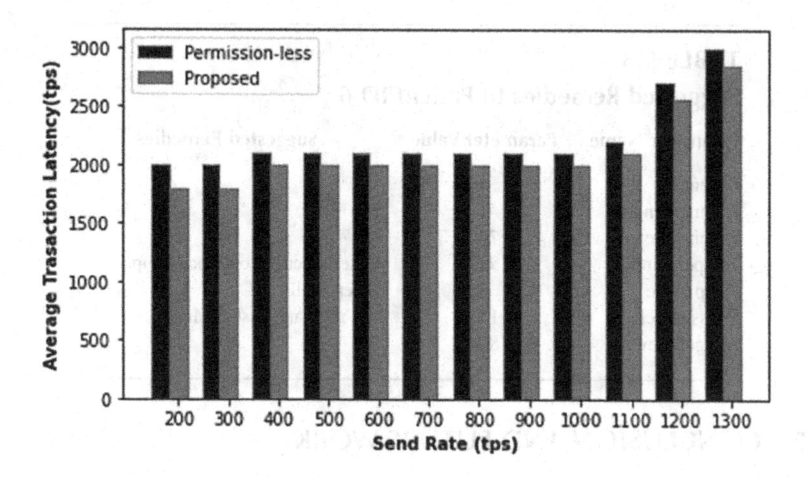

FIGURE 5.5 Average transaction latency comparison.

integrity to user data. Thus, the blocks are immutable and linked to each other. Hence, it is impossible to alter the content in the future. Similarly, the blocks are stored in the permissioned network, providing high confidentiality to patient data. This is because only authorized users are able to access and read the data. Hence, confidentiality is maintained by the individual patients. The proposed technique provides data availability to authorized users whenever required. If failure occurs in any node, it is possible to access data from other nodes. Thus, the data availability is ensured through the decentralized storage scheme. In a proposed technique, the minimal size abnormal data is transferred for analysis rather than the entire data. This minimal data size transmission and storage efficiently handles the blockchain scalability issues. Hence, the proposed technique provides efficient data handling in all aspects (integrity, confidentiality, availability, scalability, throughput and latency) are proven.

5.16 BLOCK ANALYSIS AND RECOMMENDER SYSTEM ANALYSIS

Depending on the final score, the abnormal data are then analyzed. The final score is greater than the average score, the recommendation system recommends the patient admitting the nearby hospital immediately. Otherwise the remedies are suggested and transferred to the patient. In the proposed system, the average score is fixed as 50 percent. If the abnormality level is above 40 percent, the patient is admitted to hospital. In Table 5.1, patients 4, 5 and 8 are immediately admitted to the hospital, whereas the others do not need to be admitted to hospital. The remedies suggested to patient 6 are given in Table 5.3.

These remedies are transferred to the patient and the health condition is monitored after these remedies are taken by the patient. If the health conditions are not normalized, it is suggested that the patient be admitted to hospital. Otherwise, the patient continues their regular work without any issues. Throughout this process, patient health monitoring is continuously monitored by the care center and provides better services to patients.

TABLE 5.3
Suggested Remedies to Patient ID 6

Parameter Name	Parameter Value	Suggested Remedies
Patient ID	5	–
Patient Gender	1	
Patient Age	70	
Temperature	99.2	Take Injection to reduce temperature
Sleep cycle	5	Sleep Well
Stress Level	15	Do Yoga and Listen Music
Oxygen Level	80	

5.17 CONCLUSION AND FUTURE WORK

The proposed health-mentor, a personalized health monitoring system using the Internet of Things (IoT) and blockchain technologies is used to monitor the patient's health conditions in an efficient way. The IoT sensors are used to collect the health conditions of the patient, and the ANN algorithm is used to classify the normal and abnormal data classification. When compared to the existing algorithm, the ANN algorithm with a 70:30 ratio produces better classification results. Hence the ANN-based classification technique is used in the proposed system. The abnormal data are predicted through the random forest algorithm and transferred the predicted data to block construction. To avoid collisions, the SHA256 algorithm is used to construct the block and transfer to the nearby hospital by using the shortest path algorithm. The blocks are analyzed by the care providers and send to the recommendation system. The recommendation system analyzes the historical data and final score of the abnormal data, predicts the suggestions and transfers the suggested remedies to patient. When compared with the existing health monitoring technique, the proposed technique provides better security, integrity and confidentiality to user data with minimal transmission time and latency. Similarly the recommendation technique provides a better solution to the patient. In future, the aim will be for the proposed technique to be implemented with dynamic data at an improved level of accuracy.

REFERENCES

1. Drew Ivan, "Moving toward a blockchain-based method for the secure storage of patient records", ONC/N/IST Use of blockchain for healthcare and research workshop. Gaithersburg, Maryland, 2016.
2. Bessem Zaabar, Omar Cheikhrouhou, Faisal Jamil, Meryem Ammi, Mohamed Abid, "Healthcare: A secure blockchain-based healthcare data management system", *Computer Networks*, 2021, pp. 1–16.
3. Abdullah Al Omar, "Mohammad Shahriar Rahman, Anirban Basu and Shinsaku Kiyomoto, "MediBchain: A Blockchain based privacy preserving platform for healthcare data", SpaCCS 2017 Workshops", *LNCS*, 10658, 2017, pp. 534–543.
4. M. Sumathi, S. Sangeetha, "Survey on sensitive data handling- challenges and solutions in cloud storage system", *Advances in Big Data and Cloud Computing*, Springer, 2019.

5. M. Sumathi, S. Sangeetha, “Sensitive data identification in cloud based online banking system”, International Conference on Communication & Security (ICCS 2017) Sastra University, Thanjavur.
6. Dahlia Sam, S. Srinidhi, R. Niveditha, S. Amudha and D. Usha, “Progressed IoT based remote health monitoring system”, *International Journal of Control and Automation*, Vol. 13, No. 2s, 2020, pp. 268–273.
7. L Vijay Anand, Murali Krishna Kotha, Nimmati Satheesh Kannan, Sunil Kumar, M.R. Meera, “Design and development of IoT based health monitoring system for military applications”, *Materials Today: Proceedings*, 2021.
8. Avrajit Ghosh, Arnab Raha, Amitava Mukherjee, “Energy efficient IoT Health Monitoring system using Approximate Computing”, *Internet of Things*, 2020, Vol. 9, pp. 1–17.
9. Samira Akhbarifar, Hamid Haj Seyyed Javadi, Amir Masoud Rahmani, Mehdi Hosseinzadeh, “A secure remote health monitoring model for early disease diagnosis in cloud based IoT environment”, *Personal and Ubiquitous Computing*, 2020.
10. V. Vedanarayanan, S. Durga, Avinash Sharma, T. Gomathi, S. Poonguzhali, L. Megalan Leo, A. Aranganathan, “Utilization of IoT for a secured clinical information transmission and remote monitoring of patients”, *Materials Today: Proceedings*, 2020, pp. 1–5.
11. Li Hong-tan, Kong Cui-hua, Balaanand Muthu, C.B. Sivaparthipan, “Big data and ambient intelligence in IoT based wireless student health monitoring system”, *Aggression and Violent Behavior*, 2021, pp. 1–9.
12. Wang Huifeng, Seifedine Nimer Kadry, Ebin Deni Raj, “Continuous health monitoring of sportsperson using IoT devices based wearable technology”, *Computer Communications*, 2020, Vol. 160, pp. 588–595.
13. Nonita Sharma, Monika Mangla, Sachi Nandan Mohanty, Deepak Gupta, Prayag Tiwari, Mohammed Shorfuzzaman, Majdi Rawashdeh, “A smart ontology-based IoT framework for remote patient monitoring”, *Biomedical Signal Processing and Control*, 2021, Vol. 68, pp. 1–12.
14. Mahesh Ashok Mahant, Vidyullatha Pellakuri, “Innovative supervised machine learning techniques for classification of data”, *Materials Today: Proceeding*, 2020, pp. 1–4.
15. Trong Thanh Han, Huong Yen Pham, Dang Son Lam Nguyen, Yuki Iwata, Trong Tuan Do, Koichiro Ishibashi, Guanghao Sun, “Machine learning based classification model for screening of infected patients using vital signs”, *Informatics in Medicine Unlocked*, 2021, pp. 1–11.
16. P.S.G. Aruna Sri, D. Lalitha Bhaskari, “Blockchain technology for secure medical data sharing using consensus mechanism”, *Materials Todays: Proceedings*, 2020, pp. 1–8.
17. Rubal Jeet, Sandeep Singh Kang, “Investigating the progress of human e-healthcare systems with understanding the necessity of using emerging blockchain technology”, *Materials Today: Proceedings*, 2020, pp. 1–11.
18. Asad Abbas, Roobaea Alroobaea, Moez Krichen, Saeed Rubaiee, S. Vimal, Fahad M. Almansour, “Blockchain-assisted secured data management framework for health information analysis based on internet of medical things”, *Personal and Ubiquitous Computing*, Springer, 2021, pp. 1–14.
19. Dheeraj Mohan, Lakshmi Alwin, P. Neeraja, K. Deepak Lawrence, Vinod Pathari, “A private Ethereum blockchain implementation for secure data handling in Internet of Medical Things”, *Journal of Reliable Intelligent Environments*, Springer, 2021, pp. 1–18.
20. Rajakumar Arul, Roobaea Alroobaea, Usman Tariq, Ahmed H. Almulihi, Fahd S. Alharithi, Umar Shoaib, “IoT-enabled healthcare systems using blockchain dependent adaptable services”, *Personal and Ubiquitous Computing*, Springer, pp. 1–15.

6 Image Analysis Using Artificial Intelligence in Chemical Engineering Processes

Current Trends and Future Directions

P. Swapna Reddy and Praveen Kumar Ghodke

National Institute of Technology Calicut, Kozhikode, India

CONTENTS

6.1 Introduction 80
6.2 Artificial Intelligence in Practice 80
  6.2.1 The Impact on Academic Research 81
  6.2.2 Impact in Industrial Practice 81
6.3 AI Principles 82
  6.3.1 Data-Driven Approach 82
  6.3.2 Knowledge-Based Approach 83
6.4 Image Analysis Using AI 83
  6.4.1 Image Analysis in Process Systems Engineering 83
  6.4.2 Image Analysis in the Petroleum Industry 85
    6.4.2.1 Machine Learning in Upstream 86
  6.4.3 Image Analysis in Wastewater Treatment 86
6.5 Real-Time Quality Monitoring System 89
6.6 Catalyst Design Using Image Processing 89
6.7 AI in Fault Detection and Diagnosis 90
6.8 Goals and Scopes of Image Analysis Using AI in Practice 92
6.9 Challenges of Image Analysis in Industry 93
6.10 Recent Trends and Future Outlook 94
6.11 Conclusion 95
References 95

DOI: 10.1201/9781003267782-6

6.1 INTRODUCTION

The present enthusiasm for artificial intelligence (AI), especially machine learning (ML), is apparent and addictive. Some intellectuals have outlined prophetic visions and expressed concerns about AI's potential to "revolutionize," if not even to take over from, humanity [1]. Interest in AI's business potential has attracted a lot of government-sponsored investment and venture capital worldwide, especially in China. For example, McKinsey estimates the business impact of AI in a variety of disciplines, forecasting the creation of trillion-dollar industries. All of this is fueled by AI's rapid, explosive, and unexpected breakthroughs over the past decade [2]. Computer vision, robotics, games-playing sectors, natural language processing systems, speech recognition, AlphaGo, Alexa, self-driving cars, and Watson are among the many incredible achievements of this period. In the 1990s expert systems and neural networks created a great deal of hype and a tendency to overstate the potential of these innovations. In the current scenario, many chemical engineers have enthused about the potential uses of AI, such as ML, applied in areas of catalyst design, petroleum refinery units, wastewater treatment, fault detection, etc. The idea appears to offer a unique solution to complex, long-standing chemical engineering challenges using both AI and ML. The application of AI in chemical engineering is a 35-year-old initiative which has achieved some notable achievements [3].

Chemical engineering, it seems clear, is currently at a critical juncture. The chemical engineering field is undergoing a transformation that brings both problems and opportunities in terms of modeling and automated decision-making [4]. The most crucial factors contributing to opportunities are low-cost high-computing performance that brings tremendous progress in molecular engineering, increasing automation and integrated operations, etc., that delivers faster goods and services to market. The processing of huge volumes of heterogeneous data in fractions of time is one significant outcome where AI, and particularly ML, would play an important role [5].

The present chapter is directed at chemical engineers and researchers in the field who are interested in the potential for AI, such as ML, Artificial Neural Network (ANN), Recurrent Neural Network (RNN), and so on. First, let's take a look back and highlight earlier initiatives that have yielded vital aspects for future development. Second, using these essential aspects, we can identify prospective current and future applications in chemical engineering. It is necessary to understand the "reality check" to realize current development and examine the prospects more precisely.

6.2 ARTIFICIAL INTELLIGENCE IN PRACTICE

Artificial intelligence (AI), which is the core branch of computer science, helps to develop smart systems and resolves problems in a manner comparable to the human intelligence system. The main aim of AI applications to any system is to enhance computer algorithms that are relevant to human knowledge, such as learning, problem solving, reasoning and perception [6]. Further, the growing production technologies over recent decades have contributed to a rise in complex processes across major sectors of industries, such as healthcare, smart cities and transportation, e-commerce, finance, and academia [7]. AI is classified into three main areas: machine learning, deep learning and data analytics. These techniques are widely used for intelligent decision-making, blockchain, cloud computing, the Internet of Things (IoT) and the so-called fourth industrial revolution (Industry 4.0).

6.2.1 The Impact on Academic Research

AI is one of the technologies which is growing rapidly due to its unique features to learn and adapt a system based on the available data and to make a decision. The expert systems that have been developed on the basis of AI created beliefs that started to affect many facets of engineering and, specifically, process engineering work. Expert systems were followed by an explosion in the use of neural networks, and along with the fuzzy systems these three technologies spearheaded the thrust of a rejuvenated artificial intelligence to play a practical role in engineering practice.

In the present scenario "Intelligent systems" has moved from the fringe to the mainstream in various activities of process engineering; the monitoring and analysis of process operations, fault diagnosis, supervisory control, feedback control, scheduling and planning of operations, simulation, process and product design. This evolution also signifies another important development; the early battles that pitted researchers and developers of engineering applications from artificial intelligence, operations research, systems and control theory, and statistics, against each other, have proved to be pointless.

6.2.2 Impact in Industrial Practice

The number of industrial deployments of AI-based diagnostic systems have been rising over the past four decades at a remarkable rate. Some of the more noteworthy are listed below:

- ➢ FALCON (DuPont [8], [9]): A combined effort by DuPont, Foxboro, and University of Delaware, this diagnostic system was installed at DuPont's adipic acid plant in Victoria, TX.
- ➢ PDIAS (Idemitsu Kosan Co. and Mitsubishi Heavy Industries Ltd. [10]): A fault diagnosis system for a 30,000 bbl/day catalytic cracking unit. This has already proven to be effective in assisting plant operators in determining the source of plant disturbances.
- ➢ BZOEXPERT: A support system for failure detection and diagnosis in wastewater treatment plants. It also explains how the procedure behaves during the diagnosis.
- ➢ DIAD-Kit/BOILER (Mitsubishi Kasei Corp., MIT): A cooperative effort that led to the installation of an on-line performance monitoring and diagnosis system on a cogeneration power plant at Mitsubishi's Kyushu plant site.
- ➢ Diagnosis of compressors (Exxon Chemical Co.): Exxon developed a knowledge base on the Crystal expert system shell which could analyze the vibration data from three compressors for specific faults.
- ➢ ASPEX [Watts] is a rule-based system, developed to diagnose faults in activated sludge processes.

- ➢ REACTOR [Nelson] is a rule-based system for the diagnosis and treatment of nuclear reactor accidents.
- ➢ Calgon Corp. [10] has developed an expert diagnostic advisor with over 1000 rules for the diagnosis of cooling-water treatment processes.
- ➢ ECHOS (Toyo Engineering Co. [10]): Toyo has developed the ECHOS (the Ethylene Cracking Heater Operation and Maintenance Supporting System). This system has been in use since the spring of 1987 at Osaka Petrochemical Industries' 300,000 m.t./yr ethylene plant in Takaishi. This on-line, real-time system records operating data and tries to advise operators on what caused any anomalies and what action they should take.

AI's significance is rising continually with time due to the integration of AI-based systems characterized by intelligence, adaptability and intentionality [11].

6.3 AI PRINCIPLES

In the development of decision support systems based on AI principles, the three solution approaches of data-driven, analytical, and knowledge-based have been identified. Among them, the data-driven approach and the knowledge-based approach are gaining in importance because of their wide applicability. A schematic of a typical expert system is given in Figure 6.1.

6.3.1 Data-Driven Approach

The process-monitoring techniques that have been most effective in practice are based on models constructed almost entirely from process data. The early and accurate fault identification and diagnosis of industrial processes can help to reduce manufacturing costs while also reducing downtime. The most popular data-driven process monitoring approaches include principal component analysis (PCA), Fisher

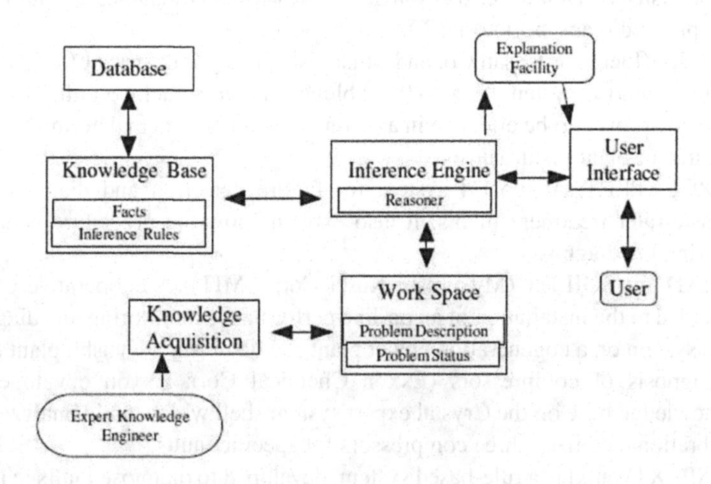

FIGURE 6.1 Typical expert system architecture [12].

discriminant analysis, partial least-squares analysis (PLS), and canonical variate analysis. Of these, PCA and PLS have been increasingly adopted for feature extraction from historical databases developed from process operations [13].

6.3.2 Knowledge-Based Approach

Heuristics and reasoning, which entail ambiguous, conflicting, and nonquantifiable information, are incorporated into knowledge-based methodologies as applied in automated reasoning systems [14]. Artificial intelligence technologies that are linked to knowledge-based methodologies and used in the process industries for monitoring, control, and diagnostics include expert systems, fuzzy logic, machine learning and pattern recognition.

6.4 IMAGE ANALYSIS USING AI

Image analysis is the process of extracting usable information from images. Modern image analysis is widely employed in many areas of research and development because it provides for quick, accurate, and reliable quantitative analysis. Images acquired by imaging hardware are processed in several steps, using a variety of image-processing algorithms to extract quantitative characteristics. Image analysis has been used in a variety of applications with great effectiveness, ranging from simple densitometric evaluation to animal phenotyping and biomass analysis. Further, the use of machine learning techniques in image analysis has become an increasingly common method.

6.4.1 Image Analysis in Process Systems Engineering

The area of process systems engineering (PSE) has existed in various forms for more than five decades, usually under the labels of process designs and process control ([15, 16, 17]). In the growing technology, the main concerns are: In future, how will a process be regulated and operated? Will smart and intelligent systems take the role of traditional control room operators? What level of intelligence will be embedded in processing equipment? Will the significance of optimization grow?

In addition, there are megatrends that are driving the economy, such as the aging infrastructure (many plants were built in the 1970s), the shift in consumer patterns (more individualized products), environmental concerns (pollution, climate change and global warming) and the skills gap (there are very few people still working that actually know how to build a plant). These are among the challenges that necessitate additional training and education, interaction among groups and other means for the maintenance of theoretical knowledge. On the other hand, digital transformation has made the main focus the shift from business-to-consumer (mobile phone apps, entertainment, communication, banking) to business-to-business apps. Here, technologies such as machine learning, image analysis, big data, time-sensitive networking, blockchains, cloud computing, and 5G communication are being considered for a range of applications. This has shown great opportunities for software vendors and almost every larger software company (e.g. Google, Amazon, Microsoft, IBM) is now

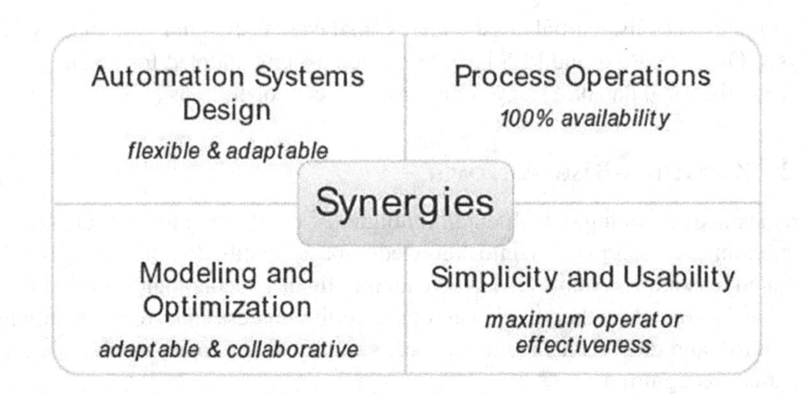

FIGURE 6.2 Synergies between the traditional PSE areas [18].

offering platforms that can be used to host process data analytics. Upcoming trends can also be observed in daily technology news, which report the increasing use of open source software, the need for standardized components, companies building up strategic directions for their digitalization and cases where traditionally isolated problems are mixed and solved together. Software companies are also approaching process industries with the aim of proving their capabilities in process analytics and decision-making. Figure 6.2 shows the different areas of PSE where the scope of digitalization can improve the performance of the process.

Machine learning is an emerging area that is now attracting a lot of attention in a variety of fields ([19, 20]). Lee et al. [21] discuss the potential of deep learning for the efficient training of neural networks with a large number of hidden layers, which, in turn, allow for hierarchical feature learning of the input data for PSE application. Furthermore, the authors also discussed the potential of reinforcement learning for handling operational problems. Sahinidis and his group ([22, 23]) worked on machine learning and optimization with the development of the system ALAMO, which allows the selection of a rich set of functions through the use of global optimization techniques for mixed integer nonlinear programming (MINLP), such as BARON. One of the challenge in machine learning is the development of hybrid models that combine basic physical principles with data-driven models based on neural networks as described by Venkatasubramanian [4].

Another area of research is in the process design where progress has been the incorporation of molecular design for the simultaneous design of materials such as solvents and a corresponding process. Claire Adjiman and her group have been developing mathematical models based on combinatorial search and optimization to accomplish this objective ([24–26]). In the area of product design Gani [27] and co-workers have been developing ProCAPD; Kalakul et al. [28]], a chemical product design simulator, based on computer-aided methods for design analysis of single molecular products (solvents, refrigerants, etc.); mixtures and blends (gasoline, jet fuel, lubricants); and liquid formulated products (cosmetics, detergents, paints, insect repellents). Further, in the area of process synthesis Chen et al. [29] are developing superstructure optimization methods for process flowsheets using Generalized

Disjunctive Programming algorithms in Python combined with global optimization methods. Many research groups are also putting in significant research efforts, such as the RAPID project conducted by AIChE for process intensification (RAPID, 2018) and the IDAES project led by NETL (IDAES, 2018). Tula et al. [30] have also developed the synthesis tool ProCAFD which can address intensification problems.

Further, optimization under uncertainty is an area that has been receiving increased attention, given the fact that design, planning and the scheduling of process systems often involve significant uncertainties ([31] and [32]). Apap and Grossmann [33] developed a multistage stochastic programming for not only exogenous uncertain parameters (e.g. prices, demands) but also endogenous uncertain parameters that are decision-dependent (e.g. outcomes in oil drilling and clinical trials). Another important computational challenge concerns de-composition methods [34] for global optimization under uncertainty, which has been addressed by Paul Barton and his group ([34]). Also, given that so much more operational data are recorded and accessible, the systematic modeling of probabilities for scenario trees [35] and actual probability distribution functions are new and interesting areas of research [36].

Another major challenge for PSE companies is: How to ensure that the existing process can continue production after overly high investments and unrealistic payback times? Among the guidelines suggested are: To ensure the step-by-step introduction of Internet of Things (IoT) concepts (protect the investments); To ensure transparent and controllable plant upgrades (this requires also good control of data flows); To use standards to deploy the full and long-term potential of digitalization (where possible); and To keep data protection and data integrity high (consider different aspects of cyber security).

6.4.2 Image Analysis in the Petroleum Industry

Systems for reservoir engineering, oil field exploration, drilling, and production engineering are all parts of the petroleum industry. If conventional fuel prices continue to increase, corporations/industries must innovate new technology, intensify the design process, and optimize & monitor operations to improve overall efficiency and expand their capacities. The simplest method to conserve efficiency and productivity is to optimize cumulative extraction through effective and smart technologies, such as inflow control devices (ICD), ResFlow ICD, ResInject injection ICD, and or inflow control valves (ICV) as well as downhole sensor systems. Quick decisions are necessary to improve automated controls in major oilfields and reservoir engineering. The Smart Oilfield and a comprehensive oilfield technological infrastructure achieved by digitizing instrumentation systems are both necessary advances [37].

Sircar et al. (2021) had reviewed three different machine learning algorithms to anticipate multiphase flowing bottom hole pressure [37]. Real field data from such an open literature library was used to build and test the model, to validate the accurate procedure of the burst header packet (BHP) obtained using ML models. Many suggested models were employed to verify the usefulness of the dataset built using ML. Hazlet et al. (2021) examined the directional well drilling rate of penetration precision and computational performance of ML methods [38]. For an Iranian carbonate oil deposit, Hassanvand et al. (2018) employed an ANN to determine the strength of rock uniaxial parameters. Few authors reviewed the analysis of cloud computing-based smart-grid technologies in the oil pipeline sensor network systems [39]. Thus, a large number of studies are being performed in drilling and reservoir engineering along with the implementation of ML.

6.4.2.1 Machine Learning in Upstream

Increased data-processing capabilities improve the performance of electronic instruments and scientific instruments. Computing technologies are desirable and have significance in the oil and gas industry sectors for production and exploration. Table 6.1 shows the upstream activities of petroleum refinery, tools applied, and AI approaches to each activity.

Companies are actively pursuing creative techniques to become increasingly efficient by cutting costs, simplifying production, and enhancing worker safety, as the oil and gas business is becoming increasingly competitive and unpredictable. Many industries are moving to digitization to defend themselves from market shocks, improve profitability at lower oil prices, and gain a competitive advantage as the economy grows. AI- and ML-based technologies are the way of the future, and they are fast-growing and being used across the value chain. Numerous industries have recognized the benefits of these new technologies, and AI applications will continue to develop in the future.

6.4.3 Image Analysis in Wastewater Treatment

The most common and frequently used technique of evaluating water quality is by traditional laboratory techniques. These are time-consuming, however, and necessitate professional personnel to perform the test. According to the World Health Organization (WHO), more accurate, sensitive, and reliable analytical techniques are required to assess water and wastewater quality from sewage and industrial plants.

TABLE 6.1
Petroleum Refinery Upstream Process, AI Application Tool, and AI Strategies

Upstream Process	AI Application Tool	AI Approach and Strategies
Subsurface Geological frameworks	➢ A tool for autonomously mapping the properties of reservoir rock across an oil field. ➢ A programme that collects geological information from well records. The procedure can be sped up by increasing the gradient by 100 times or more. A tool for rock typing has been developed based on pictures of rock samples gathered from wells.	None gradient optimization techniques with Interpolation techniques Deep neural network techniques Gradient boosting
Drilling	This tool can detect the drilled rock form and likely failure using real-time drilling telemetry.	Combination of AI/ML algorithms
Reservoir engineering	This tool can speed up traditional reservoir simulations.	Deep neural networks techniques
Optimization of product	A data-driven strategy for accurately estimating the efficacy of oil well-care programmes.	Feature selection based on expert judgment + gradient boosting

As a result of this predicament, numerous autonomous in-situ monitoring approaches have been developed.

The introduction of digital image processing simplifies these methods and produces better results in monitoring wastewater paramcter characteristics. It also leads to projections for the future. The applications of image-processing techniques for wastewater quality detection are explored along with their potential to improve water quality and reduce pollution. For many years, water characteristics have been used to monitor and control water quality in wastewater treatment plants. At present, computer-based imaging systems, satellite images, and embedded systems are the three primary areas of the processes. Image processing techniques, fractal dimension, statistics, and correlation between picture analysis and standard measurement are used to assess image attributes in order to identify water quality.

Image-processing techniques are divided into four phases: image capture, pre-processing (noise reduction and enhancement), image segmentation, and image analysis. Each of the processes mentioned above has been used to improve or produce a high-quality photo image. Figure 6.3 depicts a typical image processing and analysis block diagram.

Sample image attributes must be investigated and studied before image process analysis may be used to detect water quality in water treatment plants. As a result, image process analysis and wastewater characteristics/properties are efficiently utilized to monitor and regulate wastewater treatment plants, in particular, activated sludge wastewater treatment plants and other industrial plants such as tanneries and breweries. Among the basic wastewater monitoring measurements are effluent chemical oxygen demand (COD), active sludge mixed liquor suspended solids (MLSS), and Sludge Volume Index (SVI), which are required to be monitored in order to maintain water quality. Image utilization of sludge, automation procedures and wastewater characterization, and image process analysis have become effective, convenient, and faster. Using an image analysis approach, three aberrant situations can be identified includes viscous sludge bulking, filamentous bulking, and pinpoint flocs. Table 6.2 summarizes the wastewater image processing approaches employed in various research.

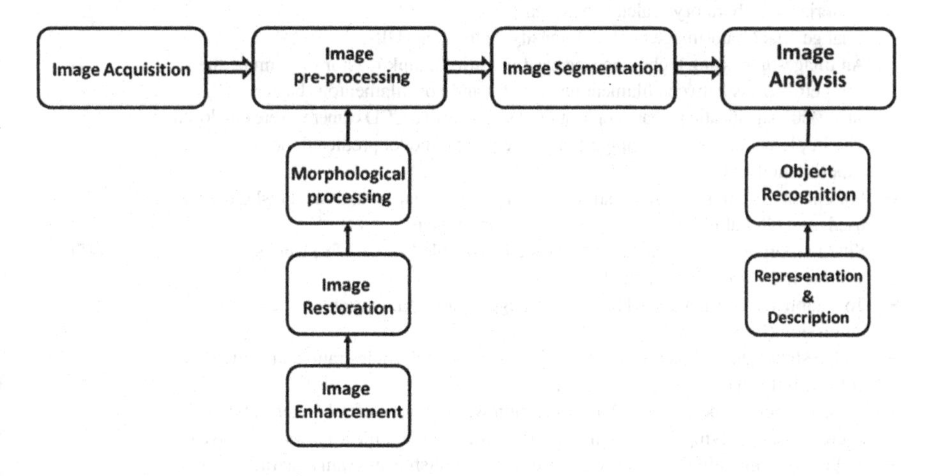

FIGURE 6.3 Typical image processing with the processes of image processing and analysis.

TABLE 6.2

Deep Literature Studies on AI-Based Research Findings in Wastewater Treatment

AI-Based Research Finding in Wastewater Treatment	References
➢ To determine the settleability and concentration level of activated sludge. ➢ Image is processed with a charge-coupled device (CCD) camera and a dark field microscope with a magnification of x50. ➢ The thresholding segmentation technique was implemented, and the processing program was performed using Microsoft Visual C++ software.	[40]
➢ The research goal was to identify the flocs and filamentous bacteria in activated sludge. ➢ Implement a fractal dimension operator for joint detection ➢ Variance and Laplacian of Gaussian operator for other detections. ➢ Bright-field microscopy with a Charge-coupled device video camera was employed. ➢ With a few irregularities due to inadequate lighting, the researcher was able to achieve the objective. ➢ The research demonstrates that texture-based picture segmentation can be performed using fractal dimensions.	[41]
➢ The importance of early identification of bulking was performed in a laboratory-based sludge simulation and found the link between sludge volume index (SVI) and filament fraction. ➢ Determine the shape, size, structure, and distribution of the floc based on the sludge's shape, size, and structure. ➢ Examine the metamorphosis of micro-flocs in the presence and absence of a hazardous substance. ➢ Bright-field microscopy and a CCD video camera are employed. ➢ To improve image contrast, histogram equalization was implemented. ➢ Both pixel-based (thresholding) and edge-based segmentation approaches were employed.	[42]
➢ A video camera with bright field microscopy having x100 was employed. ➢ With VisilogTM image processing software, the histogram enhancement approach and Threshold algorithms are used for segmentation. ➢ It has developed an automated system that monitors the properties of filamentous bacteria in a laboratory scaled experiment. ➢ Changes the Gram-image to Hue, Intensity, Saturation (HIS).	[43]
➢ An image-processing technique was used in a stirred tank reactor to examine the competitiveness between filamentous bacteria and non-filamentous bacteria. ➢ At x1000 magnification, phase contrast microscopy with a CCD cameras were employed. ➢ The implementation of the suggested mathematical model predicted the experimental results. ➢ For enhancement and segmentation, histogram equalization and the threshold were used, and Global Lab image 2.10 was used for analysis.	[44]
➢ Sludge morphological features such as total suspended solids (TSS) and sludge volume index were investigated (SVI). ➢ To establish a link between TSS and total aggregate area, partial least square regression (PLS) is used. ➢ It also shows that filamentous bacteria per suspended particles ratio and sludge volume index have a strong relationship (SVI). ➢ A Zeiss microscope with x100 magnification was employed for filaments, while for aggregates, an Olympus stereo microscope with x40 magnification was employed. ➢ MATLAB simulation software was used to accomplish histogram equalization and threshold augmentation in both cases.	[45]

(Continued)

TABLE 6.2 (Continued)

AI-Based Research Finding in Wastewater Treatment	References
➢ Based on microbial floc, image analysis was used to characterize water quality. ➢ Morphological studies are performed to understand the relation between the factors like compactness, diameter, porosity, roundness, and fractal dimensions. ➢ For image acquisition, bright field microscopy was employed with a photonic microscope and a Nikon CCD camera. ➢ Image Pro-Plus version 4.5 was used to accomplish histogram equalization and median filtering upgrades.	[46]

A key component of a wastewater quality detection system is early quality detection. The process operates as an early warning if any of the quality indicators have exceeded the prescribed limits. In order to detect wastewater quality, a vision system gathers all images of visible wastewater and further process for analysis. Many of the treatment plants use the image collection and analysis of images to monitor the active sludge of wastewater treatment plants. A system that can continually monitor wastewater quality without relying on human involvement is required to improve wastewater quality monitoring. The system can be automated utilizing a vision-based system with online analysis, reporting, and action, or it can be linked to an online system. The ultimate goal is to eliminate human interaction while monitoring any wastewater systems.

6.5 REAL-TIME QUALITY MONITORING SYSTEM

The online monitoring technology allows for real-time and in-situ monitoring of the water quality rather than sending images to a lab for analysis. As a result, human labor is used less frequently, and monitoring costs are lower. A good online wastewater quality monitoring system can track multiple quality indicators, which is cost-effective and simple to use. In online monitoring technology, the images are gathered online in real time for visible quality indicators, and analyses are performed automatically. Scientists developed the world's first fully online digital system, which comprised both monitoring and analysis systems. It consists of a magnetic pump and CCD video cameras for online quality measurements. This system detects particle size using a particle size analyzer, and images are analyzed with a digital image analyzer to calculate parameters such as diameter. Finally, particle sizes are determined using a distribution method with nearly identical peaks that vary in size. The analyzer then seeks out a linear relationship between fractal dimension and precipitation efficiency. The video camera used to capture images of sewage water for sewage water quality. The captured online images are processed in parallel using image-processing software and method to obtain the result. A real-time online measurement tool is incredibly accurate. It can also be used to conduct off-line research.

6.6 CATALYST DESIGN USING IMAGE PROCESSING

The advances in ML have a significant impact on a wide range of domains, including materials science and chemical engineering. ML has recently had a significant impact on heterogeneous catalysis research. It can be used to uncover insights or generate quick predictions about target qualities. The growing amount of data in materials databases is driving advances in the field of materials discovery and development. New catalysts are required for sustainable chemical production, alternative energy, and pollution mitigation applications. Making novel heterogeneous catalysts with good performance is a difficult task. The catalyst's performance depends on a number of factors, including particle size, composition, support, particle morphology, and an atomic coordination system. Different processes, such as Ostwald ripening, particle disintegration, surface oxidation, and surface reconstruction, might affect the characteristics of these catalysts under reaction conditions. Many heterogeneous catalysts are complexly disordered, making atomic-level characterization difficult even by modeling and simulations [47].

Computational modeling employing quantum mechanical (QM) methodologies such as density functional theory (DFT) might help speed up catalyst screening by discovering active sites and structure– activity relationships. However, due to the high computational cost of QM approaches, only a limited number of catalyst spaces can be investigated. Recent research in combining ML with QM models and experiments promises to advance rational catalyst design [48]. As a result, it is time to emphasize the ability of ML technologies to speed up heterogeneous catalyst research. In this perspective, studies on ML can help with heterogeneous catalyst design and discovery. Early investigations correlated catalytic properties and reaction conditions with measured catalytic performance using neural networks, but the number of systems investigated was limited.

Recently, ML has been used to identify heterogeneous catalysts. The method was used to predict properties of catalysts, such as stability, activity, and selectivity. Decision trees, kernel ridge regression, neural networks, support vector machines, principal component analysis, and compressed sensing are all examples of ML algorithms. The study aims to uncover how ML is influencing heterogeneous catalysis research. Homogeneous catalysis research was also aided by ML, which has many similarities (and differences) with ML studies for heterogeneous catalysis. It observed the usefulness of ML paired with QM calculations to speed up the search for effective catalysts. The use of ML-derived interatomic potentials for accurate and rapid catalyst simulations and ML's potential to assist in the discovery of descriptors of catalyst performance in huge datasets was evaluated by many authors [5, 49].

To conclude, despite the growing popularity of ML in a variety of fields, its application in catalysis is still in its infancy. Catalysts are often created and manufactured through trial and error with chemical intuition, which is a time-consuming and expensive resource. The automated machine learning approach has been found to help in developing better models, the comprehension of the catalytic mechanism, and the development of novel catalytic designs. This has been made possible by the development of cutting-edge algorithms and theory, the widespread availability of experimental data, and inexpensive processing costs.

6.7 AI IN FAULT DETECTION AND DIAGNOSIS

Diagnostic activity comprises of two important components: a priori domain knowledge and search strategy. The basic a priori knowledge that is needed for fault diagnosis is a set of failures and the relationship between the observations (symptoms) and the failures. A diagnostic system may have them explicitly (as in a table look-up), or it may be inferred from some source of domain knowledge. A priori domain knowledge may be developed from a fundamental understanding of the process using a knowledge of first principles. Such knowledge is referred to as deep, causal or model-based knowledge [50]. On the other hand, it may be gleaned from past experience with the process. This knowledge is referred to as shallow, compiled, evidential or process history-based knowledge.

Advanced supervision, fault identification and fault diagnosis methods are becoming increasingly essential for many technological and industrial processes in order to ensure reliable and safe performance. Fault detection and diagnosis have been carried out for various chemical processes, including the Tennessee Eastman process (TEP) ([51, 52]), reactor system ([53, 54]), distillation column ([55–57], [58], [59, 60]), bearing faults [61], crude and gas mixture pipelines ([62, 63]), industrial gas turbine [64], heating furnace [65], water-cooled centrifugal chiller [66] and biochemical wastewater treatment plant [67]. Further, fault detection and diagnosis can be carried out by first-principle, data-driven, or knowledge-based approaches [68]. First-principle approaches require the construction of a mathematical model based on theoretical knowledge. This approach often fails because of the complexity of the resulting mathematical model. On the other hand, the knowledge-based approach relies on having prior understanding or knowledge of the relationships between faults and model parameters or states. It is also challenging to apply this approach to large-scale systems because of the effort and skills required to construct these complex fault models ([69, 70]).

Data-driven fault detection methods can be categorized into two main types, namely, supervised and unsupervised learning approaches. The supervised approach includes ANN [71], support vector machine (SVM) [72], Bayesian network (BN), etc. On the other hand, among the examples of the unsupervised learning approach are principal component analysis (PCA) ([13, 73]), partial least square (PLS) [73], independent component analysis (ICA) [74], etc. Fault diagnosis is considered a classification problem, in which a specific type of fault is to be determined based on the data shown (similarly for determining either single or multiple classes of faults), and supervised learning approaches that are commonly utilized include SVM, decision tree (DT), K-nearest neighbor (KNN), etc.

The increasing complexity of industrial systems and their related performance requirements have induced the need to develop new approaches for their supervision. ANN is the most commonly used data-driven approach for fault detection in the process industries. It has gained substantial popularity due to its capacity to learn complicated and nonlinear dynamics of processes. On the other hand, in terms of fault diagnosis techniques, the ability of SVM to handle classification problems allow it to be extensively utilized for fault classification in numerous industrial applications.

Hence, there are a few possible areas where research is growing rapidly:

- Simultaneous faults: Some studies in the current research comprise of simultaneous faults, including a combination of two, three, and four faults occurring at the same time. In real systems, there is a possibility that avoiding a certain fault may result in the occurrence of another subsequent fault.

Hence, it would be more beneficial to train the algorithm with simultaneous faults occurring with different time delays to incorporate the more realistic scenarios of real systems.

- Adaptive fault detection and diagnosis algorithm: Adaptivity in any algorithm is always a desirable property. However, due to the ever-changing process conditions in real industrial chemical processes, development of such adaptive algorithms may be a challenging task. Nevertheless, the present-day exponential growth in efficient machine learning techniques may open up paths toward this idea.
- Deep learning image-based algorithm: In recent years, deep learning has grown increasingly popular in industry as an alternative method over the traditional machine learning. Convolutional neural networks (CNN), in particular, has seen successful application in process fault diagnosis ([75–77]) and industrial inspection ([78, 79]). Some recent applications in chemical process systems can also be found for control valve stiction diagnosis as reported in Basha et al. [63] and Song et al. [80]. These deep learning features extraction methods are exceptional in extracting the key features in the data, and that may facilitate in identifying and/or diagnosing various fault types more effectively. A drawback of deep learning methods, however, such as CNNs, is that they generally require massive amounts of data to train, although CNNs can be partially retrained to take full advantage of the data that are available in the domain of application.
- Online predictive monitoring for fault detection/diagnosis: Researchers have proposed an integrated framework based on the use of a convolutional neural network (CNN) and principal component analysis (PCA) has been proposed for stiction detection and severity identification. The CNN is used to extract features from the data representative of control valve behavior, while PCA is used to generate statistical process control charts from these features for the automated monitoring of control valve stiction. Their results illustrate the promising capability of the CNN-PCA framework for possible online predictive monitoring on the current and future health of their valves. In fact, as highlighted in Dalzochio et. al, [81] one of the key concepts highlighted in the fourth industrial revolution (Industry 4.0) is predictive maintenance. Predictive maintenance may help in reducing faults and/or unplanned downtimes; however, challenges still exist in the implementation of the data-driven machine learning techniques for online monitoring or failure prediction. Few aspects have to be considered, such as the varying dynamic operating conditions, obtaining the right training data and the generalization capability of a given machine learning technique to satisfy all scenarios in a plant [81].

6.8 GOALS AND SCOPES OF IMAGE ANALYSIS USING AI IN PRACTICE

The hype around AI has led to claims that radiologists will be rendered obsolete. However, it is currently unclear if AI will eventually replace radiologists [82]. If these applications generate abrupt alterations to the context's integrity and relational dynamics, they could cause substantial ethical and legal difficulties in healthcare. Maintaining confidence and trustworthiness is a primary goal of governance, which is essential for promoting collaboration among all stakeholders and ensuring the responsible development and application of AI in radiology. Radiologists, it is believed, should take a more active role in ushering medicine into the digital age. Professional responsibilities in this regard include investigating the clinical and social value of AI, addressing technical knowledge gaps to facilitate ethical evaluation, assisting in the recognition and removal of biases, overcoming the "black box" barrier, and brokering a new social contract on informational use and security [83]. A much closer integration of ethics, regulations, and good practices is required to ensure that AI governance accomplishes its normative goals.

The following areas have been identified in which AI can be implemented to automate the systems.

- ➢ Image segmentation, lesion detection, quantification, classification, and comparison with historical pictures are all automated processes.
- ➢ Creating radiology reports, especially with the use of natural language processing and generation.
- ➢ Semantic error detection in reports/magazines etc.
- ➢ Data mining.
- ➢ It enhanced business intelligence solutions that enable real-time dashboarding and alerting.

Policy and professional interventions may be necessary to manage job displacement, establish new employment and roles within healthcare (such as medical data scientists), and to reduce transition friction as repetitive, low-discretion, and tedious procedures become automated [84].

As a result, AI applications in healthcare are expected to generate substantial ethical and legal difficulties. Healthcare is now provided by licensed professionals in accredited facilities and is structured to flow from a professional to care recipients through regulated routes. AI applications are part of a rapidly growing number of new technologies that allow laypeople to access a vast pool of knowledge, interact with others with expertise and/or experience, and, in future, to derive accurate diagnoses and develop effective healthcare regimens without the assistance of a clinician. AI applications may potentially wreak havoc on present healthcare procedures. Clinical medicine is increasingly reflecting a shift-based approach in many health systems. While a doctor remains at the core of attribution of ethical and legal responsibilities, payment arrangements, and organizational design, AI applications may play a larger role in integrating healthcare. Further, AI may limit a patient's ability to exercise his or her right to privacy and confidentiality because ML analysis necessitates the recording of the patient's personal information.

6.9 CHALLENGES OF IMAGE ANALYSIS IN INDUSTRY

For decades, the topic of image processing has been the focus of intense research and development efforts. Many outstanding and successful applications have resulted from rapid technical advancements, particularly in terms of processing power and network transmission bandwidth. Images are now pervasive in our daily lives. Digital TV (e.g., broadcast, cable, and satellite TV), Internet video streaming, digital cinema, and video games are all examples of applications that have profited substantially. Imaging technologies are used in a wide range of applications, including digital photography, video conferencing, video monitoring and surveillance, and satellite imaging, as well as in more distant domains such as healthcare and medicine, distance learning, digital archiving, cultural heritage, and the automotive industry.

In this chapter, a few major research challenges for future image and video systems are addressed in order to accomplish breakthroughs that match end users' escalating expectations. Image processing is a vast and diverse field, with numerous successful applications in both consumer and business markets. Many technical obstacles remain, however, in order to push the boundaries of imaging technology even further. On the one hand, there is a constant push to improve the quality and realism of image and video content, while, on the other hand, there is a push to be able to successfully read and comprehend the large and complicated amount of visual data. However, there are numerous other intriguing topics, such as those relating to computational imaging, information security, and forensics, or medical imaging. Image processing, psychophysics, optics, communication, artificial intelligence, computer vision, and computer graphics will all play a role in key advances. Multidisciplinary collaborations involving researchers from both industry and academia are crucial moving ahead to achieve these discoveries.

6.10 RECENT TRENDS AND FUTURE OUTLOOK

The emergence of a largely bottom-up, data-driven strategy for knowledge acquisition using deep reinforcement learning, convolutional networks, and statistical learning has made it much easier to solve image recognition problems. However, it is unclear whether all of these tools are required to implement AI in chemical engineering. First, massive volumes of data are necessary for ML techniques to perform appropriately and understand chemical engineering applications. Second, in contrast to gameplay, vision, and voice, our systems are regulated by fundamental physics and chemistry (and biology) rules and principles, which should be compensation for the lack of "big data" acquisition. As a result, before deploying deep neural networks or ML, one should consider combining first principle knowledge with data-driven models to develop hybrid models more quickly and consistently.

ML in catalyst design and discovery have barriers and there is a necessity for systematic preservation and access to catalytic reaction data. For example, the development of testbeds for computer vision was critical to the breakthroughs made with deep neural networks. The chemical engineering community also requires open-source ML software in order to create catalyst design applications. In general, it appears that continued improvement will necessitate such community-based approaches. The Stanford Catalysis-Hub Database and the Atomistic ML Package are two of the most promising recent open-source projects. To do so, it will be necessary to develop domain-specific representations search engines for chemical entity extraction systems. While this is more difficult and requires more than a cursory understanding of AI approaches, previous proof-of-concept contributions can serve as a starting point. Recent contributions to automated reaction network generation and reaction synthesis planning are promising breakthroughs in this area.

6.11 CONCLUSION

The advances in the field of AI and ML and their applications in the oil and gas industry, catalyst design and discovery, process systems engineering, petroleum industry, wastewater treatment, and fault detection and diagnosis are presented in this chapter. According to the literature analysis, the oil and gas industry is well positioned to profit from ML due to its ability to process large amounts of data and perform computations quickly. Throughout this study, a variety of monitored learning approaches have been defined and described. The advances in image analysis and wastewater characteristics that have been made thus far have resulted in considerable positive outcomes and increases in overall quality monitoring performance. The approach is expected to accelerate substantially in the near future, becoming the hallmark of a high-level computational tool in chemical engineering.

REFERENCES

1. Caudai C, Galizia A, Geraci F, Le Pera L, Morea V, Salerno E, et al. AI applications in functional genomics. *Comput Struct Biotechnol J* 2021;19:5762–90. https://doi.org/10.1016/j.csbj.2021.10.009.
2. Alexander A, Jiang A, Ferreira C, Zurkiya D. An intelligent future for medical imaging: A market outlook on artificial intelligence for medical imaging. *J Am Coll Radiol* 2020;17:165–70. https://doi.org/10.1016/j.jacr.2019.07.019.
3. Li L, Rong S, Wang R, Yu S. Recent advances in artificial intelligence and machine learning for nonlinear relationship analysis and process control in drinking water treatment: A review. *Chem Eng J* 2021;405:126673. https://doi.org/10.1016/j.cej.2020.126673.
4. Venkatasubramanian V. The promise of artificial intelligence in chemical engineering: Is it here, Finally? 2018. https://doi.org/10.1002/aic.16489.
5. Moses OA, Chen W, Adam ML, Wang Z, Liu K, Shao J, et al. Integration of data-intensive, machine learning and robotic experimental approaches for accelerated discovery of catalysts in renewable energy-related reactions. *Mater Reports Energy* 2021;1:100049. https://doi.org/10.1016/j.matre.2021.100049.
6. Paschen U, Pitt C, Kietzmann J. Artificial intelligence: Building blocks and an innovation typology. *Bus Horiz* 2020;63:147–55. https://doi.org/10.1016/J.BUSHOR.2019.10.004.
7. Allam Z, Dhunny AZ. On big data, artificial intelligence and smart cities. *Cities* 2019;89:80–91. https://doi.org/10.1016/j.cities.2019.01.032.
8. Chester D. L. DEL and PSD. An expert system approach to on-line alarm analysis in power and process plants. *Comput Engng* 1984;1:345.
9. Venkatasubramanian V. and Dhurjati P. An object-oriented knowledge base representation for the expert system FALCON. *Found Comput Aided Process Oper* 1987:701.
10. Basta N. Expert systems. *Chem Engng*, 1988.
11. Ghahramani M, Qiao Y, Zhou M, O'Hagan A, Sweeney J. AI-based modeling and data-driven evaluation for smart manufacturing processes; AI-based modeling and data-driven evaluation for smart manufacturing processes 2020. https://doi.org/10.1109/JAS.2020.1003114.

12. Tzafestas G.S., Verbruggen H.B. Artificial intelligence in industrial decision making control, and automation: An introduction. 1995.
13. Yoon S, Macgregor JF. Principal-component analysis of multiscale data for process monitoring and fault diagnosis. *Inst Chem Eng AIChE J* 2004;50:2891–903. https://doi.org/10.1002/aic.10260.
14. Luo X, Zhang C, Jennings N.R. A hybrid model for sharing information between fuzzy, uncertain and default reasoning models in multi- agent systems. *Int J Uncertainty, Fuzziness Knowledge-Based Syst* 2002;10:401–50. https://doi.org/10.1142/S0218488502001557.
15. Sargent RWH. Introduction: 25 years of progress in process systems engineering. *Comput Chem Eng* 2004;28:437–9. https://doi.org/10.1016/J.COMPCHEMENG.2003.09.032.
16. Grossmann IE, Westerberg AW. Research challenges in process systems engineering. *AIChE J* 2000;46:1700–3. https://doi.org/10.1002/aic.690460902.
17. Stephanopoulos G, Reklaitis G V. Process systems engineering: From Solvay to modern bio- and nanotechnology.: A history of development, successes and prospects for the future. *Chem Eng Sci* 2011;66:4272–306. https://doi.org/10.1016/J.CES.2011.05.049.
18. Grossmann IE, Harjunkoski I. Process systems engineering: Academic and industrial perspectives. *Comput Chem Eng* 2019;126:474–84. https://doi.org/10.1016/J.COMPCHEMENG.2019.04.028.
19. T M. Machine Learning. 1997.
20. Kotsiantis SB, Zaharakis ID, Pintelas PE. Machine learning: A review of classification and combining techniques. *Artif Intell Rev* 2006;26:159–90. https://doi.org/10.1007/s10462-007-9052-3.
21. Lee JH, Shin J, Realff MJ. Machine learning: Overview of the recent progresses and implications for the process systems engineering field. *Comput Chem Eng* 2018;114:111–21. https://doi.org/10.1016/J.COMPCHEMENG.2017.10.008.
22. Cozad A, Sahinidis NV, Miller DC. Learning surrogate models for simulation-based optimization. *Am Inst Chem Eng AIChE J* 2014;60:2211–27. https://doi.org/10.1002/aic.14418.
23. Wilson ZT, Sahinidis NV. The ALAMO approach to machine learning. *Comput Chem Eng* 2017;106:785–95. https://doi.org/10.1016/J.COMPCHEMENG.2017.02.010.
24. Adjiman C.S., Harrison NM, Weider SZ. Molecular science and engineering: A powerful transdisciplinary approach to solving grand challenges 2017.
25. Pereira FE, Keskes E, Galindo A, Jackson G, Adjiman CS. Integrated solvent and process design using a SAFT-VR thermodynamic description: High-pressure separation of carbon dioxide and methane. *Comput Chem Eng* 2011;35:474–91. https://doi.org/10.1016/j.compchemeng.2010.06.016.
26. Jonuzaj S, Gupta A, Adjiman CS. The design of optimal mixtures from atom groups using Generalized Disjunctive Programming. *Comput Chem Eng* 2018;116:401–21. https://doi.org/10.1016/J.COMPCHEMENG.2018.01.016.
27. Gani R. Chemical product design: Challenges and opportunities. *Comput Chem Eng* 2004;28:2441– 57. https://doi.org/10.1016/J.COMPCHEMENG.2004.08.010.
28. Kalakul S, Zhang L, Fang Z, Choudhury HA, Intikhab S, Elbashir N, et al. Computer aided chemical product design – ProCAPD and tailor-made blended products. *Comput Chem Eng* 2018;116:37–55. https://doi.org/10.1016/J.COMPCHEMENG.2018.03.029.
29. Chen Q, Johnson ES, Siirola JD, Grossmann IE. Pyomo.GDP: Disjunctive Models in Python. *Comput Aided Chem Eng* 2018;44:889–94. https://doi.org/10.1016/B978-0-444-64241-7.50143-9.

30. Tula AK, Babi DK, Bottlaender J, Eden MR, Gani R. A computer-aided software-tool for sustainable process synthesis-intensification. *Comput Chem Eng* 2017;105:74–95. https://doi.org/10.1016/j.compchemeng.2017.01.001.
31. Sahinidis NV. Optimization under uncertainty: State-of-the-art and opportunities. *Comput Chem Eng* 2004;28:971–83. https://doi.org/10.1016/J.COMPCHEMENG.2003.09.017.
32. Grossmann IE, Apap RM, Calfa BA, García-Herreros P, Zhang Q. Recent advances in mathematical programming techniques for the optimization of process systems under uncertainty. *Comput Chem Eng* 2016;91:3–14. https://doi.org/10.1016/J.COMPCHEMENG.2016.03.002.
33. Apap RM, Grossmann IE. Models and computational strategies for multistage stochastic programming under endogenous and exogenous uncertainties. *Comput Chem Eng* 2017;103:233–74. https://doi.org/10.1016/J.COMPCHEMENG.2016.11.011.
34. Li X, Chen Y, Barton PI. Nonconvex generalized benders decomposition with piecewise convex relaxations for global optimization of integrated process design and operation problems. *Ind Eng Chem Res* 2012. https://doi.org/10.1021/ie201262f.
35. Calfa BA, Agarwal A, Grossmann IE, Wassick JM. Data-driven multi-stage scenario tree generation via statistical property and distribution matching. *Comput Chem Eng* 2014;68:7–23. https://doi.org/10.1016/J.COMPCHEMENG.2014.04.012.
36. Rossi F, Mockus L, Manenti F, Reklaitis G. Assessment of accuracy and computational efficiency of different strategies for estimation of probability distributions applied to ODE/DAE systems. *Comput Aided Chem Eng* 2018;44:1543–8. https://doi.org/10.1016/B978-0-444-64241-7.50252-4.
37. Sircar A, Yadav K, Rayavarapu K, Bist N, Oza H. Application of machine learning and artificial intelligence in oil and gas industry. *Pet Res* 2021. https://doi.org/10.1016/j.ptlrs.2021.05.009.
38. Hazbeh O, Aghdam SK, Ghorbani H, Mohamadian N, Ahmadi Alvar M, Moghadasi J. Comparison of accuracy and computational performance between the machine learning algorithms for rate of penetration in directional drilling well. *Pet Res* 2021;6:271–82. https://doi.org/10.1016/j.ptlrs.2021.02.004.
39. Hassanvand M, Moradi S, Fattahi M, Zargar G, Kamari M. Estimation of rock uniaxial compressive strength for an Iranian carbonate oil reservoir: Modeling vs. artificial neural network application. *Pet Res* 2018;3:336–45. https://doi.org/10.1016/j.ptlrs.2018.08.004.
40. Akhavan-Tafti H, Schaap AP, Arghavani Z, Desilva R, Eickholt RA, Handley RS, et al. CCD camera imaging for the chemiluminescent detection of enzymes using new ultrasensitive reagents. *J Biolumin Chemilumin* 1994;9:155–64. https://doi.org/10.1002/bio.1170090309.
41. Li Y, Wang T, Wu J. Capture and detection of urine bacteria using a microchannel silicon nanowire microfluidic chip coupled with MALDI-TOF MS. *Analyst* 2021;146:1151–6. https://doi.org/10.1039/D0AN02222E.
42. Mesquita DP, Dias O, Amaral AL, Ferreira EC. Relationship between sludge volume index and biomass structure within activated sludge systems. *XVII Congr. Bras. Eng. Quim.*, vol. I, 2008, p. 7.
43. Saleh MD, Eswaran C, Mueen A. An automated blood vessel segmentation algorithm using histogram equalization and automatic threshold selection. *J Digit Imaging* 2011;24:564–72. https://doi.org/10.1007/s10278-010-9302-9.
44. Dias PA, Dunkel T, Fajado DAS, de León Gallegos E, Denecke M, Wiedemann P, et al. Image processing for identification and quantification of filamentous bacteria in in situ acquired images. *Biomed Eng Online* 2016;15:64. https://doi.org/10.1186/s12938-016-0197-7.

45. Wang Z, Kawamura K, Sakuno Y, Fan X, Gong Z, Lim J. Retrieval of Chlorophyll-a and Total suspended solids using Iterative Stepwise Elimination Partial Least Squares (ISE-PLS) regression based on field hyperspectral measurements in irrigation ponds in Higashihiroshima, Japan. *Remote Sens* 2017;9:264. https://doi.org/10.3390/rs9030264.
46. Mesquita DP, Amaral AL, Ferreira EC. Activated sludge characterization through microscopy: A review on quantitative image analysis and chemometric techniques. *Anal Chim Acta* 2013;802:14–28. https://doi.org/10.1016/j.aca.2013.09.016.
47. Sirsam R, Hansora D, Usmani GA. A mini-review on solid acid catalysts for esterification reactions. *J Inst Eng Ser E* 2016. https://doi.org/10.1007/s40034-016-0078-4.
48. Burnham AK, Zhou X, Broadbelt LJ. Critical review of the global chemical kinetics of cellulose thermal decomposition. *Energy & Fuels* 2015;29:2906–18. https://doi.org/10.1021/acs.energyfuels.5b00350.
49. Goldsmith BR, Esterhuizen J, Liu JX, Bartel CJ, Sutton C. Machine learning for heterogeneous catalyst design and discovery. *AIChE J* 2018;64:2311–23. https://doi.org/10.1002/aic.16198.
50. Milne R. Strategies for Diagnosis. 1987. https://doi.org/10.1109/TSMC.1987.4309050.
51. Yu J, Zhang C. Manifold regularized stacked autoencoders-based feature learning for fault detection in industrial processes. *J Process Control* 2020;92:119–36. https://doi.org/10.1016/J.JPROCONT.2020.06.001.
52. Liu D, Shang J, Chen M. Principal component analysis-based ensemble detector for incipient faults in dynamic processes. *IEEE Trans Ind Informatics* 2021;17:5391. https://doi.org/10.1109/TII.2020.3031496.
53. Zio E, Baraldi P, Popescu IC. A fuzzy decision tree method for fault classification in the steam generator of a pressurized water reactor. *Ann Nucl Energy* 2009;36:1159–69. https://doi.org/10.1016/J.ANUCENE.2009.04.011.
54. Nawaz M, Zabiri H, Taqvi SAA, Idris A. Improved process monitoring using the CUSUM and EWMA-based multiscale PCA fault detection framework. *Chinese J Chem Eng* n.d.;29:253–65.
55. Chetouani Y. Detecting changes in a distillation column by using a sequential probability ratio test. *Syst Eng Procedia* 2011;1:473–80. https://doi.org/10.1016/J.SEPRO.2011.08.069.
56. Yahya Chetouani. Using artificial neural networks for the modelling of a distillation column. *Int J Comput Sci Appl* 2007.
57. Manssouri I, Chetouani Y and El Kihel B. Using neural networks for fault detection in a distillation column. *Int J Comput Appl Technol* 2008.
58. S. A. Taqvi, L. Tufa +2 authors F. Uddin. Fault detection in distillation column using NARX neural network. *Neural Comput Appl* 2018.
59. Taqvi SA, Tufa LD, Zabiri H, Maulud AS, Uddin F. Multiple Fault Diagnosis in Distillation Column Using Multikernel Support Vector Machine. *Ind Eng Chem Res* 2018;57:14689–706. https://doi.org/10.1021/acs.iecr.8b03360.
60. Taqvi Syed A., Tufa Lemma Dendena, Zabiri Haslinda, Mahadzir Shuhaimi, Maulud Abdulhalim Shah, Uddin Fahim. Artificial neural network for anomalies detection in distillation column. *Model Des Simul Syst* 2017.
61. Rajakarunakaran S, Venkumar P, Devaraj D, Rao KSP. Artificial neural network approach for fault detection in rotary system. *Appl Soft Comput* 2008; 8:740–8. https://doi.org/10.1016/J.ASOC.2007.06.002.
62. Muhammad Mujtaba S, Alemu Lemma T, Ali Ammar Taqvi S, Ntow Ofei T, Kumar Vandrangi S. Leak detection in gas mixture pipelines under transient conditions using Hammerstein model and adaptive thresholds. *Processes* n.d. https://doi.org/10.3390/pr8040474.

63. Basha Shaik N, Rao Pedapati S, Ali Ammar Taqvi S, Othman AR, Azly Abd Dzubir F. A feed- forward back propagation neural network approach to predict the life condition of crude oil pipeline. *Processes* n.d. https://doi.org/10.3390/pr8060661.
64. Abbasi Nozari H, Aliyari Shoorehdeli M, Simani S, Dehghan Banadaki H. Model-based robust fault detection and isolation of an industrial gas turbine prototype using soft computing techniques. *Neurocomputing* 2012;91:29–47. https://doi.org/10.1016/J.NEUCOM.2012.02.014.
65. Schubert U, Kruger U, Arellano-Garcia H, de Sá Feital T, Wozny G. Unified model-based fault diagnosis for three industrial application studies. *Control Eng Pract* 2011;19:479–90. https://doi.org/10.1016/J.CONENGPRAC.2011.01.009.
66. Zhao Y, Xiao F, Wang S. An intelligent chiller fault detection and diagnosis methodology using Bayesian belief network. *Energy Build* 2013;57:278–88. https://doi.org/10.1016/J.ENBUILD.2012.11.007.
67. Zhang X, Hoo KA. Effective fault detection and isolation using bond graph-based domain decomposition. *Comput Chem Eng* 2011;35:132–48. https://doi.org/10.1016/J.COMPCHEMENG.2010.07.033.
68. Qin SJ. Survey on data-driven industrial process monitoring and diagnosis. *Annu Rev Control* 2012;36:220–34. https://doi.org/10.1016/J.ARCONTROL.2012.09.004.
69. Chiang H.L., Russell L.E., Braatz D.R. Fault Detection and Diagnosis in Industrial Systems. 2001.
70. June W, Zhang J, Martins E, Morris AJ. Pm~edlngr of the AmwIWn Conlml Conhnnre Fault Detection and Classification through Multivariate Statistical Techniques 1995. https://doi.org/10.1109/ACC.1995.529351.
71. Chen J, Liao CM. Dynamic process fault monitoring based on neural network and PCA. *J Process Control* 2002;12:277–89. https://doi.org/10.1016/S0959-1524(01)00027-0.
72. Chiang LH, Kotanchek ME, Kordon AK. Fault diagnosis based on Fisher discriminant analysis and support vector machines. *Comput Chem Eng* 2004;28:1389–401. https://doi.org/10.1016/J.COMPCHEMENG.2003.10.002.
73. Ku W, Storer RH, Georgakis C. Disturbance detection and isolation by dynamic principal component analysis. *Chemom Intell Lab Syst* 1995;30:179–96. https://doi.org/10.1016/0169-7439(95)00076-3.
74. Lee JM, Yoo CK, Lee IB. Statistical monitoring of dynamic processes based on dynamic independent component analysis. *Chem Eng Sci* 2004;59:2995–3006. https://doi.org/10.1016/J.CES.2004.04.031.
75. Lee KB, Cheon S, Kim CO. A convolutional neural network for fault classification and diagnosis in semiconductor manufacturing processes. *IEEE Trans Semicond Manuf* 2017;30:135. https://doi.org/10.1109/TSM.2017.2676245.
76. Chen Z, Gryllias K, Li W. Mechanical fault diagnosis using convolutional neural networks and extreme learning machine. *Mech Syst Signal Process* 2019;133:106272. https://doi.org/10.1016/J.YMSSP.2019.106272.
77. Janssens Olivier (UGent), Slavkovikj Viktor (UGent), Vervisch Bram (UGent), Stockman Kurt (UGent), Loccufier Mia (UGent), Verstockt Steven (UGent) RV de W (UGent) and SVH (UGent). Convolutional neural network based fault detection for rotating machinery. *J Sound Vib* 2016.
78. Weimer Daniel, Bernd Scholz-Reiter MS. Design of deep convolutional neural network architectures for automated feature extraction in industrial inspection. *CIRP Ann - Manuf Technol* 2016.
79. Wen L, Li X, Gao L, Zhang Y. A new convolutional neural network-based data-driven fault diagnosis method. *IEEE Trans Ind Electron* 2018;65. https://doi.org/10.1109/TIE.2017.2774777.

80. Mowbray M, Savage T, Wu C, Song Z, Cho BA, Del Rio-Chanona EA, et al. Machine learning for biochemical engineering: A review. *Biochem Eng J* 2021;172:108054. https://doi.org/10.1016/j.bej.2021.108054.
81. Dalzochio Jovani, Kunst Rafael, Pignaton Edison, Binotto Alecio, Sanyal Srijnan, Favilla Jose, Barbosa J. Machine learning and reasoning for predictive maintenance in Industry 4.0: Current status and challenges. *Comput Ind* 2020.
82. Zhang K, Liu X, Shen J, Li Z, Sang Y, Wu X, et al. Clinically applicable AI system for accurate diagnosis, quantitative measurements, and prognosis of COVID-19 pneumonia using computed tomography. *Cell* 2020;181:1423–1433. https://doi.org/10.1016/j.cell.2020.04.045.
83. Sharpless NE, Kerlavage AR. The potential of AI in cancer care and research. *Biochim Biophys Acta - Rev Cancer* 2021;1876:188573. https://doi.org/10.1016/j.bbcan.2021.188573.
84. Calisto FM, Santiago C, Nunes N, Nascimento JC. Introduction of human-centric AI assistant to aid radiologists for multimodal breast image classification. *Int J Hum Comput Stud* 2021;150:102607. https://doi.org/10.1016/j.ijhcs.2021.102607.

7 Automatic Vehicle Number Plate Text Detection and Recognition Using MobileNet Architecture for a Single Shot Detection (SSD) Technique

Ahmed Mateen Buttar
and Muhammad Arslan Anwar
University of Agriculture, Faisalabad

CONTENTS

7.1 Problem Statement 102
7.2 Objective of the Study 102
7.3 Introduction 102
7.4 Review of the Literature 103
7.5 Methodology 104
7.6 Data Collection 105
7.7 Automatic Number Plate Detection Process 105
7.8 Installing and Setup Python Libraries 107
7.9 Download TF Model Pretrained Model Form TensorFlow Model Zoo and Install TFOD 108
7.10 Getting Number Plates Data 108
7.11 Training the Object Detection Model 109
7.12 Detecting Plates from an Image 110
7.13 Real-Time Detection Using WebCam 111
7.14 Applying OCR 112
7.15 Results After Detection Process 112

DOI: 10.1201/9781003267782-7

7.16 Results and Discussions 113
7.17 Comparative Analysis 115
7.18 Conclusion 116
7.19 Future Work 116
References 117

7.1 PROBLEM STATEMENT

The purpose of this research is to record and investigate the vehicles which enter the University of Agriculture Faisalabad. Vehicles are detected and their number plates are recognized and saved in the system for security purposes. This system about to improve security on confidential areas and traffic violence areas.

If any car is stolen then the system will detect the car on roads and assist the traffic police in identifying the vehicle. This proposed system will be of use to the police, who will be able to apprehend the thief through the identification of the vehicle number plate. It will also be a very useful system for parking areas.

7.2 OBJECTIVE OF THE STUDY

- To increase the effectiveness of real-time detection and identification of vehicle number plates. To record the vehicles which enter through the main gate of the University of Agriculture.
- To capture the number plates and save them.
- To improve the security of the university.
- To test the accuracy of the system to detect and recognise car plates and conduct a cost–benefit analysis.

7.3 INTRODUCTION

The automatic number plate detection and recognition system is also known as the license plate detection system. Licence Plate Detection and Recognition (LDPR) is an issue on which several academics are working. Plate detection is a huge issue that has been the subject of research across the world. This system is being developed to improve the security of transport vehicles. These vehicles can be identified through highways, toll plazas, motorways and also parking. This detection algorithm works on the detection of vehicles from the main entrance of any university, park or any industrial sides. Many applications, such as those involving image processing, computer vision, and electronic payment systems, rely on licence plates. License plates assist in the recovery of stolen cars, as well as enhanced security and the prevention of accidents [1]. LPDR is a critical technology for attaining traffic system encouragement. The LPDR difficulties include high-quality image capture, which is critical for detecting the character from an image. It's tough to offer reliable findings if the image is not captured in a straightforward manner.

There are a number of steps that have to be gone through in order to get up and running with automatic number plate recognition, as shown in Figure 7.1.

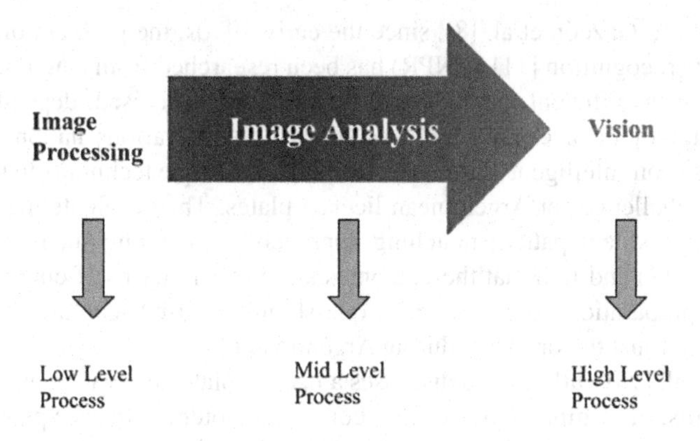

FIGURE 7.1 Process of a digital processing system. [2]

- Install and setup
- Getting number plate data
- Training a OD model
- Detecting number plates applying OCR to text
- Output ROIs and results

Having set up the environment and everything we need to get up and running I will then run the automatic number plate detection system. My system is slightly different to traditional methods of object detection or exact number plate text detection and recognition. It has to be slightly different in that respect, thus I made use of some kaggle data to train the algorithm and the object detection model to be able to detect our license plate in order to actually check that it identifies the region of interest and that it shows up as a license.

7.4 REVIEW OF THE LITERATURE

In Zheng and He [3], an improved K-means method is described for cutting characters out of licence plate pictures. Despite the fact that numerous commercial [4] LPR devices exist, the current accuracy is hampered by factors such as low lighting and moving vehicles. The K-means algorithm-based method produced improved picture segmentation results after testing and comparing other image segmentation methodologies [5]. By filtering SIFT key points, [6] the K-means method was improved to include automated cluster number determination. This efficiently recognises the local maxima that represent various clusters in the picture after modification. The process is complete when the licence plate image is clear. The experimental findings demonstrate a high accuracy of picture segmentation and a considerably greater recognition rate when using [7] OCR software [8]. After removing all undesirable [9, 10] non-character regions, the recognition rate rose to around 94.03 percent, up from about 86.6 percent before their suggested approach. Accordingly, it is clear that through the use of the method the LPR's overall recognition accuracy has increased.

According to Gazcón et al. [8], since the early 1990s, the problem of automated number plate recognition [11] (ANPR) has been researched from a number of different perspectives. Efficient methods have recently been devised, depending on the characteristics of licence plate representations used in various nations [12]. This article focuses on Intelligent Template Matching, an unique technique to tackling the ANPR [13] challenge for Argentinean licence plates. They evaluate the findings to certain other resilient pattern matching approaches (including such convolutional neural networks) and find that the outcomes are better in terms of classified performance and preparation time. The technique should be used with any licence plate perception, not just the one used during Argentina [14].

Lalimi et al. [15] outlines and discusses a licence plate detecting method. In order to achieve this, they improve the level of contrast at potential licence plate locations, propose a "region-based" filtering method for smoothing the uniform and background areas of an image, use the support vector machine algorithm and structural filtering to extract the running parallel and candidate regions, and ultimately segment the plate territory by considering the vertical edges and candidate regions. In reality, the uniqueness and strength of their licence plate identification system lies in the use of region-based filtering in the final two stages, which reduces run time and improves accuracy: Geometrical characteristics were used in conjunction with morphological filtering. The experimental findings demonstrate that their suggested approach performs well in a variety of circumstances. Their system is trustworthy since the average accuracy for diverse scenarios is above 92 percent, and it is also practical because of the low cost of computing [16].

7.5 METHODOLOGY

First of all it is necessary to understand something about vehicle number plate detection and automatic vehicle number plate detection. In essence, number plate detection pictures are uploaded to the system and the system then detects and recognizes number plates from the vehicle image. In the case of automatic number plate text detection, however, the process is one of real-time detection and recognition. The camera is operating and taking pictures of vehicles [17]. When a vehicle or motor car in the range of the camera then the Automatic Vehicle Number plate Detection and recognition (or AVNPR) system automatically detects the vehicle and its number plate region from the image. It neglects other objects and focus solely on the area of the number plate [18].

This research covers the method of Automatic Vehicle Number Plate Detection and Recognition (AVNPR). Some other researchers are working on number plate detection with MobileNet Architecture for Single Shot Detection using the CNN Network. We use the TensorFlow Object Detection Technique for detection and use easyOCR software to convert image to text. Figure 7.2 shows how the system can send images as an input to model to detect the number plate. We also do this work in real-time number plate detection. Here we need a single, high-quality camera; a normal mobile camera can be sufficient for our purposes. So the camera covers the real time scene [20], when any vehicle enter the camera's region; this optimized approach

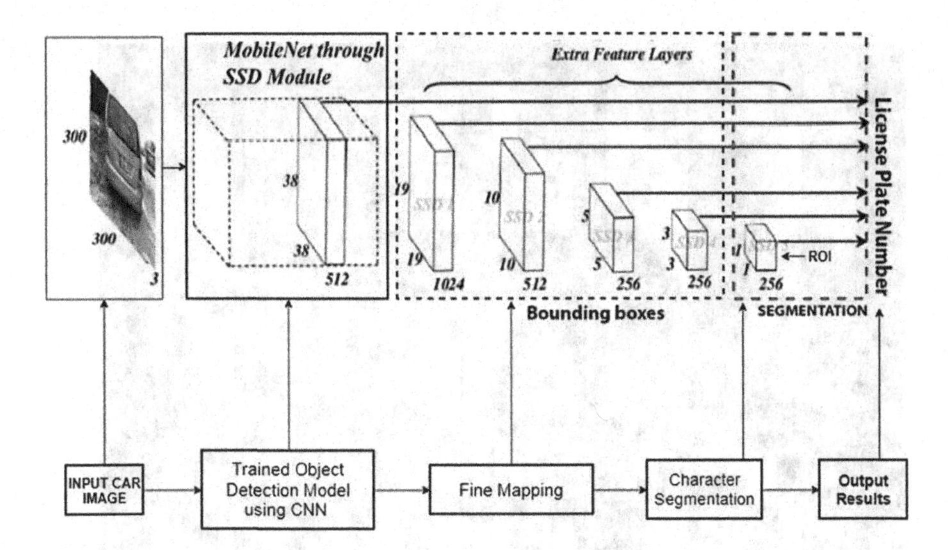

FIGURE 7.2 Architecture for AVNPR. [19]

then detects the vehicle and focuses on its number plate region. After detection, the number plate is bordered and this displays its accuracy label on the number plate in real time. This process also needs GPU, but if you can train the model, then it can work well without a GPU. If we use a GPU, then both detection and recognition process performance will increase.

7.6 DATA COLLECTION

This is the first stage of my research. Initially, we require permission from the University of Agriculture to take data from CCTV footage from the main entrance door of the university. I wrote an application letter to DSA Directorate Students Affairs to give me permission to capture the data from main door or parking. Accordingly, this process takes around two months for the data collection and permission process. I take videos from the main entrance gate of the university. Videos have a duration of around 20 minutes, although some videos take approximately 30 minutes. Vehicle images taken from these videos are then used to test the model (Figure 7.3 and Tables 7.1-7.2).

7.7 AUTOMATIC NUMBER PLATE DETECTION PROCESS

The Automatic Vehicle Number Plate Detection and Recognition (AVNPR) [21] process has to be designed and coded in Python using Jupyter Notebook. The following steps are required to detect the number plate from the image and real time from a direct recording camera which is used as [20] CCTV footage [22].

1. Installing and setting up Python Libraries
2. Obtaining the Number Plate Dataset

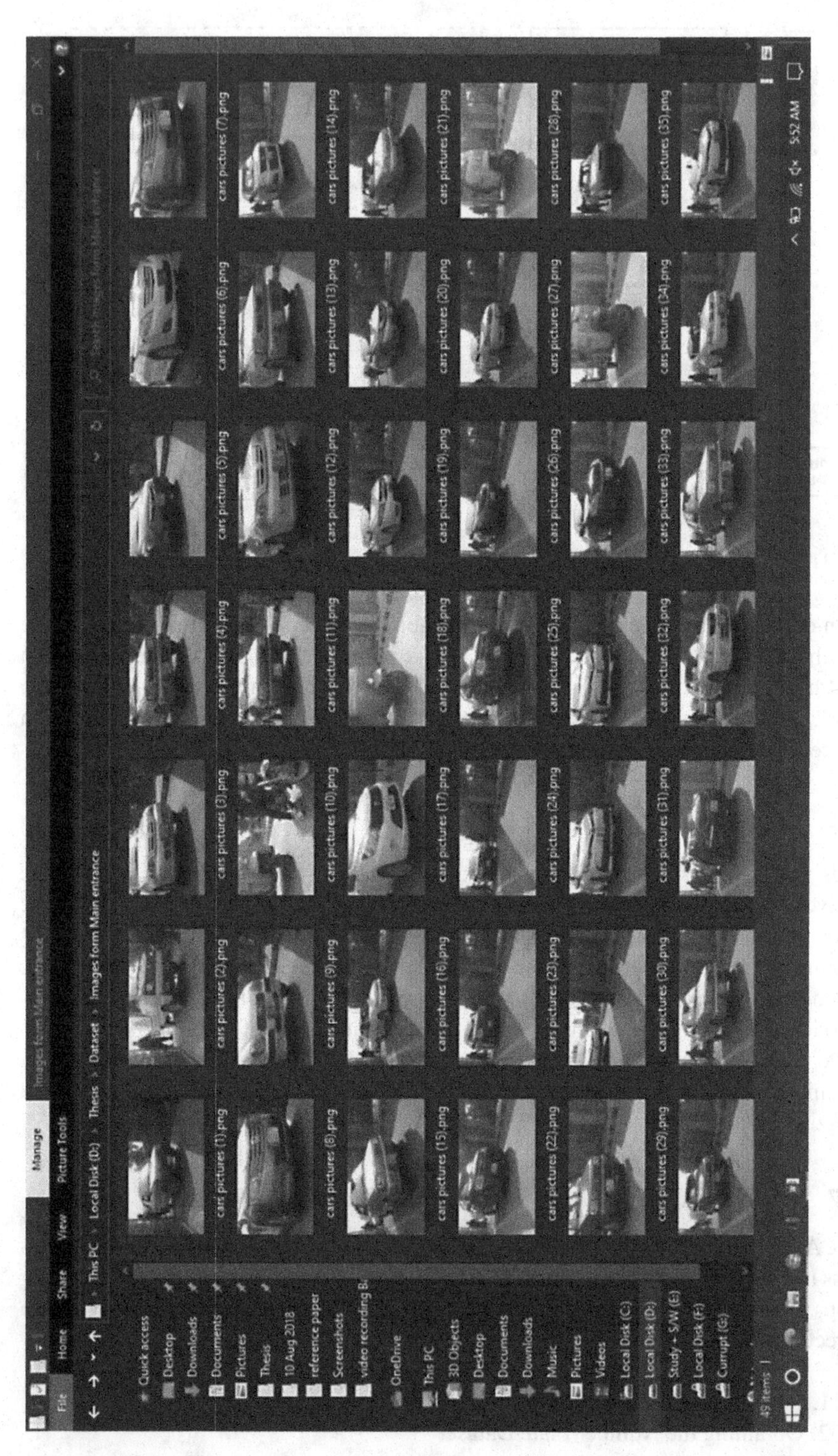

FIGURE 7.3 Data collection from main gate of the University of Agriculture.

TABLE 7.1
Software Requirements

	Software	Purpose
Operating Systems	Microsoft Windows 8,10	As an Operating System
	Jupyter Notebook	Development Tool
	TensorFlow Object Detection Model	Training Model and detect the number plates
	EasyOCR	Recognize the text
	Anaconda	Jupyter platform
	Python 3.8.8 64bit	
	Python 3.9.6 64bit	
	Microsoft Visual Studio Code	Xml file editing
Software	iVCam 6.2	Capture the view from mobile camera
	MS Office 2016	For Documentation writing, make a presentation slides etc.
	MS Edge 10	For Documentation reading

TABLE 7.2
Hardware Requirements

	Hardware	Specification
Laptop or PC	Processor	Intel Corei5-2410M CPU @ 2.30GHz
	Ram	8GB
	Camera	Cell phone camera to capture images or any other digital camera can be use
	GPU	6GB NVIDIA GeForce GTX 1660 Ti

3. Training the Object Detection Model
4. Detecting Number Plates
5. Apply OCR for Text
6. Output ROIs and Results

7.8 INSTALLING AND SETUP PYTHON LIBRARIES

There are two key dependencies I need to install to get up and running for this. Setup Tensorflow Object detection and !pip install easyocr. Create virtual environment help to avoid dependency conflicts and keep work isolated. Create a new virtual environment in command prompt as shown in Figure 7.4.

Install dependencies and add virtual environment to the Python Kernel. Ipykernal allows one to use the virtual environment inside Jupyter Notebook in Figure 7.5.

We use kernel, environment and runtime interchangeably. These represent where your code is running. After the installation in ipykernel this message is shown (installed kernelspec anprsys in C:\ProgramData\jupyter\kernals\anprsys. Following this, open Jupyter Notebook using this command: (anprsys) D:\ANPR> jupyter notebook. Jupyter Notebook will then open in Google Chrome and then we need to

```
D:\ANPR>python -m venv anprsys
D:\ANPR>.\anprsys\Scripts\Activate
(anprsys) D:\ANPR>
```

FIGURE 7.4 Code for creating virtual environment.

```
(anprsys) D:\ANPR>python -m pip install --upgrade pip
(anprsys) D:\ANPR>pip install ipykernel
(anprsys) D:\ANPR>python -m ipykernel install --user --
name=anprsys
```

FIGURE 7.5 Code to allow virtual environment inside Jupyter Notebook.

change kernel. Go to the menu bar, click on Kernel. From the dropdown menu then change the kernel to anprsys.

7.9 DOWNLOAD TF MODEL PRETRAINED MODEL FORM TENSORFLOW MODEL ZOO AND INSTALL TFOD

In the model zoo [23], I have a variety of models; some are quicker, while others are more precise. Some of the models can even identify the pixels that make up an item using segmentation. Detecting boundary boxes is adequate for our purpose. !git clone github.com/tensorflow/models. This cloning takes a few seconds, depending on internet speed.

7.10 GETTING NUMBER PLATES DATA

These data are not necessary for TFOD model training. So I use the kaggle dataset for model training. I used a pre-annotated dataset to train the object detection model. This provides 433 images which can be used to train the model [24]. We then clone the kaggle dataset link to the model. Download the dataset and take images and xml files of annotated data. The kaggle account is needed to download the images dataset. Thus, a zip file of the images is downloaded. Next, the user need to extract this folder and then obtain the images. Their [25] annotation of each image is available to be accessed by the model training deep learning mechanism using the [26] TensorFlow Object Detection Model. The annotation of image car0.png is shown in the following Figure 7.6. This is an XML file containing information about the region of the number plate.

411 images are used to train the model and the remaining 22 images are used to test the model. Train and Test is a method for determining the accuracy of your system.

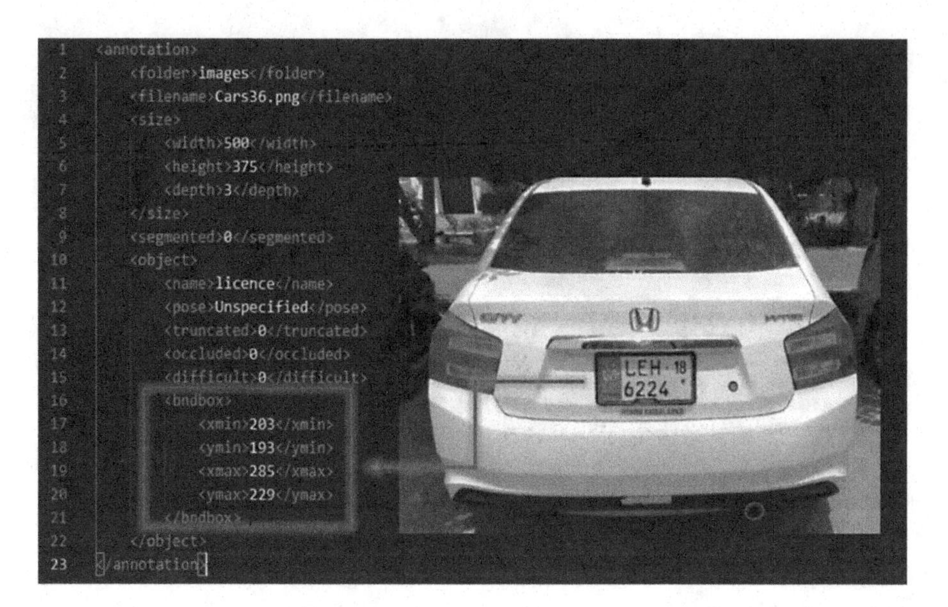

FIGURE 7.6 Annotation of car image.

It's called Train and Test since this dataset is split into two parts: a training set and a testing set. Training data account for 80 percent of the whole dataset, whereas testing accounts for 20 percent. The model has been trained using the trained model. The model is tested using the testing set. The term "train the model" refers to the process of developing a model. Testing the model entails determining its correctness [27].

In the above annotation is the corresponding car image. This image has its own particular annotation. What this particular image basically said is that <xmin>, <ymin> and <xmax>,<ymax> are actually the corresponding boundaries around the interested region of the picture. This indicates the number plate region in number plate [28]. Now I need to split up these annotations and images into training and testing portions. So the training partition is to object detection model is trained on the testing partition [29]. [30] So I train it on one particular bit of data and then I test it on a completely separate bit of data. I make two folders: 'test' and 'train'. I then choose the cars' images from the dataset. I select car0 to car411 images and cut them. These are then pasted into the 'train' folder. The other remaining images are then cut and pasted into the test folder. The same procedure is then carried out with the annotation folder. I cut car0 to car411 annotation .xml files and pasted them into the 'train' folder. The remaining xml files are then cut and pasted into the 'test' folder.

7.11 TRAINING THE OBJECT DETECTION MODEL

In order to detect plates we first need to train the object detection model. This will allow us to detect plates from a picture and in real time. In Figure 7.7 show this training phase.

API is the abbreviation for the Application Programming Interface. Developers can use an [31] API to access a collection of common operations without having to

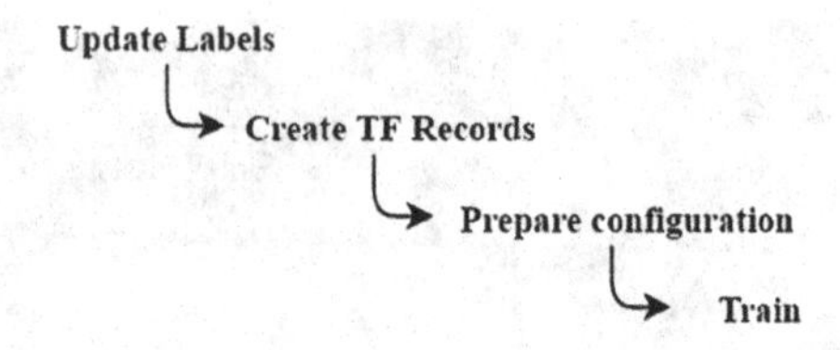

FIGURE 7.7 Phases of training.

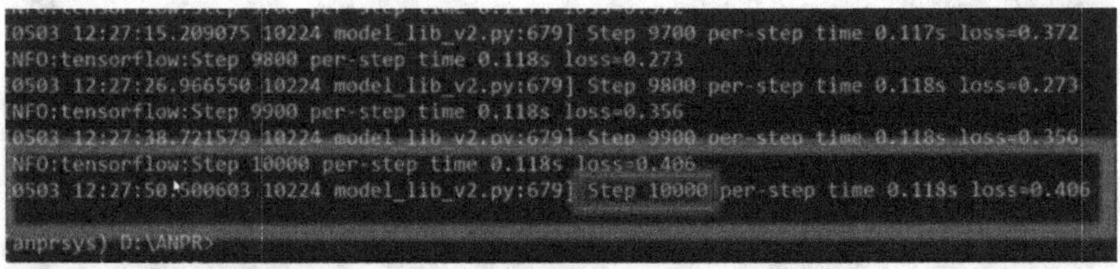

```
0503 12:27:15.209075 10224 model_lib_v2.py:679] Step 9700 per-step time 0.117s loss=0.372
NFO:tensorflow:Step 9800 per-step time 0.118s loss=0.273
0503 12:27:26.966550 10224 model_lib_v2.py:679] Step 9800 per-step time 0.118s loss=0.273
NFO:tensorflow:Step 9900 per-step time 0.118s loss=0.356
0503 12:27:38.721579 10224 model_lib_v2.py:679] Step 9900 per-step time 0.118s loss=0.356
NFO:tensorflow:Step 10000 per-step time 0.118s loss=0.406
0503 12:27:50.500603 10224 model_lib_v2.py:679] Step 10000 per-step time 0.118s loss=0.406

anprsys) D:\ANPR>
```

FIGURE 7.8 Model training steps in Command Prompt.

build code from scratch. Consider an API to be similar to a restaurant's menu, which gives a list of foods as well as descriptions for each. Continuing the analogy, when we tell the restaurant what meal we want, they prepare it and serve it to us. The TensorFlow object detection API provides a platform for building a [32, 33] deep learning network that can detect objects. Model Zoo is a framework in which pre-trained models are already available. This comprises a set of models that have been [34] pre-trained on the Open Images Dataset. If we are just interested in categories in this dataset, these models can be utilised for inference. They may also be used to initialise your models before training on a new dataset.

I need to train the model longer and I use 10,000 steps for bootstrapping to get more accurate results. This parameter represent how long the model will train for. The higher the number of steps the longer the model will train [35]. Accordingly, it takes some time to train the model due to 10,000 steps for training. In Figure 7.8 I get the command and then run in command prompt so here I could see all details like a steps 100 per-step time 0.386s loss = 1.024. If you don't have CV2 on your PC then install it first; otherwise use the command prompt give us error to install CV2. When model trained as 10,000 steps and time = 0.118s loss = 0.406.

7.12 DETECTING PLATES FROM AN IMAGE

I leverage the model to detect from an image. This vehicle number plate detection is based on MobileNet Architecture based on Single Shot Detection SSD. I put the file name into my code and try to detect the number plate. When I run this detection code then model detected the vehicle number plate region and bordered with green line and display also its accuracy for detection. By now the model is perfectly able to detect any type of number plates. This model can detect the number plate from any angle. In following Figure 7.9 shows that the vehicle number plate region is detected and its labeling is licence with 92 percent accuracy for detection. This

FIGURE 7.9 Vehicle number plate detection.

accuracy depends upon image clarity and camera's focus, which will affect just how fast the model can detect the number plate from any picture. This detection is done by the Tensorflow-trained model and I test more images on this model; their results are shown in Figure 7.9. We can give many images to model for detection at the same time. These results are based on training the model with the kaggle dataset of vehicle images. The training dataset contain all type of images, including multiple angles, shadow images, full bright images, and some blurred images. There are many types of number plates that occurs all across the world. Thus, the dataset also mixes the different types of number plates. Other results of number plates are shown in Figure 7.9.

7.13 REAL-TIME DETECTION USING WEBCAM

For real-time detection I use my laptop webcam. Thus, I open car images on my mobile and run real-time code in Jupyter Notebook. Then a small window of camera view is opened and detects the view. I show the mobile screen which contains the car image with a number plate. Then it detected real time on reasonable angles of the mobile screen. Now real-time detection is working well. It should be noted that the laptop webcam is not well suited to real-time detection because it has low resolution quality and a relatively small screen. Thus, I use my Android mobile camera as a webcam. This has 16 megapixel resolution and good image capturing quality. First of all I connect my Android phone via an application called "iVcam", then I configure this and disconnect my laptop camera from the LCD panel of my laptop.

So when I run module of real-time detection then camera window opened and now I need to launch the app iVcam on Mobile set. Then the mobile camera was connected to the PC as a webcam. Either both of the devices are connected to the same network or both devices connect using a micro USB cable. If I used my mobile webcam wireless then real-time detection is working. When I connect with the USB cable, however, then both the performance and the detection speed increase. The image quality depended on the Wi-Fi speed. In line with my findings, I suggest the use of the USB cable for better results.

7.14 APPLYING OCR

Jaided_AI, a firm that specialises in optical character recognition (OCR) services, develops and maintains the EasyOCR package. Python and the PyTorch libraries are used to implement EasyOCR. If you have a CUDA-capable GPU, the underlying PyTorch deep learning software may dramatically speed up both text detection and OCR speed.

My recommendation for getting started with EasyOCR is to use my configuration opencv. In your virtual environment, make sure to install opencv-python rather than opencv-contrib-python. Furthermore, having both of these programs installed in the same environment may have unexpected implications. If you have both installed, pip is unlikely to protest, so use the pip freeze command to double-check. Of course, the aforementioned instruction covers both OpenCV packages; just make sure you install the proper one. And, in my opinion, you should create a separate Pythonvirtual-environment for EasyOCR on your machine. The application of EasyOCR and ROI is shown in Figure 7.10. Cloning the EasyOCR [36].

7.15 RESULTS AFTER DETECTION PROCESS

As I mentioned above, I used 411 images and the annotation files to train the model in 10,000 steps for bootstrapping the model. Of course, the training time increases if we increase the number of training steps. If we use 2000 steps then the training time is short, whereas if we use either 10,000 or 20,000 steps then it will take longer. If the laptop or the PC does not have its own GPU then the training model time is between 8 and 9 hours. If we use a GPU-based system at least 2 GB or 5 GB GPU is

```
image = image_np_with_detections
scores = list(filter(lambda x: x> detection_threshold, detections['detection_scores']))
boxes = detections['detection_boxes'][:len(scores)]
classes = detections['detection_classes'][:len(scores)]

width = image.shape[1]
height = image.shape[0]

height

232

# Apply ROI filtering and OCR
for idx, box in enumerate(boxes):
    roi = box*[height, width, height, width]
    region = image[int(roi[0]):int(roi[2]),int(roi[1]):int(roi[3])]
    plt.imshow(region)
```

FIGURE 7.10 Appling EasyOCR and Region of Interest ROI.

FIGURE 7.11 Tested images and detected.

beneficial for training the model in just a few minutes. GPU improves the efficiency of the model, but detection can be done either with or without GPU. I test 50 images, which are collected from university parking. Most images are detected very well. Some images are not detected, however; these are those which have a background color that is the same as the car's body color and have no border on numberplate. Images taken from both different sides and different angles are also detected easily. Some images are in clear vision, some are in shadow and some blurred images are shown in Figure 7.11.

7.16 RESULTS AND DISCUSSIONS

I tested 50 images on real-time detection with webcam. The average accuracy of these images are 91 percent. This average result calculated all types of images. The images have been taken at a variety of different distance from the camera. The images are all taken at different angles. Some images are not clear, but they are still identified. Table 7.3 shows the average accuracy of automatic numberplate detection. When an image or vehicle passes though the webcam for real-time plate detection, the result is

TABLE 7.3
Accuracy of Detection Process

Detection Ways	Average Accuracy (%) (closed camera)	Average Accuracy (%) (distance from camera)
Number Plate Detect from Still Images	92	89.24
Detection from real time webcam	93	89

more accurate when the vehicle is close to the camera. The accuracy is inversely proportional to the distance, with lower distance increasing accuracy. The TensorFlow detection method gives us reasonable and satisfying results for detection. Accuracy is main point that needs to be improved. The results are also shown in Figure 7.12.

I trained the model with 411 images with annotations for 10,000 steps for bootstrapping. Accordingly, next time I will train the model with 800 images and 20,000 steps for bootstrapping. This training time is too long since an increased number of steps are involved. If we use high-processing machines with 6 to 8 GB GPU and a dedicated graphics card provided by a specialist manufacturer such as NVIDIA or AMD then the training time is reduced, as shown in Figure 7.13.

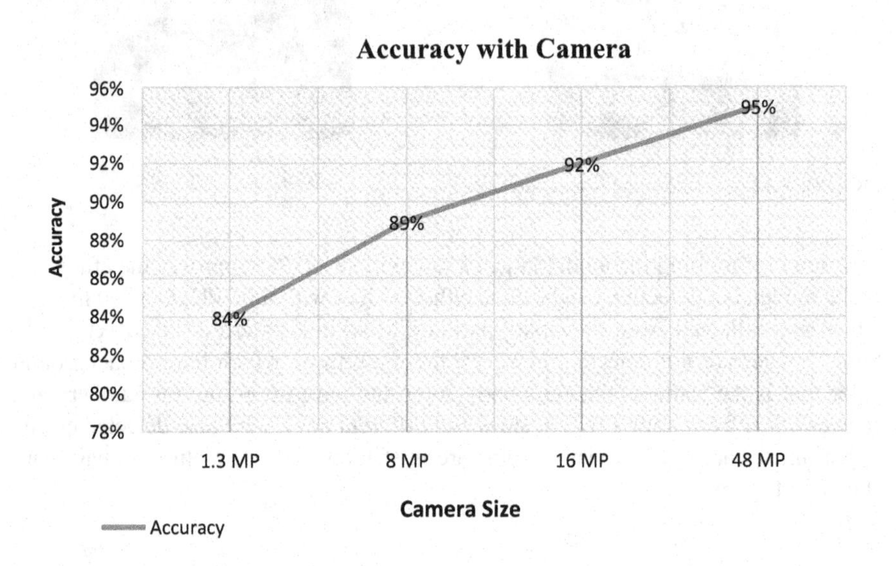

FIGURE 7.12 Graph for accuracy with camera.

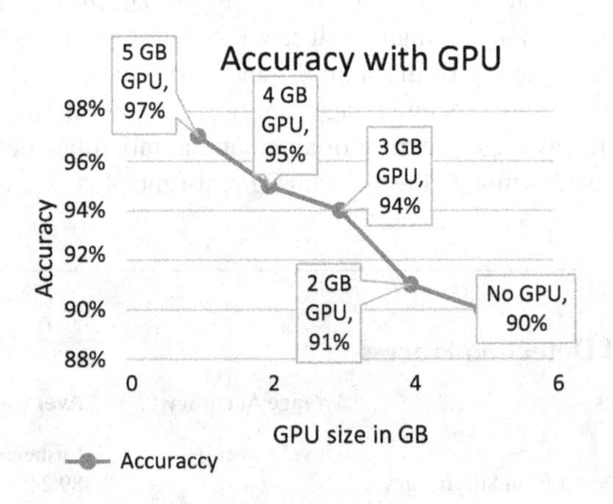

FIGURE 7.13 Graph for accuracy with GPU.

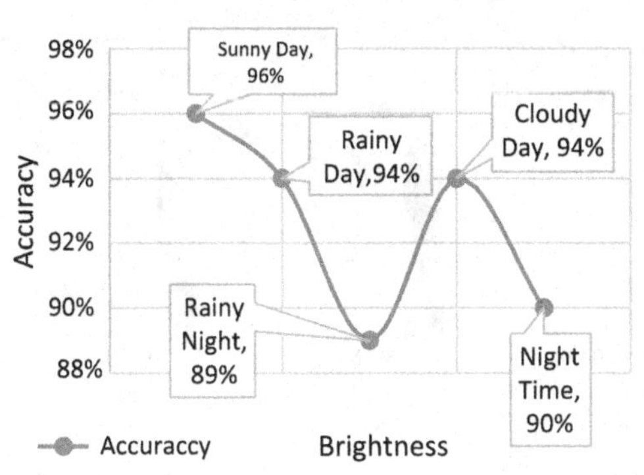

FIGURE 7.14 Graph for accuracy with daylight.

The level of light also has an effect on accuracy, but this is not sufficient to disturb our results. I have tested this system in a variety of intensities of light: these include full daylight, cloudy day, dusk, rainy night time, sunny day. The variation of light described in the above graph in Figure 7.14. As might be expected, the accuracy improves with better light. When processing at night time I use light to make the number plates visible and readable. LED white light is used at night time for the sake of clarity.

7.17 COMPARATIVE ANALYSIS

In Table 7.4 show the results, with our Optimized purposed system we achieved a maximum accuracy rate of 96 percent, and a minimum accuracy rate of 86 percent for Automatic Vehicle Number Plate Detection and Recognition AVNPR system at real-time detection and also from images data also show in graph Figure 7.15. Our optimized system yields better results than the other approaches because it used MobileNet Architecture for Single Shot Detection SSD based on YOLOv3 efficient for Convolutional Neural Networks for vision-based systems.

TABLE 7.4
Compare Our Optimized Proposed System with Others

Detection Techniques	Accuracy Max. (%)	Accuracy Min. (%)
AVNPR (TFOD) SSD	96	86
SPANS [37]	83	80
VLP (ANN) [38]	95	85
VLP (R-CNN) [39]	92	75

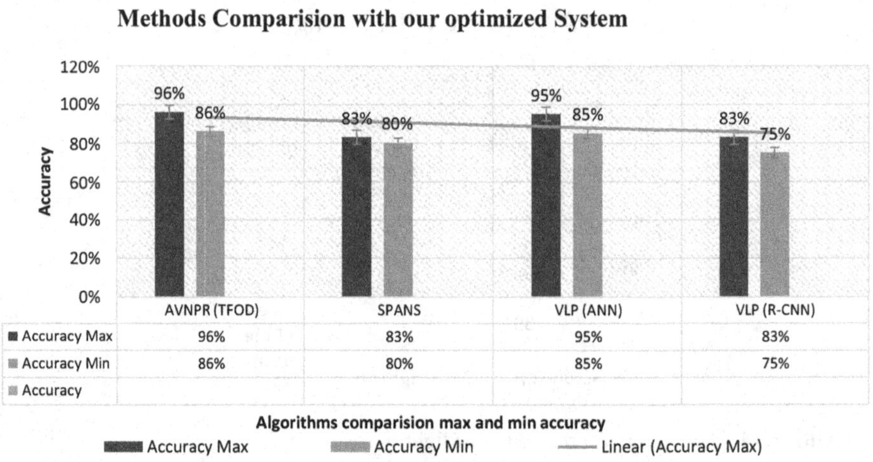
Methods Comparision with our optimized System
120%
100%
80%
60%
40%
20%
0%
Accuracy
96%
86%
83%
80%
95%
85%
83%
75%
AVNPR (TFOD)
SPANS
VLP (ANN)
VLP (R-CNN)
Accuracy Max 96% 83% 95% 83%
Accuracy Min 86% 80% 85% 75%
Accuracy
Algorithms comparision max and min accuracy
Accuracy Max
Accuracy Min
Linear (Accuracy Max)

FIGURE 7.15 Results with other algorithms.

7.18 CONCLUSION

This purpose-built system is able to detect the vehicle number plate or license plates automatically on real-time coverage. It has used AVNPR based on Single Shot Detection using YOLOv3 with CNN approaches. This is done with automatically detects the vehicles and focus on their number plates and show real-time detections accuracy and detect the region of the car's number plate. After the detection, the system recognized the detected number plate and extracted its letters or numbers in the form of text. This system automatically keeps the records in a .CSV file and also take images of detected number plates in the computer system. This system is useful for parking lots where its automatically detects plates. I get the results with optimized purposed system achieving 96 percent accuracy of automatic vehicle detection and recognition. This accuracy is achieved with high-speed GPU and good camera quality. The model training time also depends upon the amount of GPU available. lWhen I test the model on that PC, which has no GPU, then I get 84 to 96 percent accuracy in real time.

7.19 FUTURE WORK

This research will be beneficial in the long run. We linked a single webcam for real-time detection, and used the two or three webcams connected to this AVNPR system in order to detect several gates and parking areas at the same time. In the case of machine learning bootstrapping, we trained the model with 10,000 steps. In the future, we will add 20,000 steps to improve detection performance and accuracy.

REFERENCES

1. A. Agarwal and S. Goswami, "An efficient algorithm for automatic car plate detection & recognition," *Proc. – 2016 2nd Int. Conf. Comput. Intell. Commun. Technol. CICT 2016*, pp. 644–648, 2016. doi:10.1109/CICT.2016.133.
2. "Digital image processing," *Great Learning*, 2020. https://www.mygreatlearning.com/blog/digital-image-processing-explained/.
3. L. Zheng and X. He, "Character segmentation for license plate recognition by K-means algorithm," *Lect. Notes Comput. Sci. (including Subser. Lect. Notes Artif. Intell. Lect. Notes Bioinformatics)*, vol. 6979 LNCS, no. PART 2, pp. 444–453, 2011. doi: 10.1007/978-3-642-24088-1_46.
4. N.-A. Alam, M. Ahsan, M. A. Based, and J. Haider, "Intelligent system for vehicles number plate detection and recognition using convolutional neural networks," *Technologies*, vol. 9, no. 1, p. 9, 2021, doi: 10.3390/technologies9010009.
5. N. A. Borghese, P. L. Lanzi, R. Mainetti, M. Pirovano, and E. Surer, "Advances in neural networks: Computational and theoretical issues," *Smart Innov. Syst. Technol.*, vol. 37, no. JUNE, pp. 243–251, 2015, doi: 10.1007/978-3-319-18164-6.
6. I. Türkyilmaz and K. Kaçan, "License plate recognition system using artificial neural networks," *ETRI J.*, vol. 39, no. 2, pp. 163–172, 2017, doi: 10.4218/etrij.17.0115.0766.
7. A. Bhujbal and D. Mane, "A survey on deep learning approaches for vehicle and number plate detection," *Int. J. Sci. Technol. Res.*, vol. 8, no. 12, pp. 1378–1383, 2019.
8. N. F. Gazcón, C. I. Chesñevar, and S. M. Castro, "Automatic vehicle identification for Argentinean license plates using intelligent template matching," *Pattern Recognit. Lett.*, vol. 33, no. 9, pp. 1066–1074, 2012, doi: 10.1016/j.patrec.2012.02.004.
9. S. Sanjana, S. Sanjana, V. R. Shriya, G. Vaishnavi, and K. Ashwini, "A review on various methodologies used for vehicle classification, helmet detection and number plate recognition," *Evol. Intell.*, vol. 14, no. 2, pp. 979–987, 2021, doi: 10.1007/s12065-020-00493-7.
10. L. Hou et al., "Deep learning-based applications for safety management in the AEC industry: A review," *Comput. Electron. Agric.*, vol. 14, no. 2, p. 106067, 2021, doi: 10.3390/app11020821.
11. X. Jin, R. Tang, L. Liu, and J. Wu, "Vehicle license plate recognition for fog-haze environments," *IET Image Process.*, vol. 15, no. 6, pp. 1273–1284, 2021, doi: 10.1049/ipr2.12103.
12. I. V. Pustokhina et al., "Automatic vehicle license plate recognition using optimal K-means with convolutional neural network for intelligent transportation systems," *IEEE Access*, vol. 8, pp. 92907–92917, 2020, doi: 10.1109/ACCESS.2020.2993008.
13. P. R. K. Varma, S. Ganta, B. Hari Krishna, and P. Svsrk, "A novel method for Indian vehicle registration number plate detection and recognition using image processing techniques," *Procedia Comput. Sci.*, vol. 167, no. 2019, pp. 2623–2633, 2020, doi: 10.1016/j.procs.2020.03.324.
14. S. Du, M. Ibrahim, M. Shenata, and W. Badawy, "Automatic LPR a state-of-the-art review," *IEEE Trans. Circuits Syst. Video Technol.*, vol. 23, no. c, 2013.
15. M. A. Lalimi, S. Ghofrani, and D. McLernon, "A vehicle license plate detection method using region and edge based methods," *Comput. Electr. Eng.*, vol. 39, no. 3, pp. 834–845, 2013, doi: 10.1016/j.compeleceng.2012.09.015.
16. Y. Kessentini, M. D. Besbes, S. Ammar, and A. Chabbouh, "A two-stage deep neural network for multi-norm license plate detection and recognition," *Expert Syst. Appl.*, vol. 136, pp. 159– 170, 2019, doi: 10.1016/j.eswa.2019.06.036.

17. H. Nguyen, "Real-time license plate detection based on vehicle region and text detection," *J. Theor. Appl. Inf. Technol.*, vol. 98, no. 3, pp. 488–504, 2020.
18. D. Bhardwaj and S. Mahajan, "Review paper on automated number plate recognition techniques," *Int. J. Emerg. Res. Manag. &Technology*, vol. 6, no. 15, pp. 2278–9359, 2015, [Online].
19. "Real-time vehicle detection with mobilenet SSD," 2020, [Online]. Available: https://www.edge-ai-vision.com/2020/10/real-time-vehicle-detection-with-mobilenet-ssd-and-xailient/.
20. R. Panahi and I. Gholampour, "Accurate detection and recognition of dirty vehicle plate numbers for high-speed applications," *IEEE Trans. Intell. Transp. Syst.*, vol. 18, no. 4, pp. 767–779, 2017, doi: 10.1109/TITS.2016.2586520.
21. A. Kumar Sahoo, "Automatic recognition of Indian vehicles license plates using machine learning approaches," *Mater. Today Proc.*, 2020, doi: 10.1016/j.matpr.2020.09.046.
22. Z. Yang and L. S. C. Pun-Cheng, "Vehicle detection in intelligent transportation systems and its applications under varying environments: A review," *Image Vis. Comput.*, vol. 69, pp. 143–154, 2018, doi: 10.1016/j.imavis.2017.09.008.
23. H. Yu et al., "Tensor flow model garden," 2020. https://github.com/tensorflow/models.
24. Larxel, "Car license plate detection," *Kaggle*, 2020. https://www.kaggle.com/andrewmvd/car-plate-detection.
25. L. Shantha, B. Sathiyabhama, T. K. Revathi, N. Basker, and R. B. Vinothkumar, "Tracing of Vehicle Region and Number Plate Detection using Deep Learning," *Int. Conf. Emerg. Trends Inf. Technol. Eng. ic-ETITE 2020*, no. 2018, pp. 2018–2021, 2020, doi: 10.1109/ic-ETITE47903.2020.357.
26. N. J. Crane, S. W. Huffman, F. A. Gage, I. W. Levin, and E. A. Elster, "Evidence of a heterogeneous tissue oxygenation : renal ischemia / reperfusion injury in a large animal," *J. Biomed. Opt.*, vol. 18, no. 3, pp. 035001–035007, 2003, doi: 10.1117/1.
27. M. A. Raza, C. Qi, M. R. Asif, and M. A. Khan, "An adaptive approach for multi-national vehicle license plate recognition using multi-level deep features and foreground polarity detection model," *Appl. Sci.*, vol. 10, no. 6, 2020, doi: 10.3390/app10062165.
28. S. G. Kim, H. G. Jeon, and H. I. Koo, "Deep-learning-based license plate detection method using vehicle region extraction," *Electron. Lett.*, vol. 53, no. 15, pp. 1034–1036, 2017, doi: 10.1049/el.2017.1373.
29. A. Rio-Alvarez, J. De Andres-Suarez, M. Gonzalez-Rodriguez, D. Fernandez-Lanvin, and B. López Pérez, "Effects of challenging weather and illumination on learning-based license plate detection in noncontrolled environments," *Sci. Program.*, vol. 2019, 2019, doi: 10.1155/2019/6897345.
30. W. Puarungroj and N. Boonsirisumpun, "Thai license plate recognition based on deep learning," *Procedia Comput. Sci.*, vol. 135, pp. 214–221, 2018, doi: 10.1016/j.procs.2018.08.168.
31. J. Shashirangana et al., "License plate recognition using neural architecture search for edge devices," *Int. J. Intell. Syst.*, no. January, pp. 1–38, 2021, doi: 10.1002/int.22471.
32. K. T. Islam et al., "A vision-based machine learning method for barrier access control using vehicle license plate authentication," *Sensors (Switzerland)*, vol. 20, no. 12, pp. 1–18, 2020, doi: 10.3390/s20123578.
33. A. Singh and S. C. Misra, *Identifying Challenges in the Adoption of Industry 4.0 in the Indian Construction Industry*, vol. 1198. 2021.
34. R. Laroca, L. A. Zanlorensi, G. R. Gonçalves, E. Todt, W. R. Schwartz, and D. Menotti, "An efficient and layout-independent automatic license plate recognition system based on the YOLO detector," *IET Intell. Transp. Syst.*, vol. 15, no. 4, pp. 483–503, 2021, doi: 10.1049/itr2.12030.

35. P. Shivakumara, D. Tang, M. Asadzadehkaljahi, T. Lu, U. Pal, and M. H. Anisi, "CNN-RNN based method for license plate recognition," *CAAI Trans. Intell. Technol.*, vol. 3, no. 3, pp. 169– 175, 2018, doi: 10.1049/trit.2018.1015.
36. JaidedAI, "EasyOCR," *Github*, 2019. https://github.com/jaidedAI/EasyOCR.
37. N. do Vale Dalarmelina, M. A. Teixeira, and R. I. Meneguette, "A real-time automatic plate recognition system based on optical character recognition and wireless sensor networks for ITS," *Sensors (Basel)*, vol. 20, no. 1, 2019, doi: 10.3390/s20010055.
38. C. Oz and F. Ercal, "A practical license plate recognition system for real-time environments," *Lect. Notes Comput. Sci.*, vol. 3512, pp. 881–888, 2005, doi: 10.1007/11494669_108.
39. Y. J. Choong, L. K. Keong, and T. C. Cheah, "License plate number detection and recognition using simplified linear-model," *J. Crit. Rev.*, vol. 7, no. 3, pp. 55–60, 2020, doi: 10.31838/JCR.07.03.09.

8 Medical Image Compression Using a Radial Basic Function Neural Network

Towards Aiding the Teleradiology for Medical Data Storage and Transfer

L.R. Jonisha Miriam and A. Lenin Fred

Mar Ephraem College of Engineering and Technology, Marthandam, India

S.N. Kumar

Amal Jyothi College of Engineering, Kanjirapally, India

H. Ajay Kumar

Mar Ephraem College of Engineering and Technology, Marthandam, India

Parasuraman Padmanabhan and *Balàzs Gulyàs*

Nanyang Technological University, Singapore

I. Christina Jane

Mar Ephraem College of Engineering and Technology, Marthandam, India

DOI: 10.1201/9781003267782-8

CONTENTS

8.1 Introduction 122
8.2 Methodology 123
8.2.1 Data Acquisition 123
8.2.2 Medical Image Compression/Decompression Using Neural Network Algorithms 124
8.3 Results and Discussion 127
8.4 Conclusion 134
References 134

8.1 INTRODUCTION

Image compression plays a major role in communication applications. It reduces the number of bits required to transmit information and thus the transmission cost. Image compression is categorized into two types; lossy image compression and lossless image compression. In lossy image compression, there is a significant loss of data, while in lossless image compression, by contrast, there is no loss of information. For medical image compression, the Joint Photographic Experts Group (JPEG) is a standard for continuous-tone still images and it is a lossy compression scheme where the information is not recovered accurately and it has a high compression rate [1, 2]. It has two basic compression methods, namely the Discrete Cosine Transform (DCT) method for lossy compression and the predictive method for lossless compression. Grace Chang et al. [3] summarize the lossless image compression scheme for grayscale images based on the Rice coding method. A lifting scheme and Set Partitioning in Hierarchial Tree (SPIHT) lossless image compression approach achieves high PSNR and minimum error but the processing speed is minimum [4].

Lanzarini et al. [5] propose a neural network-based image compression approach to decrease the convergence time and ensure a faster transmission rate; at the same time, however, training time increases with an increase in the size of the neural network. By using the feed-forward backpropagation algorithm [6], the decompressed image quality increases with an increase in the number of neurons. Anna Durai et al. [7] adopt the method of steepest descent in order to minimize the error. In [8–10] genetic algorithm is coupled with a backpropagation neural network, it is considered to be the simplest artificial neural network that is mainly developed for image compression and it has the disadvantage of slow convergence. From the viewpoint of analysis, the genetic algorithm approach performs better results over gradient descent-based learning but it is not applicable for image compression.

Chee Wan presents edge-preserving image compression as an asymmetric compression scheme that achieves better error rates for magnetic resonance images when compared to JPEG [11]. To enhance the pixel locality, Jan-Yie et al. [12] verifies the efficiency of Hilbert space-filling curve ordering in lossless medical image compression. Computed tomography-based medical images are taken for comparing various encoding schemes. In another study, Tripathi investigates image compression using bipolar coding with the Levenberg Marquardt (LM) algorithm as a better (and more suitable) technique. It uses bipolar activation function [13]. Human Visual System (HVS) guided neural network-based image resolution enhancement is proposed in

[14]. This chapter combines fuzzy decision rules with neural networks to balance the tradeoff between speed and quality. It also achieves better visual quality. For the progressive transmission of Digital Imaging and communication in medicine (DICOM) images, Vijideva [15] highlights a wavelet-based coder with the modified preprocessing algorithm for the backpropagation neural network which leads to good quality and good performance.

The chapter is structured as follows. Section 8.2 discusses the backpropagation neural network algorithm and the radial basis function neural network algorithm for medical image compression. Section 8.3 presents experimental results for medical-computed tomography images; finally, a conclusion is drawn in section 8.4.

8.2 METHODOLOGY

The backpropagation neural network (BPNN) is widely used in computer vision and image processing, and it uses the steepest descent approach. The radial basis function neural network (RBFNN) gains prominence in medical image processing [16]. This research work utilizes two machine learning algorithms for the compression of images; BPNN and RBFNN (Figure 8.1).

When an input image is fed into the network, it is transmitted forward through the network until it reaches the output layer. The network output is then compared with the desired output with the help of the loss function and it calculates the error value of each neuron in the output layer. The error values are then propagated backward from the output, until each neuron has an associated error value. It uses this error value to evaluate the gradient of the loss function. In the second phase, the gradient is given as input to the optimization method to update the weights.

The preprocessor unit interprets the medical image as an input and it extracts the gray pixel intensity for processing. The output of this unit is in the form of the pixel array. The spectral decomposer unit performs pyramidical decomposition for the gray coefficient which is obtained from the preprocessor unit. It extracts the spectral resolutions for the given input sample. A co-similar coefficient generator separates the same spectral coefficients after the decomposition process. Based on the redundant information, the suppression of co-similar coefficients results in first-level compression. The input unit normalizes the selected coefficients and passes them to the neural network unit. It takes the min–max value by creating a feed-forward neural network in consideration with the least mean learning algorithm. The compressor coefficient unit develops array logic where the output of neural network unit is stored. The pixel interpolation unit processes the compressed information. The recovered pixel coefficients are arranged based on the sequence order. Inverse spectral decomposer unit processes the coefficients from the pixel interpolation unit. This unit operates inverse discrete wavelet transform.

8.2.1 Data Acquisition

Real-time abdomen CT DICOM datasets are used in this research work. Each dataset comprises 200 to 300 images, out of which sample images are taken from each data set for analysis of algorithms.

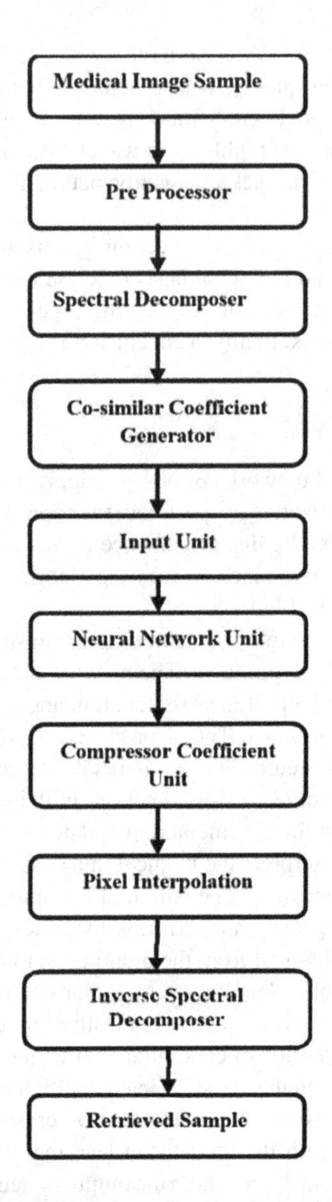

FIGURE 8.1 Medical image compression using neural network algorithm.

8.2.2 Medical Image Compression/Decompression Using Neural Network Algorithms

Steps involved in image compression utilizing BPNN are as follows

Step 1: Initially, the input medical image is transformed into matrix format (M) which is represented by Nr,Nc. Where r indicates row and c indicates column.

Step 2: With the help of the matrix format, find the value of pixel and probability of neighboring pixels for denoting the pair values.

$$P = (L_1, M_1)(L_2, M_2)\cdots(L_m, M_n) \tag{8.1}$$

where L indicates the value of pixel and M indicates the probability of neighboring pixels.

Step 3: Represent pair values using sequence order.

$$O = L_1, M_1, L_2, M_2 \cdots L_m, M_n \tag{8.2}$$

Step 4: The input to the neural network is sequence order O.

$$A_m = A_1, A_2, A_3 \cdots A_i \tag{8.3}$$

Step 5: The weight W_{nm} is calculated using the formula.

$$W_{nm} = \sum_{m=1}^{i} A_m A_m^T \tag{8.4}$$

Where $1 \le n \le k$ and A_m is the input layer.

Step 6: Hidden layer F_n is created using the formula.

$$F_n = \sum_{m=1}^{i} A_{mn} A_m \tag{8.5}$$

where $1 \le n \le$ k and Am is the input layer.

The steps involved in image decompression are as follows.

Step 1: The hidden layer is given by the formula.

$$F_n = F_1, F_2 \ldots F_k \tag{8.6}$$

Step 2: The weight W_{mn} is calculated using the formula.

$$W_{mn} = \sum_{n=1}^{k} F_n F_n^T \tag{8.7}$$

Where $1 \le n \le k$ and F_m is the hidden layer.

Step 3: The output layer is created using the formula.

$$D_m = \sum_{n=1}^{k} W_{mn} F_n \tag{8.8}$$

Where $1 \le n \le k$ and F_m is the hidden layer.

Step 4: Represent the output layer in the sequence order.

$$O = L_1, M_1, L_2, M_2 \cdots L_m, M_n \tag{8.9}$$

Step 5: Represent the sequence order in pair values. The pair value indicates pixel values and the probability of neighboring pixels.

$$O = (L_1, M_1)(L_2, M_2)\cdots(L_m, M_n) \tag{8.10}$$

Step 6: Represent the pair values into pixel values and transformed them into matrix format.
Step 7: Finally, the matrix is transformed into an image.

RBFNN was developed using a supervised learning algorithm. It has three different layers, namely input, hidden, and output layer. The input layer comprises source nodes with the same number as the input vector's dimension. It is directly connected to the hidden layer. A base function with the parameter center and width is used in the hidden unit. The hidden layers are connected with the output layer. The output layer gives the result.

The following are the fundamental computations in the RBF network.

Input layer

The input vector i is weighted by input weights w^h at the hidden unit l's input.

$$S_l = [i_1 w_{1,l}^h, i_2 w_{2,l}^h \cdots i_n w_{n,l}^h] \tag{8.11}$$

Where n denotes the input index, l is the hidden unit index, and w^h denotes the total weight between input n and hidden unit l. Hidden layer

The hidden unit l's production is estimated as follows.

$$\varphi_l(s_l) = \exp\left[-\frac{s_l - c_l}{\sigma_l}^2\right] \tag{8.12}$$

Where φ_l, c_l and σ_l denotes the triggering function, center, and width of hidden unit l respectively. Output layer

The network output m is determined using the following formula.

$$o_m = \sum_{l=1}^{L} \varphi_i(s_l) w_{l,m}^o + w_{0,m}^o \tag{8.13}$$

Where m is the output index, w^o is the output weight between hidden and output unit, and w^o is the output unit bias weight.

The output of an RBF is solely calculated using the distance between input and a given base. The image is divided into blocks, with each pixel's strength equal to the number of the outputs corresponding to the Gaussian RBFs of the RBFNN allocated to that block. Only the parameters of the RBFNs must be saved, and each sub-image must be reconstructed by adding the surfaces corresponding to the RBFs from the given RBFNN. In a quad-tree way, the image is broken into sub-image blocks, reducing the complexity. For a given Gaussian RBF, the center and dispersion coordinates must be stored as first-layer network parameters, and the amplitude must be equal to the weight corresponding to the neuron relation to the output.

The accuracy of the approximation is determined by the maximum number of RBFs in an RBFNN. If a large number of RBFs are chosen to be tested, there is a good chance that the breakdown will not go down to small blocks. For sub-images of various sizes, the number of bits used for encoding and quantization varies. On a sub-image line, the number of pixels n would be a power of two, $n = 2k$. This causes the center's x and y coordinates to be stored in only k bits. The total number of bits used to encode x and y is $k + 1$. For a linear discretization of 1/3, the dispersion will have values ranging from 1/3 to $2n/3$, requiring $k + 1$ bits to store one of the $2n$ potential values. For regular 8-bit grayscale images, the height h is stored on 8 bits. The highest number of units in an RBFN could be a power of two, and it is tested with a DC unit and up to three RBF units, each of which needs two bits to store the network's code. It is inefficient to encode a block of size 2 (4 pixels) with more than 2 units. To speed up the operation, gradient methods are used.

8.3 RESULTS AND DISCUSSION

The algorithms are developed in MATLAB 2016a and executed on the laptop with the following specifications; 1.8 GHz Dual-Core Intel Core i3 processor with 8 GB DDR3 RAM.

The medical images in DICOM format are utilized as input and the results are depicted below. The neural network-based compression approach is proposed in this work; the architecture comprises a training and a testing phase. The neural network training comprises 500 epochs, Figure 8.2 is the neural network training process. For comparative analysis, the radial basic function neural network and backpropagation neural network are used. In Figure 8.3, the first column depicts the input images, the second column depicts the compressed image and the third column depicts the decompressed image. The algorithms are tested on 9 DICOM medical images and are labeled as D1–D9. Figures 8.3 and 8.4 represents the compression results of backpropagation and radial basic function neural network algorithms. Figure 8.3 and Figure 8.4 depicts the compression results for the images D2–D5 and Figure 8.5 depicts the compression results for the images D6-D9. The performance validation was done by the following metrics.

The compression ratio judges the performance of the compression algorithm and it is defined as the ratio of input image file size to the resultant compressed image file size.

$$CR = \frac{S_{UC}}{S_C} \tag{8.14}$$

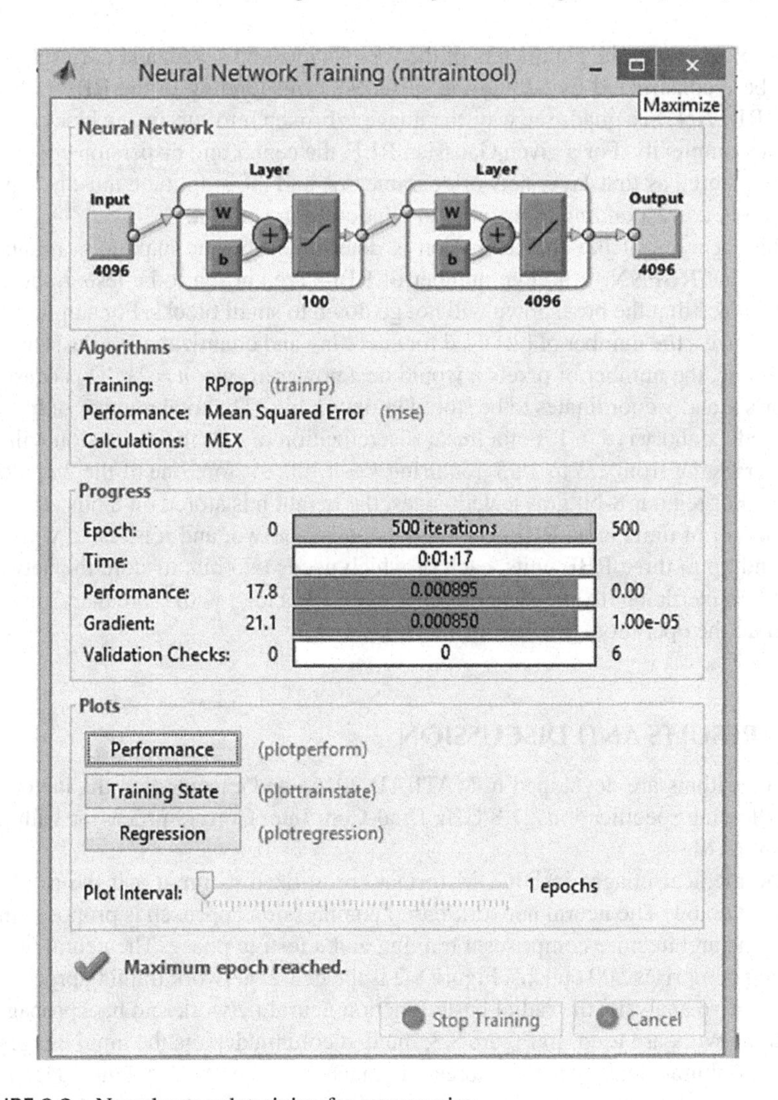

FIGURE 8.2 Neural network training for compression.

The PSNR and MSE evaluate the quality of the decompressed image. An increase in PSNR value and a low value of MSE qualifies a compression algorithm.

$$MSE = \frac{1}{N}\sum_{x=1}^{M}\sum_{y=1}^{N}(I(x,y) - I(x,y))^2 \tag{8.15}$$

$$PSNR = 10\log\frac{255^2}{MSE} \tag{8.16}$$

The (x, y) symbolizes the pixel value of the original input and $\bar{I}(x, y)$ symbolizes the pixel value of the decompressed image.

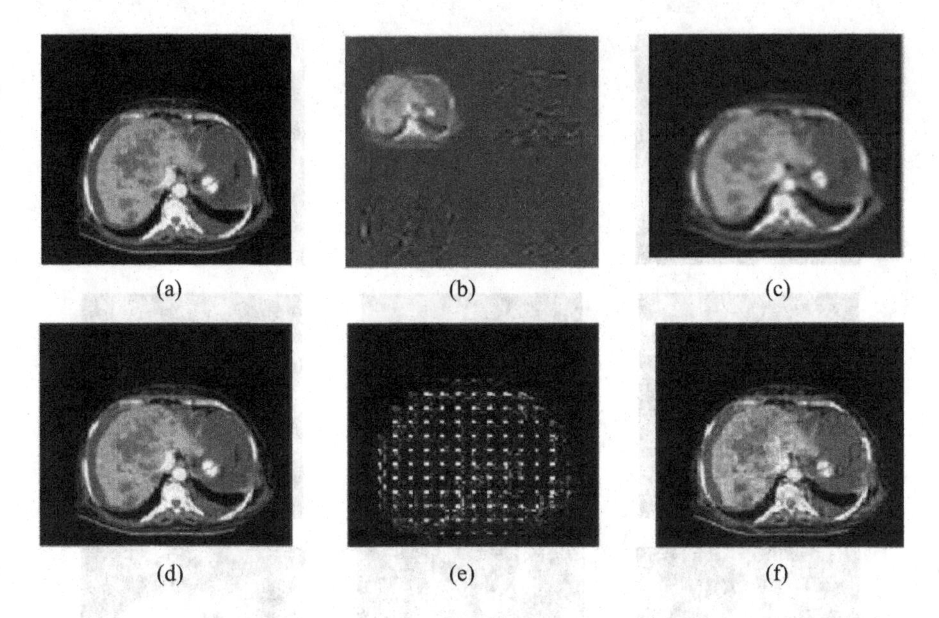

FIGURE 8.3 (a, b, c) Compression results of BPNN; (d, e, f) Compression results of RBFNN.

The PSNR and the MSE plot favor the efficiency of the RBFNN algorithm. The compression ratio plot is depicted in Figure 8.8. The compression ratio of RBFNN was found to be better when compared with the BPNN algorithm.

The RBFNN-based compression was found to be efficient when related with the BPNN-based compression. For further validation, the following metrics are used: normalized cross-correlation (NCC); structural content (SC); normalized absolute error (NAE); Laplacian mean square error (LMSE); and average difference (AD).

The NCC measures the similarity between the input image and the decompressed image. The closer the value of NCC to '1', the better is the efficiency of the compression technique

$$MSE = \frac{\sum_{x=1}^{M}\sum_{y=1}^{N} I(x,y)I(x,y)}{\sum_{x=1}^{M}\sum_{y=1}^{N} I(x,y)^2} \tag{8.17}$$

The SC also measures the degree of similarity between the input and the decompressed images. Closer the value of SC to '1', the better is the compression algorithm.

$$SC = \frac{\sum_{x=1}^{M}\sum_{y=1}^{N} I(x,y)^2}{\sum_{x=1}^{M}\sum_{y=1}^{N} I(x,y)^2} \tag{8.18}$$

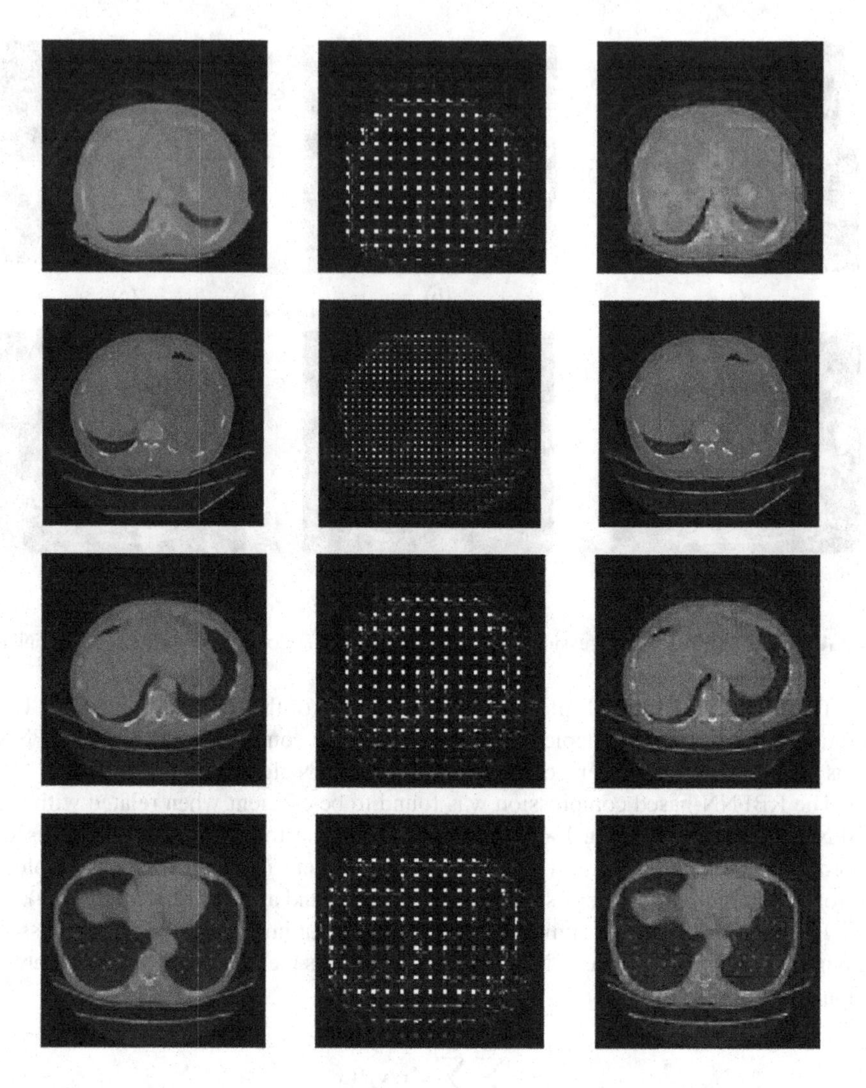

FIGURE 8.4 RBFNN compression results corresponding to the input images D2–D5.

The NAE and the LMSE are the error metric and low values prove the efficiency of the compression algorithm.

$$NAE = \frac{\sum_{i=1}^{m}\sum_{j=1}^{n}(P(i,j)-Q(i,j))}{\sum_{i=1}^{m}\sum_{j=1}^{n}P(i,j)} \tag{8.19}$$

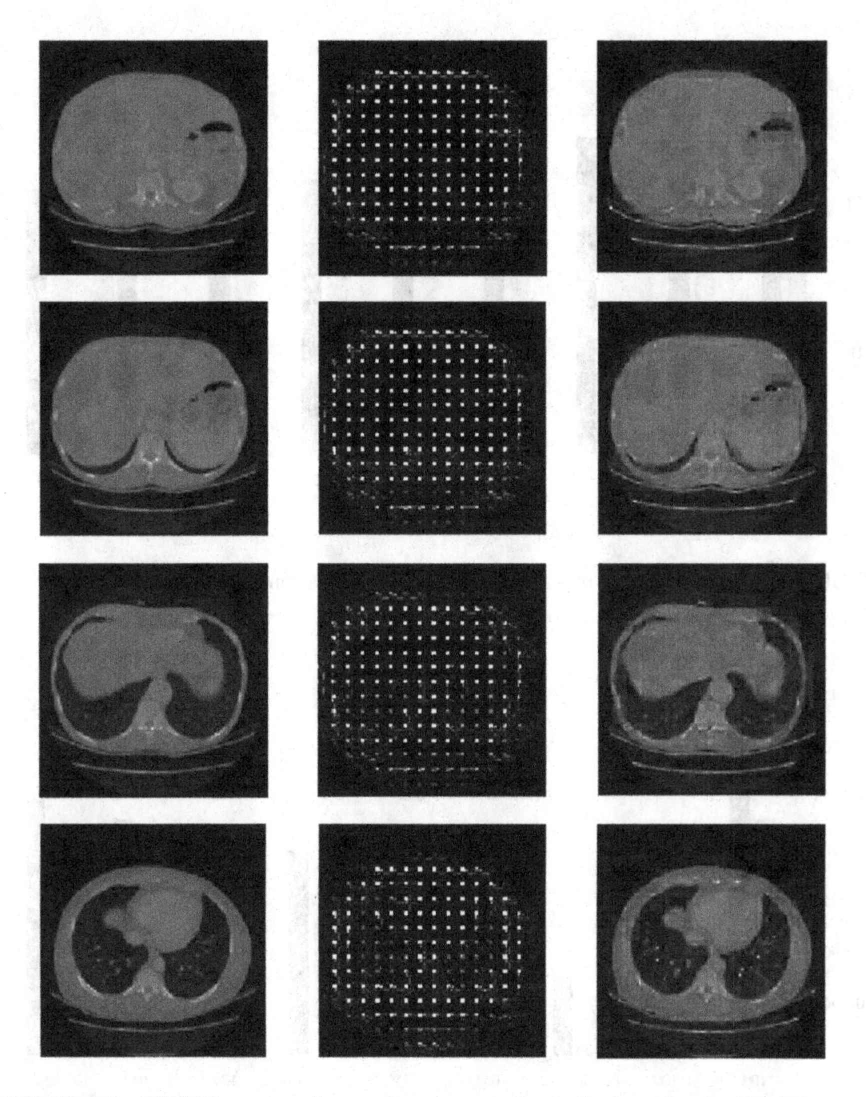

FIGURE 8.5 RBFNN compression results corresponding to the input images D6–D9.

$$LMSE = \frac{\sum_{i=1}^{m}\sum_{j=1}^{n}[L(P(i,j)) - L(Q(i,j))]^2}{\sum_{i=1}^{m}\sum_{j=1}^{n}[L(Q(i,j))]^2} \tag{8.20}$$

The AD measures the difference between the input image and the decompressed image. Low values of AD justify the ability of the compression algorithm, as shown in the Figures 8.6 and 8.7.

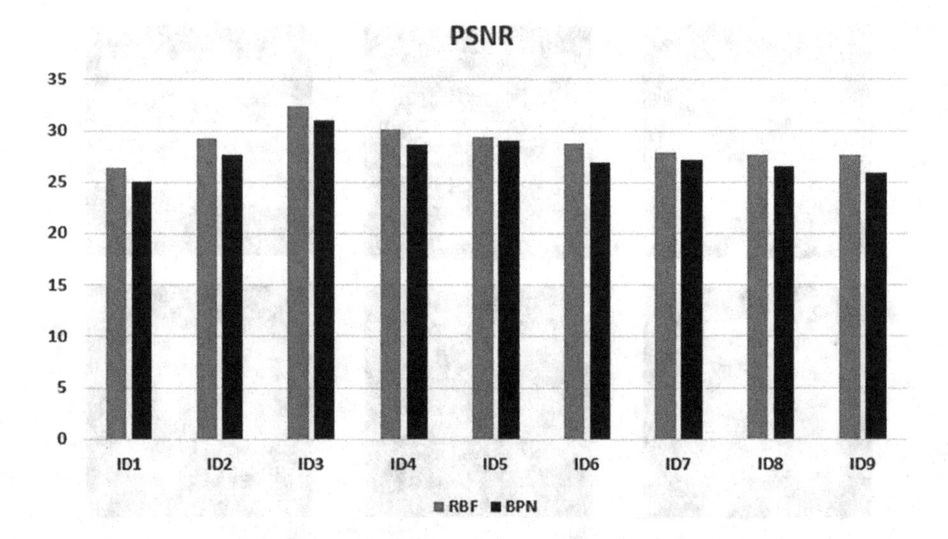

FIGURE 8.6 PSNR plot of neural network-based compression techniques.

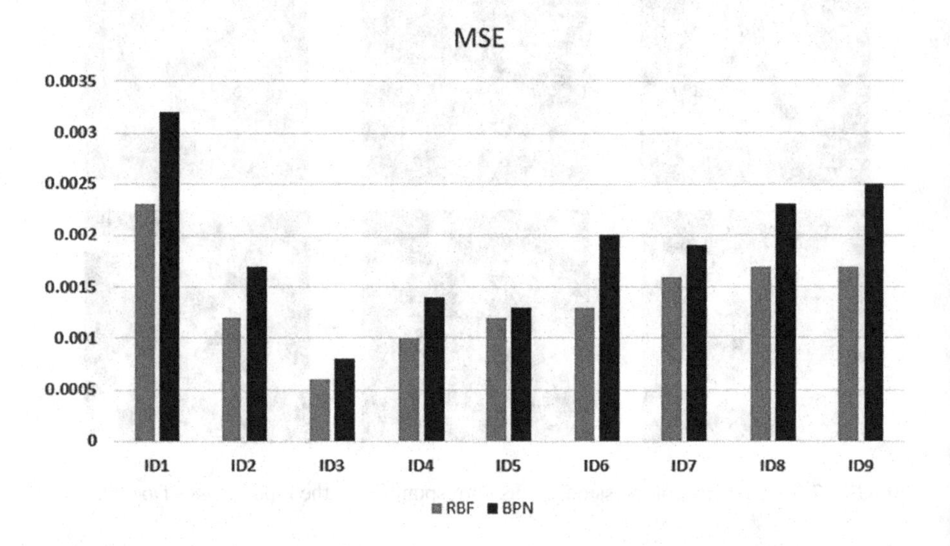

FIGURE 8.7 MSE plot of neural network-based compression techniques.

$$AD = \frac{1}{mn}\sum_{i=1}^{m}\sum_{i=1}^{n}[A(i,j) - B(i,j)] \tag{8.21}$$

The performance metrics values reveal that RBFNN-based compression is more proficient when compared with the BPNN.

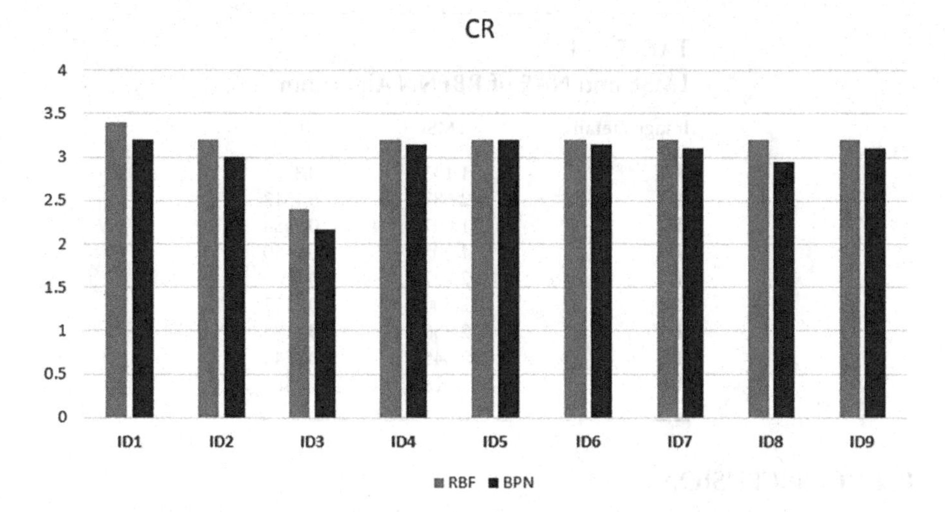

FIGURE 8.8 Compression Ratio plot of neural network-based compression techniques.

TABLE 8.1
Performance of Radial Basic Function Compression Algorithm

Image Details	File Size	Bits per Pixels	Compressed Memory Size	Space-saving	nbits/pixel
ID1	128 × 128	17	10240.00	0.71	5.00
ID2	128 × 128	16	10240.00	0.69	5.00
ID3	256 × 256	12	40960.00	0.58	5.00
ID4	128 × 128	16	10240.00	0.69	5.00
ID5	128 × 128	16	10240.00	0.69	5.00
ID6	128 × 128	16	10240.00	0.69	5.00
ID7	128 × 128	16	10240.00	0.69	5.00
ID8	128 × 128	16	10240.00	0.69	5.00
ID9	128 × 128	16	10240.00	0.69	5.00

TABLE 8.2
AD, SC, and NCC of RBFNN Algorithm

Image Details	AD	SC	NCC
ID1	0.0006	1.2104	0.8545
ID2	−0.0150	0.8792	1.0607
ID3	−0.0017	1.1203	0.9098
ID4	−0.0088	0.9010	1.0439
ID5	−0.0067	0.8994	1.0393
ID6	−0.0107	0.9163	1.0381
ID7	−0.0114	0.9104	1.0412
ID8	−0.0093	0.8881	1.0483
ID9	−0.0055	1.0683	0.9310

TABLE 8.3
LMSE and NAE of RBFNN Algorithm

Image Details	LMSE	NAE
ID1	1.1434	0.3015
ID2	2.6874	0.1312
ID3	1.0974	0.2484
ID4	1.9052	0.1550
ID5	2.4279	0.1960
ID6	2.4003	0.1127
ID7	2.3303	0.1183
ID8	2.3455	0.1745
ID9	1.1288	0.2485

8.4 CONCLUSION

Medical image storage and transfer is a crucial factor in telemedicine. This chapter proposes a system of neural network-based medical image compression. The BPNN- and RBFNN-based algorithms are outlined here for the compression of medical images. Performance metrics validation reveals that the RBFNN yields superior scores when related with the BPNN algorithm. The above Table 8.1, 8.2, and 8.3 mention the performance of the algorithm and its comparison. The future work is the development of deep learning-based medical image compression with reduced computational complexity.

REFERENCES

1. Wallace GK, "The JPEG Still Picture Compression Standard," *IEEE Transactions on Consumer Electronics*, vol. 38, pp. 17–34, Feb 1992.
2. Matsuoka R, Sone M, Fukue K, Cho K, and Shimoda H, "Quantitative Analysis of Image Quality of Lossy Compression Images," *International Society of Photogrammetry and Remote Sensing*, Sep 2013.
3. Chang SG, and Yovanof GS, "A Simple Block Based Lossless Image Compression Scheme," Thirtieth Asilomer Conference on Signals, Systems and Computer, pp. 591–595, Dec 1996.
4. Spires W, "Lossless Image Compression Via the Lifting Scheme," *International Jouranal of Engineering Sciences and Research Technology*, pp. 435–439, Apr 2015.
5. Laura L, Camacho MTV, Badran A, and Armando DG, "Image Compression For Medical Diagnosis Using Neural Networks," *British Journal of Applied Science & Technology*, pp. 510–524, 2014.
6. Yeo WK, Yap DFW, Oh TH, Andito DP, Kok SL, Ho YH, and Suaidi MK, "Grayscale Medical Image Compression Using Feed forward Neural Network," International Conference on Computer Applications and Industrial Electronics, pp. 633–638, 2011.
7. Durai SA, and Saro EA, "Image Compression With Back-Propagation Neural Network Using Cumulative Distribution Function," *International Journal of Engineering and Applied Sciences*, pp. 185–189, 2007.
8. Cottrell G, Munro P, and Zipser D, "Image Compression by Back Propagation: An Example of Extensional Programming," *Advances in Cognitive Science*, pp. 209–240, 1989.

9. Rajput GG, and Singh MK., "Modeling of Neural Image Compression Using GA and BP: A Comparative Approach," *International Journal of Advanced Computer Science and Applications*, pp. 26–34, 2011.
10. Omaima NA, "Improving the Performance of Back propagation Neural Network Algorithm for Image Compression/Decompression System," *Journal of Computer Science*, pp. 1347–1354, 2010.
11. Wan TC, and Kabuka M, "Edge Preserving Image Compression for Magnetic Resonance Images Using DANN-Based Neural Networks," *Medical Imaging*, pp. 1–17.
12. Liang JY, Chen CS, Huang CH, and Liu L. "Lossless Compression of Medical Images Using Hilbert Space-Filling Curves", *Computerized Medical Imaging and Graphics*, pp. 174–182, 2008.
13. Tripathi P, "Image Compression Enhancement using Bipolar Coding with LM Algorithm in Artificial Neural Network," *International Journal of Scientific and Research Publications*, vol. 2, pp. 1–6, 2012.
14. Lin CT, and Fan KW, "An HVS-Directed Neural Network-Based Image Resolution Enhancement Scheme for Image Resizing," *IEEE Transactions on Fuzzy Systems*, pp. 605–615, 2007.
15. Vijideva, "Neural Network-Wavelet based Dicom Image Compression and Progressive Transmission," *International Journal of Engineering Science & Advanced Technology*, pp. 702–710, 2012.
16. Lu Z, Lu S, Liu G, Zhang Y, Yang J, Phillips P. A Pathological Brain Detection System Based on Radial Basis Function Neural Network. *Journal of Medical Imaging and Health Informatics*, 2016 Sep 1;6(5):1218–22.

9 Prospects of Wearable Inertial Sensors for Assessing Performance of Athletes Using Machine Learning Algorithms

Ravi Kant Avvari and Priyobroto Basu
Department of Biotechnology and Medical Engineering, NIT Rourkela, India

CONTENTS

9.1 Introduction 137
9.2 The State of the Art in Motion Sensing 139
9.2.1 3-D Motion Capture System 139
9.2.2 Wearable IMU Sensors 139
9.2.3 Electrogoniometers 141
9.2.4 Force Plate Mechanism 141
9.2.5 Medical Imaging Techniques 141
9.3 Wearable Inertial Sensors for Sports Biomechanics 141
9.4 Machine Learning (ML) Algorithm for Precision Measurement 143
9.4.1 Kalman Filter 143
9.4.2 Extended Kalman Filter 144
9.4.3 Extended Kalman Filter Algorithm 146
9.4.4 Zero-Velocity (ZUPT) Update 146
9.4.5 Cascaded Kalman Filter 147
9.4.6 Quaternion Concept 148
9.5 Conclusion 148
References 149

9.1 INTRODUCTION

Sports biomechanics is a science which concerns the movement of living bodies, including how muscles, bones, tendons, and ligaments work in coordination to produce locomotion. The science reflects the broad interplay between classical

DOI: 10.1201/9781003267782-9

mechanics and the biological system. The biomechanics of human locomotion has become a subject of interest in order to improve the performance of the body. It has focused, in particular, on improving the endurance, agility, and performance and has been applied in the area of physiotherapy to recover from injury or disability.

The biomechanics of locomotion can be studied in two distinct ways: kinematics and kinetics. Kinematics deals with the geometry and time-dependent aspects of the body without considering the forces of motion. In human kinematics, the movement of different body parts can be measured using mechanical, magnetic or inertial means of measurement [1, 2]. Kinetics, by contrast, considers the effect of forces and torques to analyze a motion. Kinetic analysis is relevant during the estimation of muscle force and the muscle activation of a musculoskeletal body. Force plates, instrumented tools, and electromyograms (EMGs) are the notable measuring devices for muscle activation and forces [2]. Various technologies have been developed to measure human movement; examples include the use of pressure mats, force shoes, magnetic systems, floor-mounted systems, and optoelectronic systems; however, they can capture only brief periods of movement in the laboratory [3]. The introduction of wearable sensors enabled the continuation of experiments in diverse environments (indoor, outdoor, rough terrain). Biomedical sensors made of semiconductor and flexible electronics packaging technology present an exciting opportunity for the measurement of human physiological parameters in a continuous, real-time and non-intrusive manner [4]. These sensors are popularly known as Inertial Measurement Unit (IMU) and offer an alternative to the expensive Gait analysis system [5].

In sports biomechanics, IMU sensors provide real-time information on athletes' behavior and movement. These sensors placed on the body mainly consists of one triaxial accelerometer, triaxial gyroscope and triaxial magnetometer that collects raw data during the athlete's movement [6–8]. Over the years, initiatives have been taken to improve the flexibility of sportspeople in order for them to compete at an advanced level; considerable attention has been given to avoid injury through player monitoring strategies. The workload management of an athlete plays a major role in this part. Through the constant monitoring of this workload the player's performance can be judged and the risk of fatigue and injuries can be prevented [5].

Medical practitioners monitor the severity of impacts incurred by placing an IMU on the back. Turning angles or Change of Direction (COD) of athletes can be measured using IMU sensor fusion and COD detectable algorithm. The monitoring of those load angles helps in performance enhancement and the prevention of injury [7]. J. Cockcroft et al. show a potential use of wireless sensors in the estimation of dynamic acceleration and the calculation of hip angles during cycling [6]. T. Ogasawara has studied whether or not the possibility of suppressing postural tremors using IMU sensors embedded inside the bow would have potential applications in predicting the score in archery [9]. V. Bonnet et al., by contrast, focused on the lower-limb joint and torso kinematics during squat exercises in the sagittal plane [10] using only a single IMU sensor placed on the lower back. In another study, S. Ailing and Cheng Kai used embedded IMU sensors on a javelin to observe its trajectory and the parameters during its release [11]. Everyday activities can also be measured using the MEMS accelerometer, gyroscope and magnetometers [12]. These might include the hand and Foot trajectory of sportspersons participating in a range of activities, from cycling [6, 13] to golf [14], and from table tennis [15] to gymnastics [10], can be observed using these sensors. Monitoring collisions in high-impact sports such as rugby have given sports scientists valuable information for research on periodizing the impact stress in contact sport [10, 11]. Since the raw data obtained from the sensor are noisy and suffer from drift errors, they are filtered so as to obtain a better estimate of state of the system using soft computing techniques such as the Kalman filter and the extended Kalman filter.

A scientometric analysis (VOSviewer software, version 1.6.17), which emphasizes the significance of the work, is shown in Figure 9.1. The keywords to be used in this respect are sports biomechanics, gait analysis, Inertial Measurement Unit (IMU) + Gait analysis, Kalman filter (KF), Extended Kalman filter (EKF), IMU + KF, IMU + EKF, Zero-velocity update, and cascaded Kalman filter. Search sources included in this literature review have included IEEE, Elsevier, ScienceDirect, and PubMed for researched published between 2011 and 2022.

9.2 THE STATE OF THE ART IN MOTION SENSING

The technologies involved in motion sensing are the following: the 3-D motion capture system; Wearable Inertial Sensors; Electrogoniometers; Force Plate Mechanism; and Medical Imaging.

9.2.1 3-D Motion Capture System

The gait cycle is the name given to a series of rhythmical alternating movements of the trunk and limbs, causing a forward progression of the body and centre of gravity. The gait cycle is described using spatio-temporal parameters such as stance and swing, stride length, walking speed, hip and knee angles, and ground reaction forces. A 3-D motion capture system is suitable for providing this information. A number of reflective markers are placed on the subject's joints; these can be captured by infrared cameras placed at specified distances and a trajectory can be determined. The data are then processed in Visual-3D software in order to create a bone model of the lower body and generate segment-wise data. The accuracy of the measurement obtained using the calibrated camera system lies in the range of millimeters. In earlier times, the motion capture system was considered to be the gold standard in gait analysis.

9.2.2 Wearable IMU Sensors

The infrastructure required for performing gait analysis using the motion capture system is expensive and requires a specially dedicated laboratory for its application. Recent advancements in MEMS (Micro-Electro-Mechanical system) technology have led to the development of an affordable and reliable wearable IMU (Inertial Measurement Unit) system for gait analysis. This system involves a lot of sensors, including the accelerometer, the gyroscope, and the magnetometer. These sensors do not require a sophisticated laboratory and daily activities can be performed without

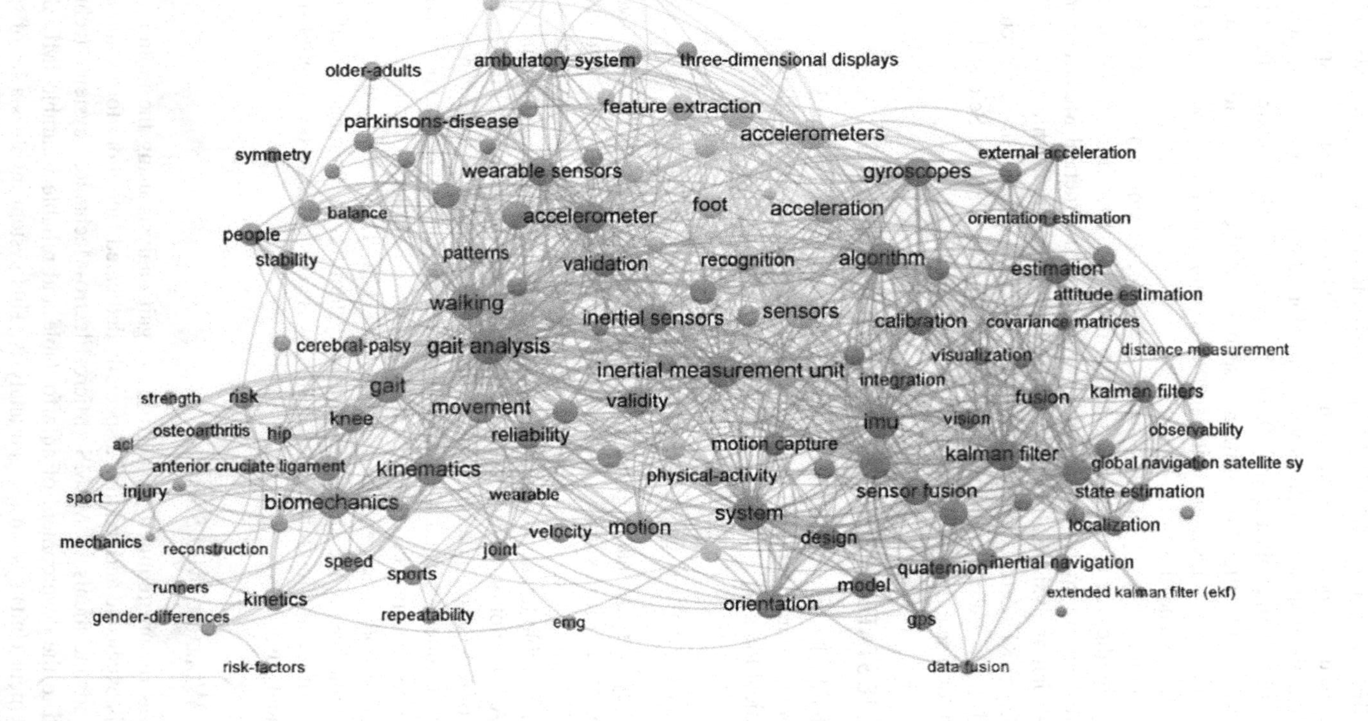

FIGURE 9.1 Scientific metric analysis of search terms of relevance to sports biomechanics.

any hindrance. Wearable sensors have helped largely in human foot trajectory detection, fall detection among elderly people, monitoring postoperative gait abnormalities and in patients suffering from various diseases such as arthritis, diabetes and Parkinson's disease. These sensors are placed, either directly or indirectly, on different body locations such as the foot, wrist, chest, or thigh, and they are attached using belts, clips, or other accessories. Despite the huge number of potential applications of this device, the inertial sensor suffers a large amount of noise and drift; hence, it is of only limited use to short time studies, where the accuracy of the measurements is of prime importance. To rectify the erroneous data different algorithms, such as Kalman filtering, extended Kalman filtering, and Zero-Order Velocity Update (ZUPT) algorithms, are employed. These are discussed in Section 9.3.

9.2.3 Electrogoniometers

The electrogoniometer is an electro-mechanical device which is used for measuring joint angles in gait analysis. The sensor converts the mechanical force into electrical signals. The angle measurement is possible in only one axis with accurate calibration. According to research, the electrogoniometer gives good angle measurement for the elbow joint, but it provides poor results for the knee joint.

9.2.4 Force Plate Mechanism

Force plates are used to measure the Ground Reaction Force (GRF) of a subject either during standing or in motion on the plates. It employs load cells connected at the ends of the plate to measure the induced force in all directions. Since the force plates remain at a fixed location on the ground, the Center of Pressure (CoP) gives erroneous results during a long track movement. Moreover, a suitable position for placing the footsteps cannot be maintained throughout, which leads to the incorrect calculation of CoP.

9.2.5 Medical Imaging Techniques

In medical imaging techniques such as ultrasound, X-rays, CT scans, and MRI imaging are commonly used to measure the geometrical structure of the human body. Such imaging modalities have been largely used for studies of relevance to injury, and in post-surgical observations.

9.3 WEARABLE INERTIAL SENSORS FOR SPORTS BIOMECHANICS

Wearable inertial sensors find a huge range of immense applications in sports such as cycling, table tennis, golf, running, javelin throwing, and also in rigorous exercises such as performing squats. It involves placing the sensor on the joints to measure kinematics details, such as hip joint angles for cycling [6], forearm angles in netball [8], and elbow angles in swimming [16]. Measures to prevent lateral ankle sprains have also been studied [17]. Table 9.1 shows various biomechanical studies using the inertial sensors, with specific references to the outcome and observations of the study.

TABLE 9.1
Sports Biomechanical Studies Using Inertial Sensors

Model	Sensor	Specification	Outcome Measure	Observations
OptimEye S5 [7]	3-φ A, G, M	Midpoint of TV, $f = 100$ Hz	COD	Successful COD detection for **athletes**
Wireless IMMSs [6]	3-φ A, G, M	On knee, $f = 75$ Hz	Hip joint angles	Hip angle detection during **cycling** accurate in sagittal plane but moderate in frontal and transverse plane.
SABEL Sense [8]	3-φ A, G, M	On forearm	Forearm shooting angle.	Variability in forearm shooting angle in **netball** sport detected successfully.
Hitoe Transmitter [9]	3-φ A	On bow, $f = 25$ Hz.	Postural tremors	Prediction of scores in **archery** by detection of postural tremors.
IMU by MTx Xsens [10]	3-φ A, G	On lower trunk	Lower limb joint angles	Angle detection during **squat** successful in sagittal plane only.
Spartan-6 series XC6SLX4 [11]	3-φ A	On javelin	Trajectory of Javelin throw.	Online trajectory determination in **javelin throw** successful.
Analog Devices [12]	3-φ A	On wrist and hip, $f = 20$ Hz	Sports activities	75–88% accuracy was observed in the identification of different sports like **bike riding, exercising, football**.
MPU 6050 [14]	3-φ A, G	On golf stick, $f = 200$ Hz	Golf swing analysis	**Golf swing movement and posture information** visualized in 3-D.
MPU 9250 [15]	3-φ A, G, M	On wrist, $f = 50$ Hz	Stroke in table tennis	Accelerometer data helpful in finding **hit or miss in table tennis**.
IMU sensor [16]	3-φ A, G	On elbow	Elbow angles in swimming.	IMU can be used to calculate the **elbow angle even in water**.
Wearable IMU [17]	3-φ A, G	On sport shoes, $f = 200$ Hz	Lateral ankle sprain hazard.	Correction system devised to provide **external electrical stimulation during ankle sprain**.
Xsens Awinda [18]	3-φ A, G, M	On pelvis and ankles, $f = 100$ Hz	Lower limb kinematics	**Knee and hip joint angles** were determined using IMU sensor and Extended Kalman filtering.
Load cell [19]	A, G	As pedal tilt	Bicycling	Mechatronic rehabilitation system for individuals suffering from **chronic or post-surgical conditions**.
IMUs by MTws [20]	3-φ A, G	On head and chin.	head and neck strength	The repeatability and sensitivity of the **head kinematics and neck strength during concussion** experimentation is not high.
Head worn IMU [21]	3-φ A	On head, $f = 60$ Hz.	Real time Gait analysis	**Foot ground contact time, contact time ratio, stride time** calculation possible from this work.
MotionFit SDK platform [13]	3-φ A	On shoe, $f = 200$ Hz	Real time cycling	Innovation of pedaling profile for **real time cycling**.
Xsens MTi-G-700 [22]	3-φ A, G, M	In googles	Vertical trajectory	Cascaded KF using MEMS IMU helps in determining **vertical trajectory in sports**.
IMU sensor [23]	3-φ A,G	On skin of the gymnasts.	IMU angles and force	High accuracy was seen during **backward somersault landings**.
MyoResearchÒ model 610 [24]	3-φ A, G, M	On the dorsal side of hands, $f = 200$ Hz.	Acceleration and orientation of pelvis, hands	High accuracy was seen while calculating the angles **during the baseball game**.
MTw, Xsens Technology [25]	3-φ A	On trunk & tibia, $f = 75$Hz	GRF	The work is unsuccessful and suggests **segmental acceleration is not whole-body acceleration**.

Abbreviations – A: accelerometer; G: gyroscope; M: magnetometer; TV: thoracic vertebrae; COD: change of direction angles; GRF: Ground reaction force.

9.4 MACHINE LEARNING (ML) ALGORITHM FOR PRECISION MEASUREMENT

Wearable IMU sensors are relatively cheaper and portable; consequently, they can be performed in both indoor and outdoor settings. Typically, sensors, such as the triaxial gyroscope and the accelerometer, are employed in the study. The angular rate of rotation is measured by a gyroscope which yields orientation angles and which, upon integration, provides a rotary position. Since the raw data suffer from noise and integration accumulates bias error and drifts the orientation from the true values, this limits the usability of the sensor for precision measurement. In order to reduce such errors, soft computing methods are employed, including the Kalman filter, the extended Kalman filter, and the Zero-Order Velocity Update (ZUPT) to provide a stable output from these sensors.

9.4.1 Kalman Filter

Kalman filtering uses a systems dynamic model, control inputs, and measurement values from multiple sensors to form the estimate of system's varying quantity (state), which is more accurate than the estimate obtained solely through measurement values. The uncertainty due to noisy sensor data can be dealt effectively by the Kalman filter ([26]; Narayan [27]). The Kalman filter produces the current state update using predicted state and sensor measurement values using the Kalman Gain factor. The Kalman Gain is calculated from covariance, which accounts for the estimation uncertainty of the prediction of the system's state. Eventually, the new state formed has a better estimated uncertainty than the previous one. The process is repeated in every time state, adjusting the Kalman Gain according to the covariance and moving towards a better state estimate ([28]; Narayan [27]).

Most modern systems are equipped with numerous sensors which provide the estimation of parameters based on a series of measurements or sensor values. For example, a GPS receiver providing location and velocity estimation (parameters), is based on the differentials in the times of the arrival of signals from various satellites. One of the biggest challenges of tracking and control systems is providing accurate and precise estimation of the hidden variables or parameters in the presence of uncertainty. In GPS receivers, the measurement uncertainty depends on many external factors, including thermal noise, atmospheric effects, slight changes in satellite positions, receiver clock precision and many more.

The Kalman filter is one of the most important and common estimation algorithms. The Kalman filter produces estimates of hidden variables based on inaccurate and uncertain measurements. The Kalman filtering algorithm is applicable for a linear system. The process and measurement model equations are,

$$a_t^- = \mathrm{F} \cdot a_{t-1}^+ + \mathrm{G} \cdot u_{t-1} + w_{t-1}$$
$$y_t = \mathrm{H} \cdot a_t + v_t$$

Where, a is the state vector, y is the measurement vector, F is the state transition matrix, H is the observation matrix, and w and v are white Gaussian process and measurement noises, respectively.

The process noise covariance matrix Q_{t-1} is defined by:

$$Q_{t-1} = \mathrm{E}(w_{t-1}w_{t-1}^T),$$

where E is the expectation operator.

Similarly, the measurement noise covariance matrix R_t is defined as

$$R_t = \mathrm{E}(v_t v_t^T)$$

After the process model, the measurement model and covariance matrices been defined as above, we can proceed with the linear Kalman filtering algorithm. The equations involved are broadly categorized in to prediction and correction blocks:

Prediction	
State Extrapolation equation	$a_t^- = \mathbf{F} \cdot a_{t-1}^+$
Covariance Extrapolation equation	$P_t^- = \mathbf{F} \cdot P_{t-1}^+ \cdot F^T + Q_{t-1}$
Correction	
Kalman Gain	$\mathbf{K} = P_t^- \cdot H^T / (H \cdot P_t^- \cdot H^T + R_t)$
State Update Equation	$a_t^+ = a_t^- + K \cdot (a_m - Ha_t^-)$
Covariance Update Equation	$P_t^+ = (\mathbf{I} - \mathbf{K} \cdot \mathbf{H}) \cdot P_t^- \cdot (I - K \cdot H)^T + K \cdot R_t K^T$

With a high Kalman Gain, the filter puts more weights on the measurement values and increases its responsiveness. With a low gain, the filter current state follows the predicted state. It smooths out the noise but decreases the responsiveness of the filter. In Figure 9.2, a Kalman filter estimation has been shown by keeping the accelerometer sensor on the wrist and obtaining the hand's trajectory. As depicted in Figure 9.2, the estimation helps in eliminating the accelerometer noise to a great extent.

9.4.2 Extended Kalman Filter

The Kalman filter can be used to model only linear transformations:

$$a_t^- = \mathrm{F} \cdot a_{t-1}^+ + \mathrm{G} \cdot u_{t-1} + w_{t-1}$$
$$y_t = \mathrm{H} \cdot a_t + v_t$$

The observation that the linear transformation of a Gaussian random variable yields another Gaussian random variable is valid upon assuming linearity for both

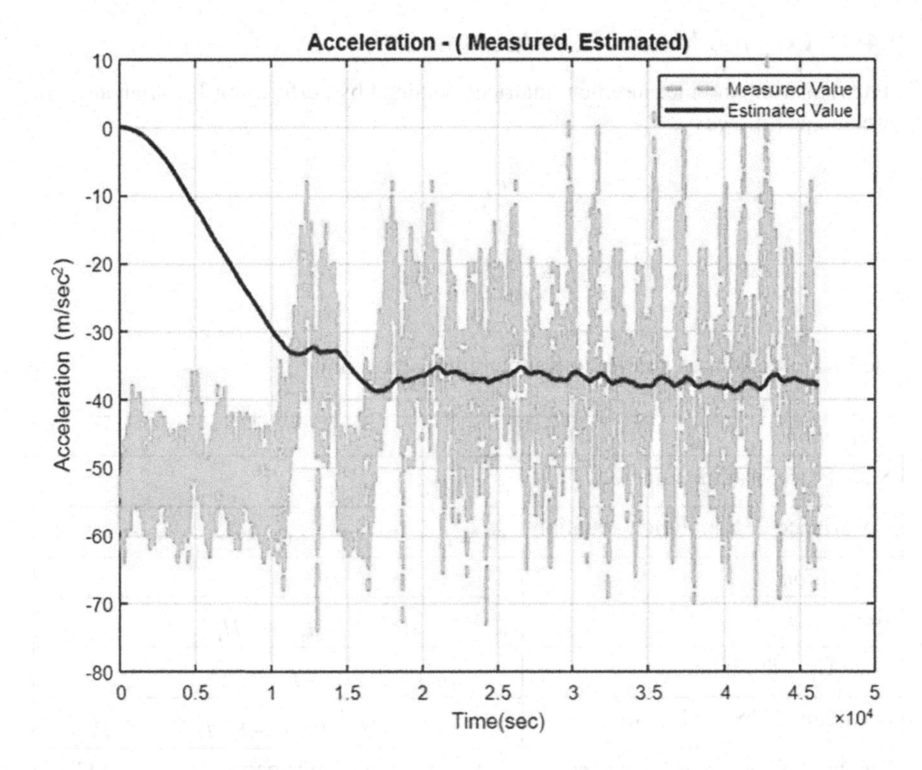

FIGURE 9.2 Kalman filter estimation of hand trajectory.

the state transition and the measurement model of a Kalman filter. When a random variable distributed over the mean is passed through a nonlinear function the resulting distribution is no longer Gaussian, which makes Kalman filtering useless over the nonlinear function domain.

The Extended Kalman filter (EKF) performs linear approximation of the nonlinear function using Taylor series expansion. EKF first evaluates the non-linear function at a mean, which is the best approximation of the distribution, and then estimates a line whose slope is around that mean. This slope is determined by the first-order derivative of the Taylor expansion as the first-order derivative gives a linear value.

Let's say we have the following models of state transition and measurement:

$$a_t = \mathrm{f}(a_{t-1}, u_{t-1}) + w_{t-1}$$
$$y_t = \mathrm{h} \cdot (a_t) + v_t$$

Where, a_t is the current state, f is the function of previous state a_{t-1} and control input u_{t-1}, h is the measurement function relating the current state a_t with the measurement y_t. w_{t-1} and v_t are process and measurement noises, respectively, having covariances Q, and R, respectively.

9.4.3 Extended Kalman Filter Algorithm

State transition and measurement matrices obtained by performing Jacobian are represented as follows:

$$F_{t-1} = \frac{\partial f}{\partial a}\Big|(\hat{a}^{+}_{t-1}, u_{t-1})$$

$$H_t = \frac{\partial h}{\partial a}\Big|\hat{a}^{-}_{t}$$

The EKF algorithm is similar to Kalman filter algorithm:

Prediction	
State Extrapolation equation	$\hat{a}^{-}_{t} = \mathrm{f}(\hat{a}^{+}_{t-1}, u_{t-1})$
Covariance Extrapolation equation	$P_t^{-} = F_{t-1} \cdot P_{t-1}^{+} \cdot F_{t-1}^{T} + \mathrm{Q}$
Correction	
Kalman Gain	$\mathrm{K} = P_t^{-} \cdot H_t^{T} / (\mathrm{R} + H_t \cdot P_t^{-} \cdot H_t^{T})$
State Update Equation	$\hat{a}^{+}_{t} = \hat{a}^{-}_{t} + \mathrm{K} \cdot (a_m - h \cdot \hat{a}^{-}_{t})$
Covariance Update Equation	$P_t^{+} = (\mathrm{I} - \mathrm{K} \cdot H_t) \cdot P_t^{-} \cdot (\mathrm{I} - \mathrm{K} \cdot H_t)^{T} + \mathrm{K} \cdot \mathrm{R} \cdot K^{T}$
The hat "^" operator, means estimate of a variable. The superscripts "+" and "–" represent "a posteriori" and "a priori" respectively.	

9.4.4 Zero-Velocity (ZUPT) Update

MEMS-based IMU sensors are used extensively today to estimate the trajectory of the foot. However, the large amount of drift present in the sensors should be removed beforehand. The zero-velocity update method (ZUPT) is used to minimize the drift in IMU sensors [29].

Gait cycle is a series of rhythmical, alternating movements of the trunk and limbs which results in the forward progression of the center of gravity and the body. A gait cycle is divided into a 60 percent stance phase and a 40 percent swing phase. There are four typical events occurring in one complete gait cycle: Heel Strike, Foot Flat, Midstance, Push Off or Heel Off and Toe-Off. The swing phase occurs from the instant the toe leaves the ground until the heel strikes. The stance phase extends from heel strike to midstance up to the detachment of the foot through gradual rolling. During a fraction of time in the midstance the velocity and acceleration of the heel are exactly zero. Without applying this correction, the sensor drift increased for a few meters in four to five steps. Therefore, ensuring the ZUPT correctly distinguishes between swing phase and stance phase is the key to eliminating drift error to a greater extent [30].

The stance phase detection is almost a pattern recognition process, where the stance phase is observed in two circumstances [29]:

i. Acceleration should be close to g, since the x-axis and the y-axis of a calibrated accelerometer is 0 and that of the z-axis is close to 9.81.
ii. Angular velocity will be 0, since for a calibrated gyroscope all the axis are 0.

The single detection threshold method is the conventional stance detection method of the ZUPT algorithm. In this method, the IMU acceleration (a_t) and angular velocities (w_t) are compared with the ZUPT acceleration threshold (σa_t) and ZUPT angular velocity threshold (σw_t) respectively (T. [30]).

A stance phase is detected when:

$$a_t \leq \sigma a_t \text{ and } w_t \leq \sigma w_t$$

The threshold is the primary and most important data of the entire ZUPT method. The threshold value of each activity like slow walking, fast walking, and running are derived by summarizing the gait data changes from collected walking data [30].

Different experimental threshold values obtained from IMU sensors are as follows

Normal walking range	$a_{\max} - a_{\min} < 3g$
Fast walking range	$3g < a_{\max} - a_{\min} < 4g$
Range of climbing stairs	$4g < a_{\max} - a_{\min} < 7g$
Range of striding/jumping	$a_{\max} - a_{\min} > 7g$

Where, a_{max} and a_{min} represent the magnitude of maximum acceleration and magnitude of minimum acceleration respectively and g is the gravitational acceleration. The angular threshold value (σw_t) is 0.6 rad/s. If the angular velocity (w_t) is less than 0.6 rad/s, the foot is in stance phase. The disadvantage of the ZUPT algorithm is that it only utilises the single threshold method for stance detection which is not an efficient way. The zero velocity is sometimes detected prior to its occurrence; sometimes it misses to detect the phase. Due to these reasons advanced ZUPT stance detectors like double threshold method for stance detection are used nowadays [30].

9.4.5 Cascaded Kalman Filter

The standard Kalman filter can be used for both angle estimation and the gyroscope bias. For estimating the gyroscope bias, however, the disturbance covariances need to be known; otherwise the Kalman filter will be unstable [31]. An efficient algorithm using two coupled Kalman filters is hence proposed to come up with a solution [32]. The first Kalman filter estimates the attitude of the rigid body and external acceleration using the accelerometer and the gyroscope, while the second filter estimates the gyroscope bias [31, 32].

9.4.6 Quaternion Concept

The Kalman filter has found its prominence in state space estimation, in tracking purposes in space research, and in the study of neural networks and sensor fusion. The quaternion concept has been demonstrated in spacecraft orientation tracking, where the focus lies in establishing mapping between the coordinate system on a reference frame x € R^3 and a local frame y € R^3 on the spacecraft's body frame, such that $y = Bx$, where x is the reference frame vector, y is the body frame vector and B is the attitude matrix or rotation matrix. The functional expression of quaternions is more mathematically tractable owing to the fat that it has a lower number of constraints. Tracking orientation in 3-D space and for the training of quaternion-valued neural network for time series prediction are the major applications of quaternion representation [33].

Algorithm of Quaternion Kalman filter: The Kalman filter equations in the quaternion domain assuming quaternion state a_t € $H^{n \times 1}$ are

1. *Prediction:*	
a. State Extrapolation Equation	$\hat{a}_t^- = Fa_{t-1}^+ + Bu_t + w_t$
b. Covariance Extrapolation Equation	$P_t^- = F \cdot P_{t-1}^+ \cdot F^H + Q$
2. *Correction:*	
c. Kalman Gain	$K - P_t^- \cdot H^H / (H \cdot P_t^- \cdot H^H + R)$
d. State Update Equation	$\hat{a}_t^+ = \hat{a}_t^- + K \cdot (z - H \cdot \hat{a}_t^-)$
e. Covariance Update Equation	$P_t^+ = (I - K \cdot H) \cdot P_t^-$

Where, F € $H^{n \times n}$ is the state transition matrix, B € $H^{n \times n}$ is the control input matrix for the control input u € $H^{n \times 1}$ and w € $H^{n \times 1}$ is the state noise. The state a_t cannot be observed directly, but can be measured through z € $H^{m \times 1}$ which relates to the state via the relation. $Z = H \cdot a_t + v_t$ where H € $H^{m \times n}$ is the observation matrix and v € $H^{m \times 1}$ is the measurement noise. Q and R are process noise covariance and measurement noise covariance, respectively.

9.5 CONCLUSION

The principles of biomechanics help us to learn the right techniques and the correct postures for maximum efficiency. It also provides us with the knowledge of the forces responsible for injuries in sports and taking preventive measures beforehand. Earlier useful technologies for analysis include force plates, electrogoniometers, and motion capture systems which were either unreliable or expensive. The advent of the wearable sensor has paved the way for research involving much less investment and easy-to-wear devices to increase the detection sensitivity and an improved signal-to-noise ratio. The miniaturization of IC chips and the reduction in power consumption will enable those sensors to be placed on athletes' clothing, such as in the heel of sports shoes to ensure the online measurement of the gait. A large amount of drift and noise contaminates the sensor output, which can be rectified using soft computing algorithms. The Kalman filter, the extended Kalman filter, and the Zero-Order Velocity Update (ZUPT) provide a stable output from these sensors via drift compensation, noise elimination, and improved estimation in constrained environments. In sports, sensors have shown remarkable results in terms of the measurement of gait associated with various activities. Inventions like multi-sensor fusion and constrained optimization could pave the way for the improved estimation of the inertial sensing.

REFERENCES

1. Nihat Özkaya, Margareta Nordin, David Goldsheyder, and Dawn Leger. *Fundamentals of biomechanics*. Vol. 86. New York: Springer, 2012.
2. C. Wong et al., "Wearable Sensing for Solid Biomechanics: A Review", *IEEE Sensors Journal*, vol. 15, no. 5, May 2015, Page: 2747–2760.
3. M.D. Akhtaruzzaman et al., "Gait Analysis: Systems, Technologies, and Importance", *Journal of Mechanics in Medicine and Biology*, vol. 16, no. 7, 2016, Page: 1630003 (45 pages).
4. Anna E. Saw et al., "Monitoring Athletes Through Self-report: Factors Influencing Implementation", *Journal of Sports Science and Medicine*, vol. 14, 2015, Page:137–146.
5. Dhruv R. Seshadri, et al., "Wearable Sensors for Monitoring the Physiological and Biochemical Profile of the Athlete", *Digital Medicine*, vol. 2, 2019, Page:72.
6. J. Cockcroft, J.H. Muller, C. Scheffer, "A Novel Complimentary Filter for Tracking Hip Angles During Cycling Using Wireless Inertial Sensors and Dynamic Acceleration Estimation", *IEEE Sensors Journal*, vol. 14, no. 8, August 2014, Page: 2864–2871.
7. M. Meghji, A. Balloch, D. Habibi, I. Ahmed, N. Hart, R. Newton, J. Weber, R. Waqar, "An Algorithm for the Automatic Detection and Quantification of Athletes' Change of Direction Incidents Using IMU Sensor Data", *IEEE Sensors Journal*, vol. 19, no. 12, 2019, Page: 4518–4527.
8. J.B. Shepherd, G. Giblin, G-J Pepping, D. Thiel, and D. Rowlands, "Development and Validation of a Single Wrist Mounted Inertial Sensor for Biomechanical Performance Analysis of an Elite Netball Shot", *IEEE Sensors Letters*, vol. 1, no. 5, October 2017, 1–4.
9. T. Ogasawara, H. Fukamachi, K. Aoyagi, S. Kumano, H. Togo, and K. Oka, "Archery Skill Assessment Using an Acceleration Sensor", *IEEE Transactions on Human-Machine Systems*, vol. 51, no. 3, June 2021, Page: 221–228.
10. V. Bonnet, C. Mazza, P. Fraisse, A. Cappozzo, "Real-time Estimate of Body Kinematics During a Planar Squat Task Using a Single Inertial Measurement Unit", *IEEE Transactions on Biomedical Engineering*, vol. 60, no. 7, July 2013, Page: 1920–1926.
11. Song Ailing and Chen kai, "Design and Fabrication of Intelligent Training Javelin based on Embedded Technique", *IEEE Sensors Letters*, vol. 5, no. 4, April 2021.
12. M. Ermes, J. Parkka, J. MantyJarvi and I. Korhonen, "Detection of Daily Activities and Sports with Wearable Sensors in Controlled and Uncontrolled Conditions", *IEEE Transactions on Information Technology in Biomedicine*, vol. 12, no. 1, January 2008, Page: 20–26.

13. James Y. Xu et al., "Integrated Inertial Sensors and Mobile Computing for Real-Time Cycling Performance Guidance via Pedalling Profile Classification", *IEEE Journal of Biomedical and Health Informatics*, vol. 19, no. 2, March 2015, Page: 440–445.
14. Y. J. Kim et al., "Golf Swing Analysis System with a Dual Band and Motion Analysis Algorithm", *IEEE Transactions on Consumer Electronics*, vol. 63, no. 3, August 2017, Page: 309–317.
15. X. Sha et al., "Accurate Recognition of Player Identity and Stroke Performance in Table Tennis Using a Smart Wristband", *IEEE Sensors Journal*, vol. 21, no. 9, May, 2021, Page: 10923–10932.
16. B. Guignard et al., "Validity, Reliability and Accuracy of IMU to measure angles: application in swimming", *Sports Biomechanics*, July 29, 2021, Pages: 1–33 doi: 10.1080/14763141.2021.1945136.
17. Daniel T.P. Fong et al., "Using a Single Uniaxial Gyroscope to Detect Lateral Ankle Sprain Hazard", *IEEE Sensors Journal*, vol. 21, no. 3, February, 2021, Page:3757–3762.
18. L. Wicent, N.H. Lovell and S.H. Redmond, "Estimating Lower Limb Kinematics Using a Lie Group Constrained Extended Kalman Filter with a Reduced Wearable IMU Count and Distance Measurements", *Sensors*, vol. 20, no. 6829, 2020, Page: 1–28.
19. R.G. Ranky et al., "Modular Mechatronic System for Stationary Bicycles Interfaced With Virtual Environment for Rehabilitation", *Journal of Neuro Engineering and Rehabilitation*, vol. 11, 2014, Page: 93.
20. M. Nazarahari, J. Arthur, H. Rouhani, "A Novel Testing Device to Assess the Effect of Neck Strength on Risk of Concussion", *Annals of Biomedical Engineering*, vol. 48, no. 9, September 2020, Page: 2310–2322.
21. Tong-Hun Hwang, Julia Reh, Alfred O. Effenberg, and Holger Blume, "Real-Time Gait Analysis Using a Single Head-Worn Inertial Measurement Unit", *IEEE Transactions on Consumer Electronics*, vol. 64, no. 2, 2018, Page: 240–248.
22. S. Zihajehzadeh, Tien Jung Lee, Jung Keun Lee, Reynald Hoskinson, and Edward J. Park, "Integration of MEMS Inertial and Pressure Sensors for Vertical Trajectory Determination", *IEEE Transactions on Instrumentation and Measurement*, vol. 64, no. 3, 2014, Page: 804–814.
23. E. J. Bradshaw et al., "Agreement between force and deceleration measures during backward somersault landings", *Sports Biomechanics*, April 20, 2020, Pages: 1–9. doi: 10.1080/14763141.2020.1743348
24. N.G. Punchihewa et al., "Identification of Key Events in Baseball Hitting Using Inertial Measurement Units", *Journal of Biomechanics*, vol. 87 (2019), Page: 157–160.
25. S. J. Callaghhan, "The relationship between inertial measurement unit derived 'force signatures and ground reaction forces during cricket pace bowling", *Sports Biomechanics*, vol. 19, no. 3, June 2020, Pages: 307–321.
26. Hamad Ahmed, Muhammad Tahir, "Improving the Accuracy of Human Body Orientation Estimation with Wearable IMU Sensors", *IEEE Transactions on Instrumentation and Measurement*, vol. 66, no. 3, March 2017, Page: 535–542.
27. Narayan Kovvali, Mahesh Banavar, Andreas Spanias, "An Introduction to Kalman filtering with MATLAB Examples", *Synthesis Lectures on Signal Processing*, vol. 6, no. 2, 2013, Page: 1–81.
28. Özkan Bebek et al., "Personal Navigation via High-Resolution Gait-Corrected Inertial Measurement Units", *IEEE Transactions on Instrumentation and Measurement*, vol. 59, no. 11, November 2010, Page: 3018–3027.
29. Z. Wang et al., "Stance-Phase Detection for ZUPT-Aided Foot-Mounted Pedestrian Navigation System", *IEEE/ASME Transactions on Mechatronics*, vol. 20, no. 6, December 2015, Page: 3170–3181.

30. T. Zhao et al., "Pseudo-Zero Velocity Re-Detection Double Threshold Zero-Vèlocity Update (ZUPT) for Inertial Sensor-Based Pedestrian Navigation" *IEEE Sensors Journal*, vol. 21, no. 12, June, 2021, Page: 13772–13785.
31. David M. Bevly et al., "Cascaded Kalman Filters for Accurate Estimation of Multiple Biases, Dead- Reckoning Navigation, and Full State Feedback Control of Ground Vehicles", *IEEE Transactions on Control Systems Technology*, vol. 15, no. 2, March 2007, Page: 199–208.
32. M.A. Zaved et al., "Cascaded Kalman Filtering-Based Attitude and Gyro Bias Estimation With Efficient Compensation of External Accelerations", *IEEE Access*, vol. 8, 2020, Page: 50022–50035.
33. C. Jahanchachi et al., "A Class of Quaternion Kalman Filters", *IEEE Transactions on Neural Networks and Learning Systems*, vol. 25, no. 3, March 2014, Page: 533–544.

10 Long Short-Term Memory Neural Network, Bottleneck Distance, and Their Combination for Topological Facial Expression Recognition

Djamel Bouchaffra, Faycal Ykhlef, and Assia Baouta
Centre for Development of Advanced Technologies, Baba Hassen, Algeria

CONTENTS

10.1 Introduction 154
10.2 Some Mathematical Background 157
 10.2.1 A Brief Introduction to Homology Theory 157
 10.2.2 Barcodes and Persistence Diagrams 158
 10.2.3 Distance Functions 159
10.3 A Methodology for Facial Expression Recognition 160
 10.3.1 Global View of the Proposed Design 160
 10.3.2 Barcode Extraction for Facial Expressions 160
 10.3.3 Facial Expression Classification 160
 10.3.4 Classification Based on the Bottleneck Distance 161
 10.3.5 Classification Based on LSTM 161
 10.3.6 Classification Based on a Combination of Bottleneck and LSTM 162
10.4 Experiments and Results 165
 10.4.1 Data Collection 166
 10.4.2 Evaluation Standards 166
 10.4.3 Classification Results 166
  10.4.3.1 Classification Based on Bottleneck Distance 166

DOI: 10.1201/9781003267782-10

10.4.3.2 Classification Based on LSTM Recurrent Neural Network 167
10.4.3.3 Classification Using a Combination of Bottleneck and LSTM Classifiers 167
10.5 Conclusion and Future Work 168
Acknowledgments 168
References 169

10.1 INTRODUCTION

Research on facial expressions and physiognomy has been around since the early Aristotelian era (4th century BCE). The study of physiognomy is the evaluation of a person's character from their outer appearance, namely the face [1]. However, in recent years this interest in physiognomy has waned significantly and in recent years it has been superseded by facial emotions. The foundational work on facial emotions that constitutes the underpinning of the current research's thrust dates back to the 17th century. One of the original seminal works on facial emotion analysis, and a work which had a considerable impact om the science of automatic facial expression recognition was performed by Charles Darwin in 1872. In his work, Darwin proposed a treatise that constitutes the fundamental principles of those emotions exhibited by humans as well as animals [2]. Because it is difficult to separate out each facial emotion, he proposed the taxonomy of various semantically similar emotions into homogeneous clusters. The sets of emotion categories formed are classified thus: cluster 1: {aversion, defiance, disgust, guilt, pride}; cluster 2: {surprise, astonishment, dread, horror}; and cluster 3: {self-awareness, shame, introversion, modesty}.

The physical facial deformations assigned to each cluster have been described thoroughly by Darwin in the same treatise. One can cite: "the closure of the mouth when a human is in a cogitation state", "the contraction of the muscles around the eyes when being in affliction", and "the depression of the corners of the mouth when one is in despair". Similar research in the study of facial emotions and human expressions that is worthy of mention is the research which has been conducted by the psychologist Paul Ekman and his colleagues since the 1970s. However, the rapidly increasing power of computer science-related fields such as artificial intelligence, computer graphics, computer vision, pattern recognition and behavioral science have been the causes of a paradigm shift in facial expression recognition (FER). The work of Suwa and his colleagues [3] was a great achievement in this new field. The authors developed a system that analyzes facial expressions from a sequence of images (video frames) based on twenty tracking points. In the period just after the 1990s, research on automatic FER flourished and has become a major field of artificial intelligence. Although FER can be achieved via the use of a system of multiple sensors, the proposed work focuses exclusively on facial images, since visual expressions represent the essential information channels in interpersonal communication. Traditionally, FER involves three main steps: (i) facial region detection; (ii) feature extraction; and (iii) expression classification. The process of feature extraction which conveys the most relevant information for FER success is conducted using facial component landmarks [4]. Shallow machine learning, such as support vector machine

(SVM), AdaBoost, and random forest, have subsequently been employed for recognition tasks.

In more recent years, deep learning (DL) has emerged as a powerful approach to machine learning, yielding competitive results in multiclass face expression problems and also in many other computer vision applications [5]. This accomplishment is facilitated by the current availability of big data repositories. Among several deep learning models available in the literature, the convolutional neural network (CNN) remains the most well-established network model. In CNN-based approaches, the input image is convolved through a filter collection to produce a feature map. Each feature map is further combined to a fully connected network that classifies face expressions using a Softmax function. However, both the explicability and the interpretability of DL's final decision remains barely understood. In other words, unless a deeper insight of its functionality is revealed, DL can still be considered to be a black box.

The methodology that we propose in this study consists of assigning a single emotion cluster (or a quadrant in the valence–arousal space) to the input facial expression image. This two-dimensional space contains all of the emotions that are close in the physiognomic sense (refer to Figure 10.1). It is worth underscoring that the mapping from the set of facial expression images to the set of emotions is a one-to-one mapping (refer to Figure 10.2). In other words, for each facial expression image, there is one and only one emotion assigned to it. We first extracted qualitative facial features using Topological Data Analysis (TDA) [6]. This latter area invokes the field of topology (shape properties that are unchanged by continuous deformations) and allows the design of algorithms that compute qualitative features characterizing facial expressions [7, 8]. The extraction of these qualitative features, known as Betti numbers (the number of k-dimensional holes on a face which represents a topological surface), is conducted via a computational mechanism known as persistent homology (PH) [9–11]. We computed persistence diagrams (PDs) (viewed as two multi-sets) for each Betti number that disclose stable qualitative features describing a facial expression image. Once the set of features have been extracted, we designed three facial expression classification approaches: (i) the first classifier is based on the Bottleneck

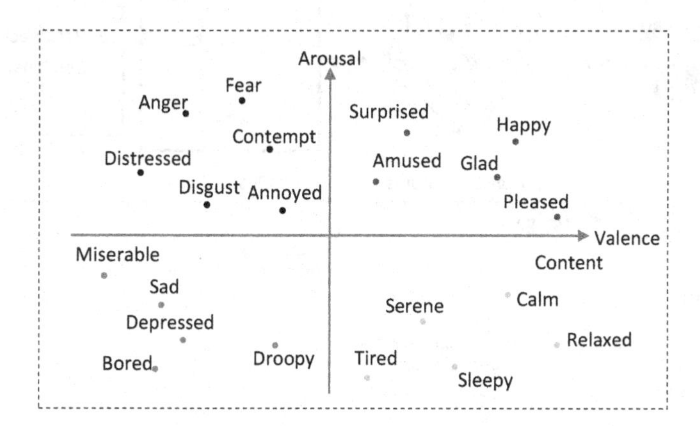

FIGURE 10.1 Two-dimensional valence-arousal space decomposed into four quadrants of emotions.

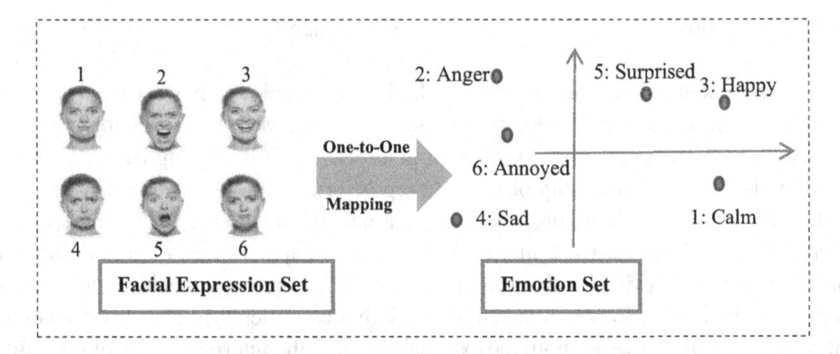

FIGURE 10.2 The passage from the facial expression set to the emotion set shows that the input for the three classifiers is a facial expression and the output is an emotion.

distance between two PDs; (ii) the second classifier hinges on a Long Short-Term Memory (LSTM) recurrent neural network classifier fed by topological clues supplemented by geometrical descriptors exhibited through a sequence of ten filtrations; and (iii) the third classifier combines both classifiers using their mutual strengths. A comparison of the performance of the three classifiers is subsequently conducted and reported. This performance is computed with respect to the capability of each classifier in assigning the correct emotion cluster to an input facial expression image.

It is worth outlining that the assignment problem of an input facial expression image onto one cluster of emotions represents a first step towards achieving full emotion recognition. The second stage, whose mission is beyond the scope of this manuscript, is needed to recognize the true emotion within one winning cluster. A global view of this research in facial expression recognition is depicted by Figure 10.3. The

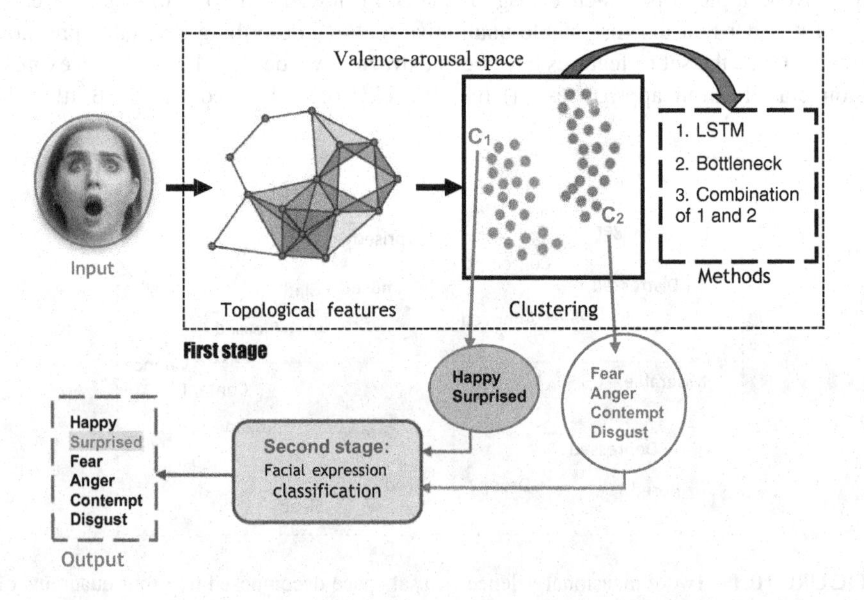

FIGURE 10.3 Global view of the expressions classification full design.

motivation behind proposing a two-stage solution to this problem is explained by the fact that the multiclass expression problem is very challenging; however, this challenge can be overcome if those emotions that are close in the valence–arousal space are first partitioned into two separate clusters and then the distances between emotions are considered. This latter information is available through the valence feature.

The organization of this manuscript is as follows: Section 10.2 covers some mathematical background relating to computational topology, including homology theory. Section 10.3 is devoted to the introduction of our methodology, including the description of the three proposed classifiers. This very section focuses on the extraction of qualitative features via persistence diagrams and the emotion classification. Experimental results for facial expression recognition using Bottleneck distance and LSTM classifiers as well as their combination are laid out in Section 10.4. Finally, a conclusion and directions for future work are presented in Section 10.5.

10.2 SOME MATHEMATICAL BACKGROUND

In this section, we provide a brief background on homology theory, barcodes, persistence diagrams (PDs) and the Bottleneck distance metric associated with a pair of persistence diagrams.

10.2.1 A Brief Introduction to Homology Theory

A topological space is a mathematical entity expressing the intuitive notions of closeness, connectivity, and continuity. Any metric space (such as ^{n}), or its subset, can be perceived as a topological space. Homology is invoked to measure the topological characteristics of a topological space. The dimension of the k-th homology vector space is known as the k-th Betti number. More precisely, it represents the rank of the k-th homology group of a topological space and it tallies the number of k-dimensional holes in the space. In order to computationally analyze a topological space, it is necessary to discretize the space and represent it using relevant pieces of information. Moreover, the discretization process of topological spaces is conducted through triangulation: Thus, the space is decomposed into a simplicial complex that represents a collection of vertices, edges, triangles, tetrahedrons, and higher-dimensional simplices. Information about the connectivity of all these simplices is expressed mathematically. The computation of Betti numbers from a simplicial complex representation associated with a topological space is conducted using a filtration process. This latter operation is based on a distance between points, where points closer than a distance *d* are connected in the simplex by an edge/face. However, a mathematical definition of a filtration is introduced by the following definition:

Definition 1

A filtration of a simplicial complex K is a nested sequence of sub complexes of K:

$$K_0 \leq K_1 \leq K_2 \leq \cdots \leq K_m = K. \tag{10.1}$$

When the simplicial complex K is filtrated, topological features can therefore be generated for each member in the sequence through the derivation of the homology group of each simplicial complex. For a family of topological spaces, or simplicial complexes, persistent homology (PH) provides a method for quantifying the dynamics of topological features (e.g., when holes appear and disappear). Such family of simplicial complexes in which simplices are formed but never erased is called a *filtered simplicial complex*. Persistent homology can be exhibited through its barcode representation: for each dimension *k*, barcodes represent a collection of horizontal intervals $[d_i \ldots d_j]$ whose left endpoint d_i represents the birth of a particular k-dimensional homology, whereas the right endpoint d_j designates its death (all within a filtration value). The number of intervals $[d_i \ldots d_j]$ found throughout a filtration value corresponds to the Betti number β_k at that value (*k*: dimension of the holes). Betti numbers are computed for each simplicial complex (by varying *d*). From each stage to the next, pairing up the births and the deaths, as described above, we obtain a set of intervals (or bars), which is called the *barcode of the filtration* [11]. Each bar represents a class in one of the homology groups and thus has a finite dimension.

10.2.2 Barcodes and Persistence Diagrams

Betti numbers are computed for each simplicial complex (by varying *d*). From each stage to the next, pairing up the births and the deaths, as described above, we get a set of intervals or bars, which is called the *barcode of the filtration*. Each bar represents a class in one of the homology groups and thus has a finite dimension. Figure 10.4 depicts an example of a cloud of 5 points representing a "house with 1 hole". The persistent homology (barcodes) discloses for $2 \leq d \leq 2.8$, one connected component ($\beta_0 = 1$), one hole ($\beta_1 = 1$) yielding a series of Betti numbers 1,1.

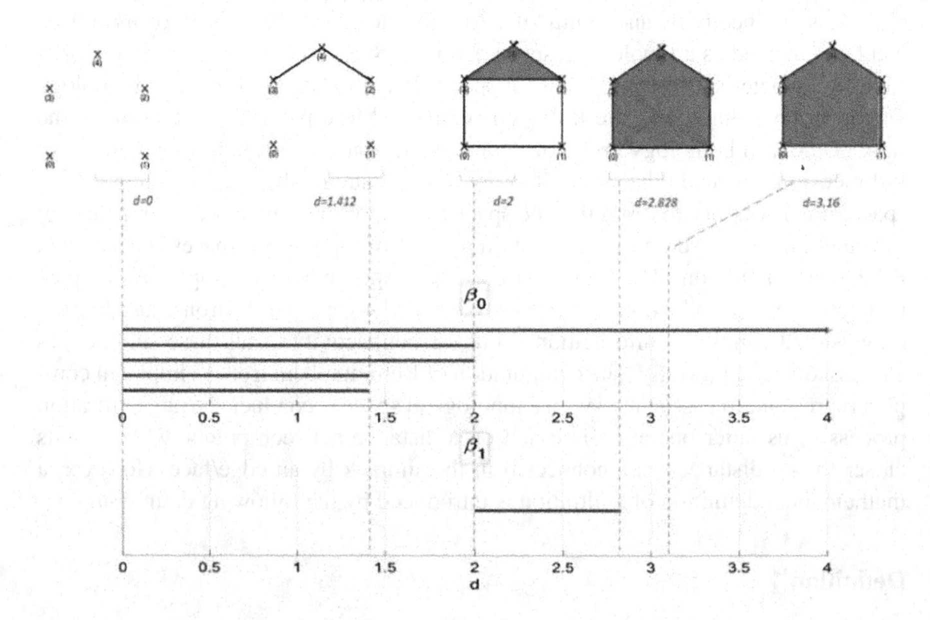

FIGURE 10.4 Simplicial complexes with their barcodes. PH discloses one connected component ($\beta_0 = 1$) and one hole ($\beta_1 = 1$) ($2 \leq d \leq 2.8$). d is the filtration value (extracted from [9]).

Definition 2

A persistence diagram (PD) is a collection of points in the plane where each point (x, y) is associated to a qualitative topological feature that emerges at scale x and disappears at scale y. We state that the corresponding feature possesses a persistent value of y–x.

In other words, a PD is a descriptor of the topological activity of a set of data. It is worth underscoring that the concept of barcode is equivalent to the concept of persistence diagram. Therefore, these two words can be used interchangeably in the pure mathematics literature.

10.2.3 Distance Functions

In problems that invoke persistent homology, distance functions on a space of data are most widespread.

Definition 3

Let $E \leq {}^{n}$, a function:
$d(x) = \text{Inf}_{y \in E}$ "$x - y$" defined by assigning every $x \in {}^{n}$ to a distance to E is called a distance function.

For computational reasons, among several possible distance functions, we adopted the bottleneck distance during our experiments.

Definition 4

Let PD(t) and PD( ) be two persistence diagrams associated with the data sets t and, the bottleneck distance between these two persistence diagrams in dimension k is defined using the following formula:

$$G_{\infty,k}(\text{PD}(t), \text{PD}(\quad)) = \inf_{\gamma} \sup_{x \in PD(t)}{}'' x - \gamma(x)''\infty, k, \tag{10.2}$$

where γ is a bijection from PD(t) to PD( ). It is the shortest distance $G_{\infty,k}$ for which there exists a perfect match between points of two diagrams PD(t) and PD( ) (using all the points on the main diagonal to disregard cardinality mismatches) such that any two matched points are at distance not exceeding the value of $G_{\infty,k}$.

The global bottleneck distance between two PDs is the weighted sum of the bottleneck distance in each dimension. This global metric is expressed as follows:

$$\sum_{k=0}^{k=n} \lambda_k w_{\infty,k}(\text{PD}(t), \text{PD}(\cdots)), \tag{10.3}$$

where $(n + 1)$ is the total number of dimensions.

10.3 A METHODOLOGY FOR FACIAL EXPRESSION RECOGNITION

We first introduce the Bottleneck distance-based classifier, the Long Short-Term Memory (LSTM) recurrent neural network classifier for this facial expression classification problem. Both classifiers incorporate topological features; however, it is worth indicating that the LSTM features were supplemented by geometric information which are: the perimeter and the area of a simplex at each filtration value.

Subsequently, we introduce a third classifier that seamlessly combines these two classifiers. In this section we show how topological features are extracted from images and how classification is performed given input patterns.

10.3.1 Global View of the Proposed Design

Our approach to the recognition of face expressions is achieved via six operations: (1) collect the database containing a set of images with 68 points landmarks (generated via active appearance models) in each image. (2) Project the set of all 68 points onto a 2D-Euclidean space. (3) Construct a filtration depicting a sequence of 10 Vietoris-Rips (VR) simplicial complexes for each image from the cloud of 2D points. (4) Extract qualitative features using homology theory within each filtration out of ten. (5) Compute the persistence diagrams (barcodes), and (6) classify input facial expression images into two selected clusters of emotions in the valence–arousal space using three classifiers. The classification phase is undertaken via three different methodologies: (i) topological classification based on the Bottleneck distance between an input face image persistence diagram and the persistence diagram of all expressions stored during a training phase; (ii) Classification based on a Long Short-Term Memory (LSTM) recurrent neural network trained through sequences of topological and geometrical features within each filtration considered as a point in time; and (iii) Combination of both classifiers using a novel multi-classification criterion. We now describe in more details the six operations that represent the foundation of our methodology.

10.3.2 Barcode Extraction for Facial Expressions

Given a cloud of points generated from an input facial expression image, a sequence of subcomplexes (filtration values) based on a VR filtration is built in order to derive a facial expression persistence barcode (refer to Figure 10.5).

A persistence diagram (PD) assigned to each face expression class is often computed. As pointed out in Section 10.2, each point (x,y) of this PD is associated to qualitative topological features that emerges at scale x and disappears at scale y.

10.3.3 Facial Expression Classification

In this section, we propose three classification techniques; the first one invokes the Bottleneck distance between a pair of PDs, the second uses a LSTM recurrent neural

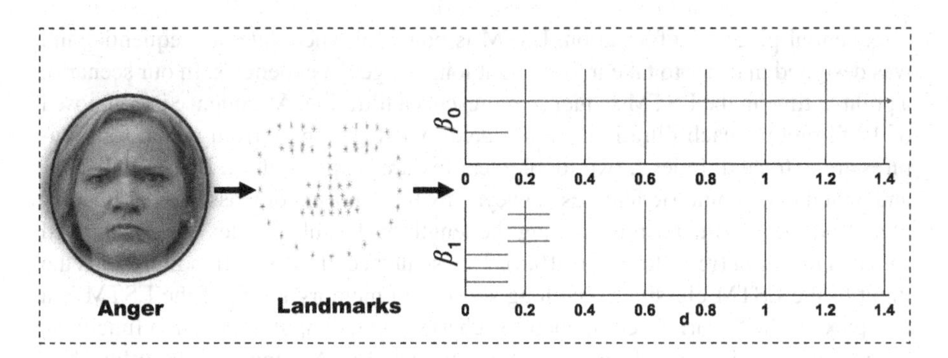

FIGURE 10.5 Barcodes for the anger facial expression: The x-axis represents the filtration values and the y-axis are the Betti numbers.

network, and the third relies on a combination scheme of Bottleneck and LSTM classifiers.

10.3.4 Classification Based on the Bottleneck Distance

We compute the Bottleneck distance between a persistence diagram (or barcode) PD(I) of an input image I and a persistence diagram of an image with a known expression class PD(m) amongst the predefined classes. The optimization criterion consists of determining the class m^* whose PD is closer (in some sense) to the PD associated to the input image. This statement can be formally stated as: determine the class m^* such that:

$$\omega^* = \underset{\omega}{\operatorname{argmin}} \sum_{k=0}^{k=n} \lambda_k w_{\infty,k}(PD(I), PD(\omega)), \tag{10.4}$$

where the weights λ_k are computed using a cross-validation scheme and are subject to: $\sum_{k=0}^{k=n} \lambda_k w_{\infty,k}$. The parameter n is the number of dimensions (Betti numbers) considered. It is noteworthy that the lack of stability of the persistence diagrams with respect to the data is a major issue within this metric-based classifier. In fact, unless we define a suitable metric function that accounts for unstable persistence diagrams, the recognition error rate using the Bottleneck distance between pairs of persistence diagrams might remain significantly high.

10.3.5 Classification Based on LSTM

The second classifier invokes the Long-Short Term Memory Recurrent Neural Network. This choice is justified by the facts that: (i) the first classifier is affected by the instability of the persistence diagrams, relies only on topological clues and lacks a training procedure; (ii) the presence of the filtration subcomplexes allows extracting a sequential piece of information. LSTM is powerful when data are sequential and was designed in order to take into account long-range dependencies. In our scenario, a point in time in the LSTM sequence represents a filtration. A sequence is composed of 10 filtrations; each filtration is associated to a subcomplex from which four features are extracted. Indeed, two topological features expressed via Betti numbers β_0 and β_1 and two geometric features conveyed by the "area encompassed by a subcomplex" and the "perimeter representing the length of the subcomplex", are computed for each filtration (refer to Figure 10.6). This sequence of information represents the input to the LSTM classifier. The long short-term memory block of the LSTM is a complex unit with various components such as weighted inputs, activation functions, inputs from previous blocks and eventual outputs [12]. A sample of these four features within a filtration sequence fed to the LSTM classifier for one facial expression image is depicted via Figure 10.6.

10.3.6 Classification Based on a Combination of Bottleneck and LSTM

a. **Motivation: Diversity Measures**

The motivation behind the combination of Bottleneck and the LSTM classifiers is justified by the diversity (also known as the orthogonality) of these two classifiers. This diversity is expressed via the following measures: Correlation Coefficient, Yule's Q-Statistic, Disagreement, Kullback–Liebler (KL) divergence (which is the distance between the class-posterior probability distributions computed from the two classifiers), and other statistics. For example, if the KL divergence value is greater than 0, then one can assert that the two classifiers are diverse; otherwise, they are dependent (or redundant). We reported in the experiment section the values of these measures that promoted a classifier ensemble strategy. We now show how to compute these diversity measures in a specific example: Let's assume we have the following predicted outputs assigned to two classifiers:

Classifier 1: + × + + × × + ×

Classifier 2: + + × × × + × +

where + and × represent the correct and incorrect classes, respectively.

The values inside Table 10.1 are expressed as:

- a = P(classifier 1 is correct and classifier 2 is correct)
- b = P(classifier 1 is correct and classifier 2 is incorrect)
- c = P(classifier 1 is incorrect and classifier 2 is correct)
- d = P(classifier 1 is incorrect and classifier 2 is incorrect).

The value a is estimated by the ratio of (the number of times classifier 1 is correct (depicted by the + sign) and classifier 2 is correct in the two sequences) and (the total sequence length). This ratio is equal to 1/8 in this example. All other values, b, c, and d, are computed the same way. In conclusion, Table 10.1 represents the probability distribution denoted by P.

Since the probability distribution P is known then one can compute a set of diversity measures in Table 10.2 [13].

The probability distribution Q in the KL divergence formula represents the classifier independence probability distribution expressed via Table 10.3.

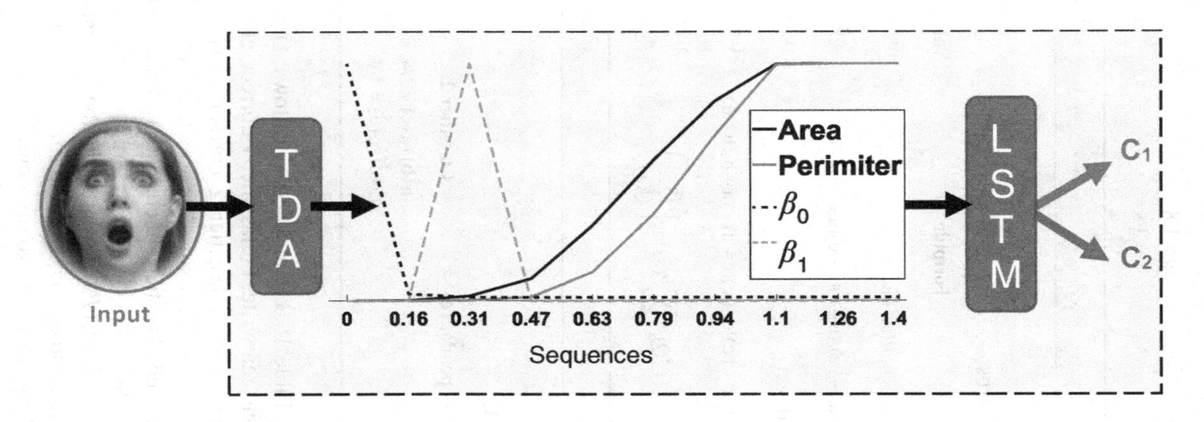

FIGURE 10.6 Feature extraction from an input facial expression image and classification into an emotion cluster C_1 or C_2 using three classifiers. Each feature within a sequence of filtration is represented by one of the four shaded graphs.

TABLE 10.1
Probability Distribution P

Distribution P (Observed)	Classifier 2 (Correct)	Classifier 2 (Incorrect)
Classifier 1 (Correct)	$a = 1/8$	$b = 3/8$
Classifier 1 (Incorrect)	$c = 3/8$	$d = 1/8$

TABLE 10.2
Values of Diversity Measures

Measure	Notation	Formula	Value	Observation
Yule's Q-Statistic	YQ	(ad – bc)/(ad + bc)	–0.8	Commit errors on different patterns
Correlation Coefficient	ρ	(ad – bc)/(a + b)(c + d)(a + c)(b + d)	–2.0	Negatively Correlated
Disagreement Measure	D	(b + c)	0.75	Disagree
Interrater Agreement	K	2(ad – bc)/(a + c)(c + d) + (a + b)(b + d)	–0.5	Disagree
KL Divergence	KLD	$D_{KL}(P''Q) = \sum_{x \in Y} P(x) \log\left(\frac{P(x)}{Q(x)}\right)$	0.18	independence

TABLE 10.3
Probability Distribution Q

Distribution Q (Predicted by Independence)	Classifier 2	Classifier 2
Classifier 1	(a+b) (a+c) = 1/4	(a+b) (b+d) = 1/4
Classifier 1	(a+c) (c+d) = 1/4	(b+d) (c+d) = 1/4

The computation of Table 10.3 is justified as follows: Using the independence assumption, one can write: P(classifier 1 correct, classifier 2 correct) = *P*(classifier 1 correct) × *P*(classifier 2 correct) = (*P*(classifier 1 correct, classifier 2 correct) + *P*(classifier 1 correct, classifier 2 incorrect)) × (P(classifier 2 correct and classifier 1 incorrect) + P(classifier 2 correct and classifier 1 correct)) = (a + b) × (c + a). The same reasoning is undertaken to fill all cells of Table 10.3. $D_K(P''\ Q)$ represents the amount of information lost when Q is used to approximate P. In the simple case, if KLD is strictly positive, which indicates that the two distributions in question convey different quantities of information. In this case, there is a loss of information in approximating the true distribution P by the Q distribution. *This means that the two classifiers with these two distributions are diverse, and therefore, there is a need to combine them.* The computation of the Kullback–Liebler divergence measure is computed as follows:

$$D_{KL}(P''Q) = a \times \log \frac{a}{(a+c)(a+b)} + b \times \log \frac{b}{(a+b)(b+d)} + c$$
$$\times \log \frac{c}{(a+c)(c+d)} + d \times \log \frac{d}{(b+d)(c+d)}$$
$$= \frac{1}{8} \times \log 0.5 + \frac{3}{8} \log 1.5 + \frac{3}{8} \times \log 1.5 + \frac{1}{8} \log 0.5 = 0.18.$$

Since the KL divergence measure is not equal to 0, one can conclude that in this example there is a loss of information in using distribution Q to approximate the true distribution P. In other words, the distribution Q is not a very good estimator of the distribution P. This also means that classifier 1 and classifier 2 are independent; this ascertainment promotes the diversity hypothesis. However, one is more interested in the complementarity (expressed via the disagreement measure D) of classifiers to conduct the classification task. In other words, if classifier 1 is incorrect therefore classifier 2 should be correct and conversely. In this example, this disagreement measure is equal to 3/8 + 3/8 = 3/4 = 75%, which indicates that both classifiers can be considered as complementary.

b. **Combination Scheme**

The idea in this section consists of combining the two diverse classifiers into one more accurate classifier that exploits the prediction powers of both classifiers. However, before performing this combination, it is necessary to convert the two classifiers' scores into a posteriori probability values. This action allows a fair comparison between these two classifiers since their scores are put into the same scale. Since the Bottleneck classifier outputs a distance d as a score, and given the fact that a high score corresponds to a small distance, therefore, we transformed this score into $(1 - d)$. We finally used the Softmax function to perform the class score conversion into a posteriori probabilities. Furthermore, it is crucial to underscore that the LSTM classifier prediction is deemed correct when this latter classifier's class probability is over a certain optimal threshold value t^*. This latter value is computed using a cross-validation procedure. Therefore, the combination scheme expressed via a function $f(x)$ can be written as follows:

$$f(x) = \begin{cases} \omega_i^* \text{ if } P_{LSTM}(\omega_i^* / x) \geq t^*, otherwise \\ \omega_j^* : class\ predicted\ by\ Bottleneck\ classifier \end{cases}. \tag{10.6}$$

10.4 EXPERIMENTS AND RESULTS

In this section, we show how the data are collected, and prepared for both classifiers: the Bottleneck-based distance and the LSTM. We also describe the evaluation metric used and report the performance obtained using each classifier separately and their performance when they are combined into a single one. Finally, we provide a comparison between the three classifiers in terms of true positive and false negative rates.

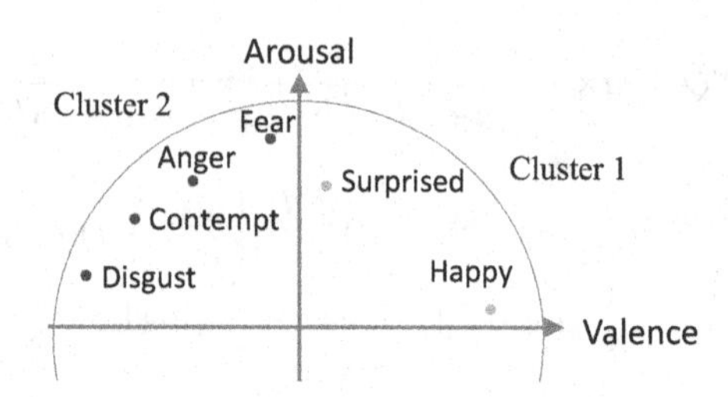

FIGURE 10.7 The two clusters of emotions considered in this study.

10.4.1 Data Collection

To assess the performance of the proposed model, we have used Cohn–Kanade Dataset (CK+) version [14]. This dataset is one of the most common benchmarks used for facial expression recognition. The entire dataset is composed of 298 images which convey six expressional states, namely: "anger", "contempt", "disgust", "fear", "happy", and "surprised". Figure 10.7 shows the two clusters of emotions considered during this study.

Furthermore, cluster 1 contains 151 facial expression images, whereas cluster 2 contains 147 images. We have selected two-thirds of the entire dataset for training and one-third for testing. However, a fraction of the training set was extracted to represent a validation set required during the classifier combination procedure.

10.4.2 Evaluation Standards

Since our dataset is balanced, the evaluation metrics we computed is the recognition accuracy (Acc), We used 70 percent of the data for training and 30 percent for testing. The accuracy is defined as:

$$Acc = \frac{TP + TN}{TP + TN + FP + FN}, \tag{10.7}$$

where TP, TN, FP, and FN, denote, respectively, True Positive cases, True Negative cases, False Positive cases, and False Negative cases.

10.4.3 Classification Results

We now lay out the results obtained from the three classifiers used during facial expression classification.

10.4.3.1 Classification Based on Bottleneck Distance

Once the PD of an input face expression image is computed, the weighted Bottleneck distance is calculated from all PDs of the labeled training sample. The value of n, which is the dimension of the homological space, is in our case equal to 1. A weight value of 0.3 was assigned to λ_0, and a weight value of 0.3 was assigned to λ_1. The predicted cluster is the one that is assigned to the PD that has the smallest distance to the input PD.

10.4.3.2 Classification Based on LSTM Recurrent Neural Network

Before we invoke the LSTM neural network in order to assign the test image to one of the two clusters, a set of four sequences as features with length equal to 10 is extracted from the input image. The number of hidden units which represents the quantity of information stored between different time steps (the hidden state) is set to 10. This number is carefully chosen to avoid over-fitting. A fully connected layer is added to the LSTM architecture and a Softmax function is invoked to output probability values assigned to the targeted classes. The maximum number of times that LSTM went through the complete training dataset, defined as the maxEpoch, is set to 70. This latter represents a hyperparameter of the LSTM learning algorithm. Finally, the re placement optimization algorithm selected for training LSTM is "Adam Solver".

10.4.3.3 Classification Using a Combination of Bottleneck and LSTM Classifiers

As explained in Section 10.3.3.3, we showed that there is a need to compute some diversity measures between the Bottleneck and the LSTM classifiers. We have computed their values from the classification Softmax scores representing probability distributions of both classifiers when considering 100 output classes. Table 10.4 depicts the results obtained.

Since the Yule's Q-Statistic is negative, this indicates that the two classifiers commit errors on different patterns. Likewise, since the KL divergence value is greater than 0, and the disagreement measure is over 50 percent, then, one can assert that the two classifiers are diverse and their combination is certainly desirable. Moreover, as pointed out in Section 10.4.1, we have used a validation set to determine the optimal threshold value t* needed during the combination scheme. The optimal threshold value that we found through this validation set is equal to 0.90. This means that if the a posteriori probability value $P_{LSTM}(m^*/x)$ is greater than 0.90 then we choose the LSTM class decision; otherwise, we choose the Bottleneck class decision as the predicted class. The classification performance based on the three classifiers: Bottleneck distance-based, LSTM and their combination are depicted by Table 10.5.

TABLE 10.4
Values of Diversity Measures in the Dataset

Measure	Value	Observation
Yule's Q-Statistic	−0.84	Commit errors on different patterns
Correlation Coefficient	−2.0	Negatively Correlated
Disagreement Measure	0.75	Disagree
Interrater Agreement	−0.5	Disagree
KL Divergence	0.068	Independence

TABLE 10.5
Classification Performance of All Classifiers

	Acc (%)	TPR (%)	TNR (%)	FPR (%)	FNR (%)
Bottleneck Distance	61.0	91.8	31.4	68.6	8.2
LSTM	85.0	71.4	98.0	2.0	28.6
Combination	**99.0**	100.0	**98.0**	**2.0**	0.0

According to these results, the LSTM classifier exhibits a higher accuracy (Acc = 85.0%) than the Bottleneck classifier (Acc = 61.0%). This is explained by the fact that LSTM is capable to learn sequential information obtained during each filtration, whereas the Bottleneck classifier is not embedded with a learning paradigm. However, it is worth underscoring that the Bottleneck classifier outperforms the LSTM classifier in TPR and FNR. This observation was vital since it indicates that their combination is necessary to improve the classification performance. In fact, the combination of these two classifiers exhibited a classification accuracy of 99.0 percent, TPR = 100 percent and FNR = 0%. Given this high classification accuracy of the correct quadrant, the emotion recognition is much easier since it will focus on this winning cluster and should rely only on the valence values that are close to the predicted emotion. However, this further examination is part of an undergoing work.

10.5 CONCLUSION AND FUTURE WORK

We have presented a novel methodology based on topological data analysis for facial expression recognition. The mission consists of assigning a facial expression to one of two emotion clusters of the valence–arousal space. Three classifiers have been proposed to address this two-class problem. The results obtained indicate that the Bottleneck-based classifier needs a stable persistence diagram to achieve its recognition goal. However, the LSTM recurrent neural network exhibits a better performance than the Bottleneck classifier since it seamlessly embeds topological and geometrical features together within a single framework. In fact, this second classifier handles perfectly sequential data provided naturally via the subcomplexes filtration. Finally, the combination of both classifiers has proven to be more effective than both classifiers considered separately since the accuracy in this combination scheme has reached the rate of 99 percent. Our next objective consists of embedding other geometrical descriptors, such as chain code or Fourier descriptors, within topology in order to conduct emotion recognition within a winning cluster. Furthermore, several levels of circularity (measured as a function of area and perimeter) in facial regions should be computed during TDA. This information allows for example discriminating a "smiley mouth" against "a surprised mouth".

ACKNOWLEDGMENTS

The authors would like to thank the Algerian Thematic Agency of Research and Health Sciences (ATRSS) for funding this research.

REFERENCES

1. L. Hartley, *Physiognomy and the Meaning of Expression in Nineteenth-Century Culture*, Cambridge University Press, 2001.
2. C. R. Darwin, *The Expression of the Expressions in Man and Animals*, John Murray, London, 1st edition, 1872.
3. N. Suwa, N. Sugie, and K. Fujimora, "A Preliminary Note on Pattern Recognition of Human Expressional Expression", *International Joint Conference on Pattern Recognition*, pages 408–410, 1978.
4. N. Munasinghe, "Facial Expression Recognition Using Facial Landmarks and Random Forest Classifier", 17th IEEE/ACIS International Conference on Computer and Information Science, Singapore, 2018.
5. A. S. Vyas, H. B. Prajapati, and V. K. Dabhi, "Survey on Face Expression Recognition using CNN", Proceedings of the 5th International Conference on Advanced Computing & Communication Systems (ICACCS), 2019.
6. G. Carlsson, "Topology and Data", *Bull. Amer. Math. Soc.* 46, 255–308, 2009.
7. D. Bouchaffra, "Nonlinear Topological Component Analysis: Application to Age-Invariant Face Recognition," *IEEE Transactions on Neural Networks and Learning Systems*, Volume 26, Issue 7, pp. 1375–1387, 2014.
8. D. Bouchaffra, "Mapping Dynamic Bayesian Networks to Alpha-Shapes: Application to Human Faces Identification across Ages", *IEEE Transactions On Neural Networks and Learning Systems (TNNLS)*, Volume 23, Issue 8, pp. 1229–1241, 2012.
9. D. Bouchaffra, & F. Ykhlef, "Persistent Homology for Land Cover Change Detection", *Oxford Research Encyclopedia of Natural Hazard Science*, 2021.
10. D. Bouchaffra, A. Baouta, F. Ykhlef, M. Khelladi, & J. Tan, "Land Cover Change Detection based on Homology Theory", In IEEE 6th International Conference on Image and Signal Processing and their Applications (ISPA) (pp. 1–4), 2019.
11. H. Edelsbrunner, and J. L. Harer, *Computational Topology: An Introduction*, American Mathematical Society, 2010.
12. R. K. Behera, M. Jena, S. K. Rath, & S. Misra, "Co-LSTM: Convolutional LSTM Model for Sentiment Analysis in Social Big Data", *Information Processing & Management*, 58(1), 102435, 2021.
13. Y. Bian, and H. Chen, "When Does Diversity Help Generalization in Classification Ensembles?" *IEEE Transactions on Cybernetics*, 2021.
14. P. Lucey, J.F. Cohn, T. Kanade, J. Saragih, Z. Ambadar, and I. Matthews, "The Extended Cohn-Kanade Dataset (CK+): A Complete Expression Dataset for Action Unit and Expression-Specified Expression", Proceedings of the Third International Workshop on CVPR for Human Communicative Behavior Analysis, San Francisco, United States of America, 94–101, 2010.

11 A Comprehensive Assessment of Recent Advances in Cervical Cancer Detection for Automated Screening

J. Jeyshri and M. Kowsigan

SRM Institute of Science and Technology, Kattankulathur, India

CONTENTS

11.1 Introduction....172
11.1.1 Cervical Cancer Monitoring and Detection Methods....172
11.2 Manual Screening Procedure....173
11.2.1 Cervical Cancer Screening and Diagnosis Procedures....173
11.3 Applications of Artificial Intelligence in Cervical Cancer Early Screening....174
11.3.1 Testing and Detection of HPV....174
11.3.2 Cervical Cytology Examination....174
11.3.2.1 Cervical Cell Segmentation....175
11.3.2.2 Cervical Cell Classification....177
11.3.2.3 AI Enhances Cervical Intraepithelial Lesion Screening Accuracy....177
11.4 Applications of Artificial Intelligence in Cervical Cancer Diagnosis....178
11.4.1 Colposcopy....178
11.4.1.1 Artificial Intelligence Improves Image Classification....178
11.4.1.2 Artificial Intelligence Aids in the Detection of High-Grade Cervical Lesions and Biopsy Guidance....178
11.4.2 MRI of the Pelvis....179
11.4.2.1 Cervical Cancer Lesions Segmentation....179
11.4.2.2 Cervical Cancer Diagnosis LNM....179
11.5 Future Directions and Limitations....180
References....181

DOI: 10.1201/9781003267782-11

11.1 INTRODUCTION

Intelligence methods are being used to troubleshoot brain tumors, uterine cancer, prostate cancer, Covid analysis, regular exercise identification, radiative feedback identification, and intellectual health evaluations of Alzheimer's patients. They have proven more successful than traditional diagnostic procedures because of advances in the healthcare industry. It has been estimated, based on proven clinical studies from global cancer statistics, that around half a million new cervical cancer cases are diagnosed, amounting to around 15 percent of all female cancer patients [1]. With an 83 percent mortality rate, this illness is mostly prevalent in impoverished nations. This is particularly true of the experience in African countries like Uganda, which has the world's 15th highest cervical cancer prevalence, with 64.9 percent of confirmed cases.

Cervical cancer is the most prevalent site of HPV infection and it is spread through sexual contact. In this example, cervical cancer is far more easily prevented by testing and identification that is easy to use than other types of cancer, and thus is crucial to achieving risk expectations. A malignant cervical development is a tumor that is cancerous. In the absence of any controlled cell division and death cause, cervical tissue cells grow and reproduce improperly [2]. When a tumor develops dangerous characteristics, the cancerous growth spreads to other areas of the body, causing infection in some places that, in most cases, may be averted by early detection. Cervical cancer deaths can be minimized if appropriate screening programmes are introduced. Various screening and diagnostic procedures rely on computer-aided designs (CAD) due to the rapid growth of modern clinical discovery and computer technological innovation.

Data mining is a method of extracting relevant information from a variety of sources. Real-world data includes flawed data, such as that which is either erroneous or incomplete [3]. The cleaning and modification of raw data to allow for a trustworthy analytic delivery can appropriately depict the conclusion in this way. The dataset is used to implement it [4, 5]. There is duplication, missing values, and noise in the cervical cancer dataset that was obtained for analysis. Due to the growing importance of health problems, data mining tools are recognized as among the most challenging and significant areas of medical study. With the insights it retrieves, the data mining system can assist progress the cervical cancer screening procedure [6, 7]. These approaches are used in the medical field not just to explore relationships and commonalities between symptoms, but also to anticipate illnesses [8–10]. Several mining approaches may be used to propose ongoing research and medical treatment; these can save lives, particularly in the case of cervical cancer. The first stage is to pre-train some data, which is an important phase in any data mining processes [11–14].

11.1.1 Cervical Cancer Monitoring and Detection Methods

The most recent World Health Organization recommendations offer three screening procedures for cervical cancer early detection: HPV testing, cytology and acetic acid inspection [15, 16]. HPV examinations and cytology are carried out on brushed-exfoliated cervix cells. HPV examination identifies strong infection around the vaginal area, whereas histopathology analysis uses the process of microscopy to recognize

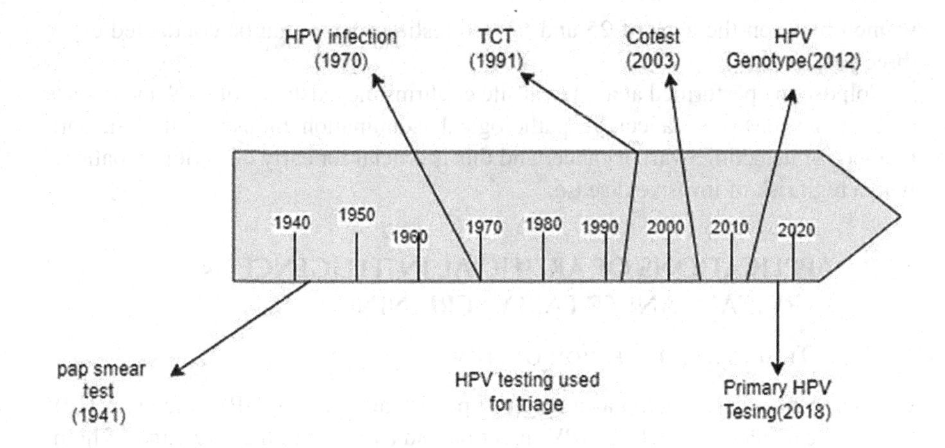

FIGURE 11.1 The growth of cervical cancer screening.

tissues extracted from the vagina that are cancerous or premalignant. The progress of cervical cancer screening technologies is seen in Figure 11.1. However, colposcopy-guided biopsy continues to be the benchmark for diagnosing cervical cancer; thereafter grading is determined using clinical evaluation.

11.2 MANUAL SCREENING PROCEDURE

A traditional Pap smear is a screening procedure that is performed manually. This is a process that uses a microscope to recognize and categorize depilation cervical cells based on their color and nucleus and cytoplasm features. Liquid-based cytology (LBC) has the potential to significantly enhance preparation processes. In comparison to the traditional specimen, this specimen is more securely held on a microscope slide, which is easier to store and handle, and offers a more uniform distribution of the sample [17]. Cervical cells are classified into several subtypes based on their aberrant nuclear size, the degree of staining, and so on. A glandular cell with atypical squamous cell carcinoma is one such example of an abnormal epithelial cell.

Colposcopy is a procedure that utilizes a specialized device to magnify the whole subjected uterus in order to carry out a complete examination of the vagina in real time. In order to detect cervical intraepithelial neoplasia (CIN), squamous intraepithelial lesions (SIL) and various stages of cancer. A biopsy assisted by a Pap smear of the suspicious location is conducted to ascertain the necessity for further treatment, namely cryotherapy; this is particularly important in people with other serious illnesses [18].

11.2.1 Cervical Cancer Screening and Diagnosis Procedures

According to the U.S. Cancer Society's most recent cancer viewing recommendations, all woman aged 25 years or older should undertake cervical cancer screening. In addition, basic HPV testing should be conducted once every five years on women between the ages of 25 and 65. Co-testing review can be conducted every three years.

Colposcopy performed at a referral site confirms the existence of CIN and detects or excludes invasive cancer. A pathological examination biopsy is the standard method for detecting ovarian cancer and this is crucial for early detection in patients with a high risk of invasive disease.

11.3 APPLICATIONS OF ARTIFICIAL INTELLIGENCE IN CERVICAL CANCER EARLY SCREENING

11.3.1 Testing and Detection of HPV

Cervical cancer can develop as a result of persistent high-risk HPV infection. HPV testing is capable of detecting HPV infection and assisting in the screening of high-risk groups [19]. HPV genotyping will make it easier to determine the risk of cancer with HPV DNA-positive results, making cervical cancer screening and therapy more feasible. This learning technology is being used to increase the accuracy and breadth of testing for screening.xsummarizes these investigations.

HPV types are related to a variety of stages of lesions. Cervical adenocarcinomas, for example, are frequently linked with HPV 18, making them harder to identify by cytology. Possessing high-risk types such as 16, 18, and 31 increases one's risk of acquiring cervical cancer. As a result, differentiating between different forms of HPV simplifies the classification and management of HPV-infected women [20]. The method identified patients at increased risk of getting CIN2/3+, and demonstrated that certain viruses containing several forms of HPV carry extra dangers. Accordingly, the researchers focused on the most relevant genetic variants [21]. However, further research is being conducted [22]. Due to the great sensitivity of HPV testing, the number of colposcopy referrals increases, which may result in more potentially hazardous treatments [23, 24]. They then created a pelvic intermediate lesions risk-grading model using a machine learning method (random forest), which accurately predicted CIN2+ with an average accuracy [25, 26]. This method successfully classified lesions into risk categories and offered useful integrated care options.

Currently, HPV typing is principally based on tests, which have a number of limitations, including a high rate of false-negative findings and an extensive demand. AI has demonstrated tremendous promise in terms of its use in detecting which stages are involved by using prognostic markers that can prove useful in identifying the tumors.

11.3.2 Cervical Cytology Examination

These cancer preventive initiatives based on cytology have significantly lowered the prevalence of cervical cancer in a number of developed countries [27]. Microscopic examination assessment for high-grade precancerous lesions are unique and less reactive, requiring careful microscope examination by skilled cytologists [28]. Each procedure is time-consuming, labour-intensive, and prone to error. In addition, its repeatability in cytology is limited, resulting in low levels of accuracy [29]. Additionally, altering the viewers' results in unclear and subjective outcomes [30, 31]. As a result, the researchers seek to create ways for autonomous picture processing that will reduce these stresses.

Introduced in 1992, PAPNET was the first commercially available automated screening system. This technology has been certified for re-screening slides that have been ruled negative by cytologists [32]. The Thin Prep Imaging System was recognized by FAD as a commercial screening tool in 2004 [33]. According to the proprietary algorithm, the system may choose the 21 most critical fields of view; if irregular cells are detected, the pathologist must manually examine the entire slide [34]. The technology enhances the level of screening sensitivity and efficiency. Later that same year (2008), an imaging system was introduced. It selected ten fields of view (FOVs) of cervical cells that most maximized efficiency. However, other assessments show that these automation systems are inefficient and unsuitable for usage in low- and middle-income nations. Furthermore, its scientific technology continues to have flaws and is still subjected to a final manual screening procedure. As a result, several academics are continuing to optimize the use of methodology in the field of cytology.

11.3.2.1 Cervical Cell Segmentation

An automated smear analysis system typically consists of five stages: picture capture, pre-processing, segmentation, feature extraction, and classification [35]. The classification and splitting stages of autonomous smear analysis, AI technology is applied, which helps to improve screening efficiency.

Cellular recognition, that is the precise identification of their internal structures, is the first step in cytological diagnostics [36]. Because the majority of cytology clinical guidelines are based on nuclear and cytoplasmic malignancy, adequate categorization is necessary for the testing of solutions. Continuous research on the use of artificial intelligence to cell segmentation has shown positive results in the segmentation of various cells [37]. It has also been used to segment cervical cells automatically, with a degree of reported success. For instance, segmenting single-cell pictures into the nucleus, cytoplasm, and background using fuzzy c-means clustering technique enables whole-cell segmentation. Using supervised learning, to separate cells with overlapping cytoplasm in cervical smear images, several researchers used a flexible form employing cytoplasmic shape segments and statistics. Experimental evidence indicates that this strategy is highly suitable and is superior to the most sophisticated methods. The models on Pap smear slide images were also investigated. This was accomplished by combining nucleus localization with a single-cell classification technique to categorize normal and pathological cells. The model's accuracy and sensitivity are 92 percent and it comprises of two levels as seen in Figure 11.2's Mask-RCNN architecture section.

AI enables real, accurate, and consistent cervical cell segmentation. Thus, the time- consuming manual segmentation procedure and subjective problems associated with categorization of abnormal cells may be eliminated.

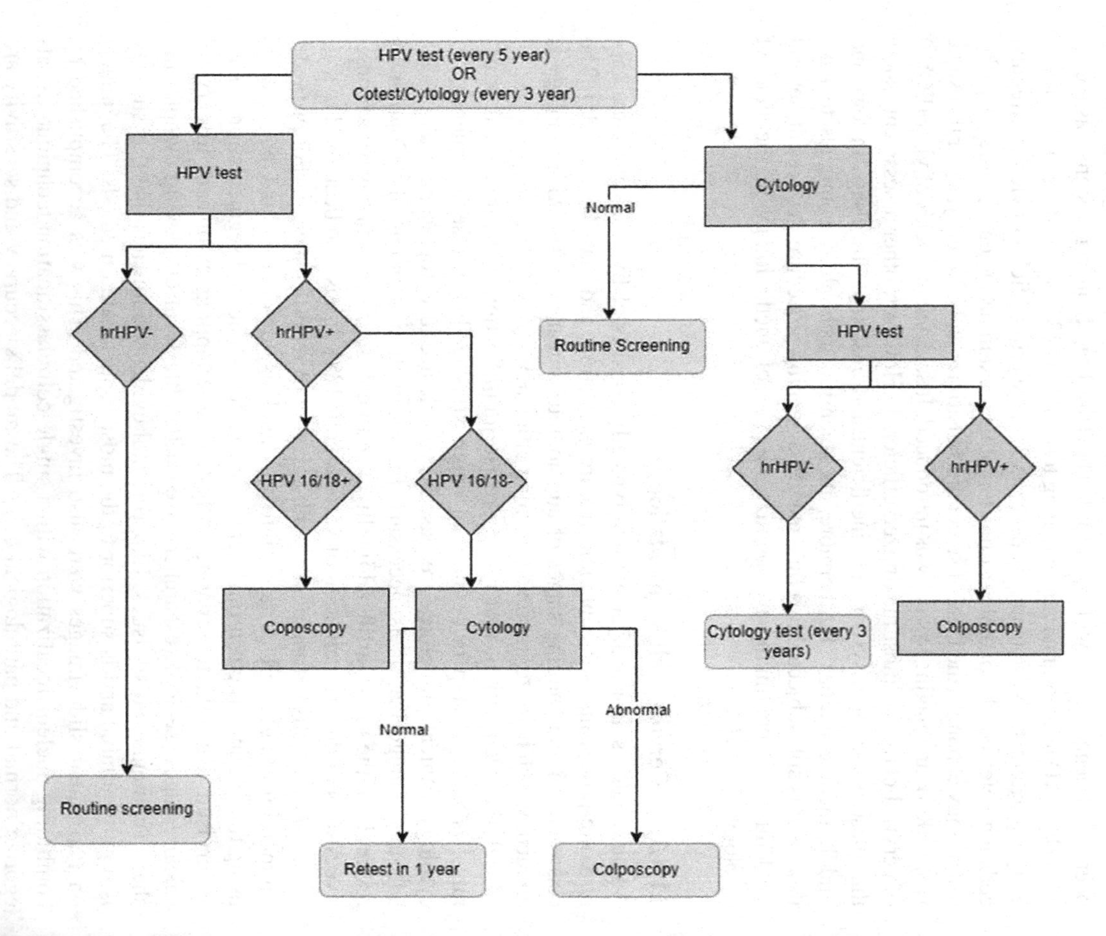

FIGURE 11.[illegible] Cervical cancer screening for every woman is recommended by the American Cancer Society.

TABLE 11.1
Usability of AI in Categorization Methods

Purpose of Study	Number of Samples Taken	Classification Methods
Identifying High-grade Lesions	605 cytology samples	SVM, RF, Decision tree
HPV Screening (Point-of-care)	13,000 samples	CNN
Forecasting the grade of cervical lesions	10,000 HPV cases 11 CIN1 samples 15 CIN2+ samples	RF, Clustering technique

11.3.2.2 Cervical Cell Classification

Cervical cancer screening requires an accurate categorization of cervical cells in smears. The low accuracy of manual classification and the high requirement for a formal qualification for the viewer restrict the application of cytology, particularly in areas with a scarcity of trained cytopathologists [38]. The application of artificial intelligence has circumvented these constraints (Table 11.1).

The past few decades have seen a lot of different ways to categorize things, and the majority of them use different ways to find different features. A growing number of scholars have proposed and accepted classification methods that do not depend solely on an accurate segmentation algorithm. For the first time, we used supervised learning to help us classify the cells in our bodies. Cell images can be automatically categorized by considering deep-level features that can be found in the images. This new method outperforms well for existing algorithms in detection accuracy, area under the curve (AUC), and specificity. When this was carried out for the first time, six different convolutional neural networks were used to identify precancerous lesions [39]. A technique for categorizing cells based on a graphical method convolutional network is used in order to investigate the possible association between cervical cell pictures and classification performance improvement. Additionally, hybrid deep feature algorithms with great accuracy were presented for the spammed dataset [40–46].

11.3.2.3 AI Enhances Cervical Intraepithelial Lesion Screening Accuracy

Several studies have revealed that an AI-assisted cytological examination can categorize cervical cells to aid in urgent care and have significantly increased the recognition accuracy of CIN when compared to abnormal biopsy as per standard practise results after developing a strong autonomous cytological detection model has been developed based on AI methodology.

According to the data shown above, AI can be used for viral detection and cytology with a high detection rate and accuracy. Much research and implementation is being carried out in this field, such as the creation of a machine learning microscope with an immersive virtual reality screen for screening [47]. They observed that it considerably increased the accuracy of recognition for low-grade squamous intraepithelial lesions (LSIL) and high squamous intraepithelial lesions (HSIL), as well as consistency across several classes and the recognition of atypical squamous cells of unknown significance. Apart from diagnostic applications, AI microscopes may be used to instruct cytopathologists in training.

11.4 APPLICATIONS OF ARTIFICIAL INTELLIGENCE IN CERVICAL CANCER DIAGNOSIS

Colposcopy-guided biopsy is used to identify cancer, which are then staged according to the universal obstetrics and gynaecology staging criteria. The use of AI in colposcopy and MRI to aid in detecting and identifying the stage of cervical cancer has yielded positive results. The CNN model offers a learning procedure for colposcopy picture categorization.

11.4.1 Colposcopy

Currently, there is a lack of balance between colposcopy and pathology, which might result in misinterpretation and missed diagnoses [48]. Colposcopy conducted by an inexperienced physician has the potential to cause injury, such as coughing up blood, or distress. Thus, it requires sufficient skill and experience to acquire competency and ensure that medical protocols are followed. However, an experienced colposcopy expert must undergo further training and a scarcity of competent workers make colposcopy difficult to utilize in cervical cancer detection.

11.4.1.1 Artificial Intelligence Improves Image Classification

DL has recently become popular in dealing with medical images. The use of DL technology in colposcopy categorization is helpful in resolving the traditional colposcopy bottleneck and boosting diagnostic performance.

11.4.1.2 Artificial Intelligence Aids in the Detection of High-Grade Cervical Lesions and Biopsy Guidance

Identifying the distinction between normal and CIN 2/3+ is among the most significant clinical aims of cancer screening. Treatment is indicated if the lesion is classed as CIN 2/3+. Mild dysplasia in CIN 1, on the other hand, is generally cured after a year of immune system and may thus be monitored more cautiously, based on creating a color and texture-based computer processing based on data from the system for cervical pictures. In our study, researchers were able to distinguish high-grade from low-grade tumors and also to comparing with normal tissues with a sensitivity of 75 percent and a specificity of 91 percent studied 9,405 women over a period of seven years in a longitudinal cohort study. The cervical pictures acquired were utilized to verify the quick R-CNN method-based model. The model's AUC for identifying CIN 2+ was 0.91, which was higher than the colposcope evaluator's assessment of the same picture and better than standard Pap smears and alternative forms of cytology. Cho et al. created a binary decision strategy to identify whether or not a cervical lesion requires biopsy [49]. In the case of CIN+ and LSIL+, the Need-To-Biopsy was described as "not being normal." Several deep learning models had a good average AUC, a sensitivity, and a specificity, indicating that the model can assist a novice clinician in deciding whether to undergo a test or to send the patient to a doctor.

The difficulty of identifying malignancy using a huge number of colposcopy pictures has been overcome as a result of AI's superior image processing capabilities. AI technology makes it easier to find lesions and perform tissue samples under colposcopy, which makes it easier to carry out and lowers the percentage of colposcopy misdiagnoses.

11.4.2 MRI of the Pelvis

In terms of cervical cancer staging prior to surgery, MRI has proved to be quite accurate. MRI is therefore the best way to find out how far the disease has spread in your body, assessing therapy response, detecting tumor recurrence, and monitoring cervical cancer patients. The major goal of an MRI scan is to detect peritumoral infiltration and lymph node metastasis (LNM).

11.4.2.1 Cervical Cancer Lesions Segmentation

The resolution of soft tissue is greater in MRI than in CT. It can examine peri-uterine invasion, uterine and vaginal involvement, and tumor size and neighbouring pelvic tissues. In diffuse-weighted imaging, researchers developed a universal network CNN to reliably find and segment cervical cancer. With a dice coefficient, sensitivity, and a positive predictive value, they claimed maximum learning efficiency during picture training. Researchers have also developed a wireless network-based computational model of a DL algorithm that can segment cervical cancer MRI images with a high degree of accuracy, one which is significantly superior to typical depth-learning algorithms. By contrast, manual segmentation is less precise and objective, and also more time-consuming, than AI segmentation. In order to predict peri-uterine invasion, researchers developed a non-invasive radiologic model based on T2-weighted imaging (T2WI) and DWI, split MRI images, and some extracted characteristics.

11.4.2.2 Cervical Cancer Diagnosis LNM

AI can also help with the early detection of cervical cancer LNM. Although computed tomography and magnetic resonance imaging (MRI) only had an accuracy of 83 percent to 85 percent in detecting lymph node involvement, their specificity was quite high, ranging between 66 percent and 93 percent. The level of staging system was updated for the first time in 2018 to add the status of surrounding tissue as a staging factor [50]. Stage IIIC cervical cancer was defined as a cancer in which lymph node involvement may be seen. Studies have seen researchers creating a DL model that use MRI prior to surgery to predict LNM in cervical cancer patients. In T1WI, the AUC utilized both intratumorally and intramuscular DL models; the hybrid model, on the other hand, through integrating tumor image data from DL mining with lesion status which was determined by MRI, has significantly enhanced the detection rate of LNM.

Over the past decade, radiology has made significant progress in creating a connection between screening and precision medicine. Radiology extracts rich information buried in medical pictures using complicated image-processing methods mixed with statistical analysis. In patients with cervical cancer MRI biomarkers evaluation was employed to increase the diagnostic grade of LNM. The combination of T2WI with a lymph node status decision tree provided the best diagnostic impact and found that T2WI- and DWI-based radiography images had a high prediction potential in the early stages of cervical cancer, in cases when pelvic LNM is used.

11.5 FUTURE DIRECTIONS AND LIMITATIONS

AI excels in both computation and image analysis. These characteristics establish it as a key player in the field of medical research, assisting physicians in decision-making, decreasing the doctor's burden, and lowering the rate of misdiagnosis. Overall, AI can improve the specificity and accuracy of screening and diagnostic programmes, overcome time constraints and limits on the number of professional and technical personnel, and avoid bias caused by subjective factors, allowing cervical cancer screening to be implemented in resource-poor areas and resulting in a significant reduction in the incidence of the disease.

The use of AI, on the other hand, poses significant difficulties. The first is the data to be employed in many ML algorithms. In order to achieve acceptable performance levels, millions of observations are required; thus, there is frequently a lack of sufficient data. Current clinical data, on the other hand, may be limited, lacking in indicators, and questionable in quality. Another important impediment to the development of computerized healthcare solutions is the administration of medical data. A future concern is the creation of not just various, but also standardized and massive databases. Bias and information security risks must also be examined, since they may lead to exaggerated conclusions and misdiagnosis. Second, AI-based models are yet to be used and popularized in clinical practise, necessitating the conduct of a series of prospective clinical investigations to confirm these findings. Third, because AI is simply an auxiliary diagnostic technique, it cannot replace physicians. AI may potentially create system paralysis, necessitating the employment of technical support. In addition, maintenance systems must be educated and implemented.

The use of AI in cervical cancer screening is a promising area, especially because its use in cervical cytology is relatively well developed. However, there are still a number of roadblocks in the segmentation technology, which is crucial for autonomous classification. All of these issues must be addressed, including overlapping nuclei segmentation, non-target cell and fragment handling, and quality control of slide-dyeing abnormalities. We also mentioned in the previous section that some of the predictive models make no use of segmentation techniques. It will eliminate a number of pre-training levels and could be a viable growth in future developments. Intelligence technology could be utilized to treat, predict prognosis, and eliminate cervical cancer, in addition to the benefits of enhanced prediction and treatment discussed in this work. In order to achieve better therapeutic decision-making, future research into both therapy and prediction will be required in future. As a result, efforts to eradicate cervical cancer will be facilitated all around the world. In addition, when the rate of cancer rises, and other uncommon disease types increases, intelligence techniques should be employed to support in the timely identification of such illnesses in the future. AI may also be used to distinguish cervical cancer from other conditions in a non-invasive manner. Further AI research will considerably improve cervical cancer prediction, increase cancer screening and predicting improvements, optimizing staging systems, and improving patient prognoses.

REFERENCES

1. Sung H, Ferlay J, Siegel RL, Laversanne M, Soerjomataram I, Jemal A, et al. Global Cancer Statistics 2020: GLOBOCAN Estimates of Incidence and Mortality Worldwide for 36 Cancers in 185 Countries. *CA: A Cancer J Clin* (2021) 71:209–49. doi: 10.3322/caac.21660
2. Brisson M, Kim JJ, Canfell K, Drolet M, Gingras G, Burger EA, et al. Impact of HPV Vaccination and Cervical Screening on Cervical Cancer Elimination: A Comparative Modelling Analysis in 78 Low-Income and Lower-Middle-Income Countries. *Lancet* (2020) 395(10224):575–90. doi: 10.1016/S0140-6736(20)30068-4
3. Schiffman M, Castle PE, Jeronimo J, Rodriguez AC, Wacholder S. Human Papillomavirus and Cervical Cancer. *Lancet* (2007) 370(9590):890–907. doi: 10.1016/S0140-6736(07)61416-0
4. Simms KT, Steinberg J, Caruana M, Smith MA, Lew JB, Soerjomataram I, et al. Impact of Scaled Up Human Papillomavirus Vaccination and Cervical Screening and the Potential for Global Elimination of Cervical Cancer in 181 Countries, 2020-99: A Modelling Study. *Lancet Oncol* (2019) 20(3):394–407. doi: 10.1016/S1470-2045(18)30836-2
5. Fontham ETH, Wolf AMD, Church TR, Etzioni R, Flowers CR, Herzig A, et al. Cervical Cancer Screening for Individuals at Average Risk: 2020 Guideline Update From the American Cancer Society. *CA Cancer J Clin* (2020) 70(5):321–46. doi: 10.3322/caac.21628
6. Redman CWE, Kesic V, Cruickshank ME, Gultekin M, Carcopino X, Castro Sanchez M, et al. European Federation for Colposcopy and Pathology of the Lower Genital Tract (EFC) and the European Society of Gynecologic Oncology (ESGO). *Eur Consensus Statement Essential Colposcopy Eur J Obstet Gynecol Reprod Biol* (2021) 256:57–62. doi: 10.1016/jejogrb.2020.06.029
7. WHO. *World Health Organization Human Papillomavirus (HPV) and Cervical Cancer, Fact Sheet*. Available at: https://www.who.int/news-room/fact-sheets/detail/human-papillomavirus-(hpv)-and-cervical-cancer (Accessed 17 September 2019).
8. Bray F, Ferlay J, Soerjomataram I, Siegel RL, Torre LA, Jemal A. Global Cancer Statistics 2018: GLOBOCAN Estimates of Incidence and Mortality Worldwide for 36 Cancers in 185 Countries. *CA Cancer J Clin* (2018) 68(6):394–424. doi: 10.3322/caac.21492
9. Pollack AE, Tsu VD. Preventing Cervical Cancer in Low-Resource Settings: Building a Case for the Possible. *Int J Gynaecol Obstet* (2005) 89 Suppl 2:S1–3. doi: 10.1016/j.ijgo.2005.01.014
10. World Health Organization. *Cervical Cancer Screening in Developing Countries: Report of a WHO Consultation*. Geneva: World Health Organization (2002).
11. Esteva A, Kuprel B, Novoa RA, Ko J, Swetter SM, Blau HM, et al. Dermatologist-Level Classification of Skin Cancer With Deep Neural Networks. *Nature* (2017) 542(7639):115–8. doi: 10.1038/nature21056
12. Maron RC, Weichenthal M, Utikal JS, Hekler A, Berking C, Hauschild A, et al. Systematic Outperformance of 112 Dermatologists in Multiclass Skin Cancer Image Classification by Convolutional Neural Networks. *Eur J Cancer* (2019) 119:57–65. doi: 10.1016/j.ejca.2019.06.013

13. Schmidt-Erfurth U, Sadeghipour A, Gerendas BS, Waldstein SM, Bogunović H. Artificial Intelligence in Retina. *Prog Retin Eye Res* (2018) 67:1–29. doi: 10.1016/j.preteyeres.2018.07.004
14. Bi WL, Hosny A, Schabath MB, Giger ML, Birkbak NJ, Mehrtash A, et al. Artificial Intelligence in Cancer Imaging: Clinical Challenges and Applications. *CA Cancer J Clin* (2019) 69(2):127–57. doi: 10.3322/caac.21552
15. Marth C, Landoni F, Mahner S, McCormack M, Gonzalez-Martin A, Colombo N. Cervical Cancer: ESMO Clinical Practice Guidelines for Diagnosis, Treatment and Follow-Up. *Ann Oncol* (2017) 28(suppl_4):72–83. doi: 10.1093/annonc/mdx220
16. Eddy DM. Screening for Cervical Cancer. *CA Cancer J Clin* (2020) 70(5):347–8. doi: 10.3322/caac.21629
17. Sawaya GF, Smith-McCune K, Kuppermann M. Cervical Cancer Screening: More Choices in 2019. *JAMA* (2019) 321(20):2018–9. doi: 10.1001/jama.2019.4595
18. Nanda K, McCrory DC, Myers ER, Bastian LA, Hasselblad V, Matchar DB, et al. Accuracy of the Papanicolaou Test in Screening for and Follow-Up of Cervical Cytologic Abnormalities: A Systematic Review. *Ann Intern Med* (2000) 132(10):810–9. doi: 10.7326/0003-4819-132-10-200005160-00009
19. Yuan C, Yao Y, Cheng B, Cheng Y, Li Y, Li Y, et al. The Application of Deep Learning Based Diagnostic System to Cervical Squamous Intraepithelial Lesions Recognition in Colposcopy Images. *Sci Rep* (2020) 10(1):11639. doi: 10.1038/s41598-020-68252-3
20. Walboomers JM, Jacobs MV, Manos MM, Bosch FX, Kummer JA, Shah KV, et al. Human Papillomavirus Is a Necessary Cause of Invasive Cervical Cancer Worldwide. *J Pathol* (1999) 189(1):12–9. doi: 10.1002/(SICI)1096-9896(199909)189:1<12:AID- PATH431>3.0.CO;2-F
21. Guo M, Gong Y, Wang J, Dawlett M, Patel S, Liu P, et al. The Role of Human Papillomavirus Type 16/18 Genotyping in Predicting High-Grade Cervical/Vaginal Intraepithelial Neoplasm in Women With Mildly Abnormal Papanicolaou Results. *Cancer Cytopathol* (2013) 121(2):79–85. doi: 10.1002/cncy.21240
22. Wong OGW, Ng IFY, Tsun OKL, Pang HH, Ip PPC, Cheung ANY. Machine Learning Interpretation of Extended Human Papillomavirus Genotyping by Onclarity in an Asian Cervical Cancer Screening Population. *J Clin Microbiol* (2019) 57(12):e00997–19. doi: 10.1128/JCM.00997-19
23. Pathania D, Landeros C, Rohrer L, D'Agostino V, Hong S, Degani I, et al. Point-of-Care Cervical Cancer Screening Using Deep Learning-Based Microholography. *Theranostics* (2019) 9(26):8438–47. doi: 10.7150/thno.37187
24. Tian R, Cui Z, He D, Tian X, Gao Q, Ma X, et al. Risk Stratification of Cervical Lesions Using Capture Sequencing and Machine Learning Method Based on HPV and Human Integrated Genomic Profiles. *Carcinogenesis* (2019) 40(10):1220–8. doi: 10.1093/carcin/bgz094
25. Yu K, Hyun N, Fetterman B, Lorey T, Raine-Bennett TR, Zhang H, et al. Automated Cervical Screening and Triage, Based on HPV Testing and Computer-Interpreted Cytology. *J Natl Cancer Inst* (2018) 110(11):1222–8. doi: 10.1093/jnci/djy044
26. Melnikow J, Henderson JT, Burda BU, Senger CA, Durbin S, Weyrich MS. Screening for Cervical Cancer With High-Risk Human Papillomavirus Testing: Updated Evidence Report and Systematic Review for the US Preventive Services Task Force. *JAMA* (2018) 320(7):687–705. doi: 10.1001/jama.2018.10400
27. Dijkstra MG, Snijders PJ, Arbyn M, Rijkaart DC, Berkhof J, Meijer CJ. Cervical Cancer Screening: On the Way to a Shift From Cytology to Full Molecular Screening. *Ann Oncol* (2014) 25(5):927–35. doi: 10.1093/annonc/mdt538

28. Perkins RB, Langrish SM, Stern LJ, Figueroa J, Simon CJ. Comparison of Visual Inspection and Papanicolau (PAP) Smears for Cervical Cancer Screening in Honduras: Should PAP Smears be Abandoned? *Trop Med Int Heal* (2007) 12(9):1018–25. doi: 10.1111/j.1365-3156.2007.01888.x
29. Stoler MH, Schiffman M. Interobserver Reproducibility of Cervical Cytologic and Histologic Interpretations: Realistic Estimates From the ASCUS-LSIL Triage Study. *JAMA* (2001) 285(11):1500–5. doi: 10.1001/jama.285.11.1500
30. William W, Ware A, Basaza-Ejiri AH, Obungoloch J. A Review of Image Analysis and Machine Learning Techniques for Automated Cervical Cancer Screening From Pap-Smear Images. *Comput Methods Programs Biomed* (2018) 164:15–22. doi: 10.1016/j.cmpb.2018.05.034
31. Bengtsson E, Malm P. Screening for Cervical Cancer Using Automated Analysis of PAP-Smears. *Comput Math Methods Med* (2014) 2014:842037. doi: 10.1155/2014/842037
32. Chivukula M, Saad RS, Elishaev E, White S, Mauser N, Dabbs DJ. Introduction of the Thin Prep Imaging System (TIS): Experience in a High Volume Academic Practice. *Cytojournal* (2007) 4:6. doi: 10.1186/1742-6413-4-6
33. Thrall MJ. Automated Screening of Papanicolaou Tests: A Review of the Literature. *Diagn Cytopathol* (2019) 47(1):20–7. doi: 10.1002/dc.23931
34. Chankong T, Theera-Umpon N, Auephanwiriyakul S. Automatic Cervical Cell Segmentation and Classification in Pap Smears. *Comput Methods Programs Biomed* (2014) 113(2):539–56. doi: 10.1016/j.cmpb.2013.12.012
35. Landau MS, Pantanowitz L. Artificial Intelligence in Cytopathology: A Review of the Literature and Overview of Commercial Landscape. *J Am Soc Cytopathol* (2019) 8(4):230– 41. doi: 10.1016/j.jasc.2019.03.003
36. Firuzinia S, Afzali SM, Ghasemian F, Mirroshandel SA. A Robust Deep Learning-Based Multiclass Segmentation Method for Analyzing Human Metaphase II Oocyte Images. *Comput Methods Programs Biomed* (2021) 201:105946. doi: 10.1016/j.cmpb.2021.105946
37. Wang P, Wang L, Li Y, Song Q, Lv S, Hu X. Automatic Cell Nuclei Segmentation and Classification of Cervical Pap Smear Images. *Biomed Signal Process Control* (2019) 48:93–103. doi: 10.1016/j.bspc.2018.09.008
38. Zhao L, Li K, Wang M, Yin J, Zhu En, Wu C, et al. Automatic Cytoplasm and Nuclei Segmentation for Color Cervical Smear Image Using an Efficient Gap-Search MRF. *Comput Biol Med* (2016) 71:46–56. doi: 10.1016/j.compbiomed.2016.01.025. ISSN 0010-4825.
39. Gautam S, Bhavsar A, Sao AK, Harinarayan KK. CNN Based Segmentation of Nuclei in PAP-Smear Images with Selective Pre-Processing. *Digital Pathol* (2018) 10581:105810X. doi: 10.1117/12.2293526
40. Cox S. Guidelines for Papanicolaou Test Screening and Follow-Up. *J Midwifery Wom Heal* (2012) 57:86–9. doi: 10.1111/j.1542-2011.2011.00116.x
41. Phaliwong P, Pariyawateekul P, Khuakoonratt N, Sirichai W, Bhamarapravatana K, Suwannarurk K. Cervical Cancer Detection Between Conventional and Liquid Based Cervical Cytology: A 6-Year Experience in Northern Bangkok Thailand. *Asian Pac J Cancer Prev* (2018) 19(5):1331–6. doi: 10.22034/APJCP.2018.19.5.1331
42. Hoda RS, Loukeris K, Abdul-Karim FW. Gynecologic Cytology on Conventional and Liquid-Based Preparations: A Comprehensive Review of Similarities and Differences. *Diagn Cytopathol* (2013) 41(3):257–78. doi: 10.1002/dc.22842
43. Nayar R, Wilbur DC. The Pap Test and Bethesda 2014. *Cancer Cytopathol* (2015) 123(5):271–81. doi: 10.1002/cncy.21521

44. Hussain E, Mahanta LB, Das CR, Choudhury M, Chowdhury M. A Shape Context Fully Convolutional Neural Network for Segmentation and Classification of Cervical Nuclei in Pap Smear Images. *Artif Intell Med* (2020) 107:101897. doi: 10.1016/j.artmed.2020.101897
45. Khan MJ, Werner CL, Darragh TM, Guido RS, Mathews C, Moscicki AB, et al. ASCCP Colposcopy Standards: Role of Colposcopy, Benefits, Potential Harms, and Terminology for Colposcopic Practice. *J Low Genit Tract Dis* (2017) 21(4):223–9. doi: 10.1097/LGT.0000000000000338
46. Miyagi Y, Takehara K, Miyake T. Application of Deep Learning to the Classification of Uterine Cervical Squamous Epithelial Lesion From Colposcopy Images. *Mol Clin Oncol* (2019) 11(6):583–9. doi: 10.3892/mco.2019.1932
47. Lal S, Das D, Alabhya K, Kanfade A, Kumar A, Kini J. NucleiSegNet: Robust Deep Learning Architecture for the Nuclei Segmentation of Liver Cancer Histopathology Images. *Comput Biol Med* (2021) 128:104075. doi: 10.1016/j.compbiomed.2020.104075
48. Piotrowski T, Rippel O, Elanzew A, Nießing B, Stucken S, Jung S, et al. Deep-Learning-Based Multi-Class Segmentation for Automated, Non-Invasive Routine Assessment of Human Pluripotent Stem Cell Culture Status. *Comput Biol Med* (2021) 129:104172. doi: 10.1016/j.compbiomed.2020.104172
49. Conceição T, Braga C, Rosado L, Vasconcelos MJM. A Review of Computational Methods for Cervical Cells Segmentation and Abnormality Classification. *Int J Mol Sci* (2019) 20(20):5114. doi: 10.3390/ijms20205114
50. Song Y, Zhu L, Qin J, Lei B, Sheng B, Choi KS. Segmentation of Overlapping Cytoplasm in Cervical Smear Images via Adaptive Shape Priors Extracted from Contour Fragments. *IEEE Trans Med Imaging* (2019)

12 A Comparative Performance Study of Feature Selection Techniques for the Detection of Parkinson's Disease from Speech

Faycal Ykhlef and Djamel Bouchaffra

Center for Development of Advanced Technologies, Baba Hassen, Algeria

CONTENTS

12.1 Introduction 185
12.2 Proposed Methodology 187
12.3 PD Features 187
12.4 Feature Selection 187
12.5 Fisher Score 188
12.6 mRMR (Minimum Redundancy Maximum Relevance) 189
12.7 Chi-Square 189
12.8 Classification 189
12.9 Assessment of Feature Selection Methods 189
12.10 Results and Interpretation 190
12.11 Conclusion and Perspectives 191
References 192

12.1 INTRODUCTION

Parkinson's Disease (PD) is a neurological illness that affects the central nervous system [1]. It is the second most common neurodegenerative disorder (after Alzheimer's disease). It affects people of all ages, but is more common among elderly people [2]. MRI data analysis and effective image processing are considered among the best PD diagnosis methods since they allow the analyses of brain's motors in a progressive manner [1]. However, this solution has the drawback of being expensive and inaccessible to a large population. A novel diagnosis method has been proposed by M.A.

DOI: 10.1201/9781003267782-12

Little [3]. This consists of using the patient's voice to detect PD symptoms [4]. This approach is very straightforward and non-expensive since it only requires the waveform of the voice acquired using only a microphone.

The diagnostic system is composed of three main stages: (i) data acquisition; (ii) feature extraction; and (iii) classification. Data acquisition consists in the collection of a set of healthy and dysphonic voices (uttered by patients suffering from Parkinsonian syndromes). These data can be one of these types: sustained vowels, isolated consonants, or continuous sentences. A set of domain-specific databases are available in the literature. One can mention a number of datasets: the Massachusetts Eye & Ear Infirmary voice disorders database [5], the Saarbrucken voice database [6] and the Parkinson dataset [3]. Feature extraction consists in computing the most relevant acoustic measurements that better characterize the phenotypes of PD. These measurements constitute the input feature vector. Advanced signal-processing techniques are used in this step. Pitch, Shimmer, Jitter and Harmonic to Noise Ratio are among the examples of PD measurements [3]. In the classification stage, the feature vector is used as input to a decision process. Several low-complexity models can be used to detect PD. One can mention: Artificial Neural Network (ANN), Gaussian Mixture Models (GMMs), Naive Bayes, Logistic Regression, Hidden Markov Models (HMMs) and Support Vector Machines (SVMs) [7, 8]. Other schemes based on deep learning can be employed when massive amounts of data are collected [9]. It was reported in the literature that when dealing with small datasets, low-complexity models will give the best general results [10]. This is explained by their small number of parameters that are learned during training. These few parameters can be optimal when training is conducted with a small dataset rather than with a large dataset. In other words, simple models exhibit better learning from small datasets than more complicated models (such as a deep neural network with several parameters) since they are essentially trying to capture less information from a small size dataset. This is compatible with the principle of Occam's Razor, which promotes low-complexity models over high-complexity models. It is worth underscoring that, when compared to other classifiers, SVM is more effective when dealing with a high-dimension dataset. This is due to the fact that SVM is based on a high-dimensional distance metric that impacts the class decision task. The decision performance and the computational cost of the methods cited above can be improved when dimensionality reduction and features selection approaches are employed [7, 11]. Several approaches for PD detection have been proposed in the literature [1–12]. The main goal of these approaches is to design a non-complex, inexpensive, and accurate diagnosis system [17, 18].

In contrast to most of the conventional approaches addressing this problem that rely on a large dataset, the study presented in this chapter aims to assess the performance of three feature selection methods for SVM-based PD detection on small datasets. The techniques we have selected are: Fisher, minimum Redundancy Maximum Relevance (mRMR) and Chi-square. We have evaluated the performance of these feature selection methods on the small size corpus known as the Parkinson dataset [13]. This latter dataset contains the measurements of 22 different features recorded by healthy and disordered speakers. We have used the SVM classifier in our investigation due to its efficiency with regard to small size samples. The metrics we used to assess the performance of these feature selection approaches are: the classification accuracy (Ca), the Receiver Operating Characteristics (ROC); and the area under the curve (AUC).

This chapter is organized as follows. In Section 12.2, the proposed methodology is presented. In Section 12.3, the global evaluation is reviewed. In Section 12.4, the experimental results are provided. Section 12.5 lays out the conclusion and offers some future perspectives.

12.2 PROPOSED METHODOLOGY

The summary of the methodology we have proposed is shown in Figure 12.1. The different stages of the system are described in the following sections.

12.3 PD FEATURES

We have used the Parkinson dataset for the evaluation of the feature selection methods. The whole of the data are publicly available on the website of the University of California at Irvine (UCI) [13]. It consists of a collection of 22 biomedical features recorded by 31 speakers. The set of features is represented in a matrix form. Each column in the matrix denotes a specific feature, and each row corresponds to one of 195 voice recordings. The 23rd row denotes the speaker status (0: healthy, 1: Parkinsonian symptoms) (Table 12.1). More information about the dataset is given in [3].

12.4 FEATURE SELECTION

The selection methodology consists in choosing the best set of features for the classification task. It aims to facilitate the learning process and reduce the computational complexity. There are three main selection methods: (i) Filters; (ii) Wrappers; and (iii) Embedded [12]. In our investigation, we have chosen to assess the performance of Filters. They select features regardless of the classification model. It basically

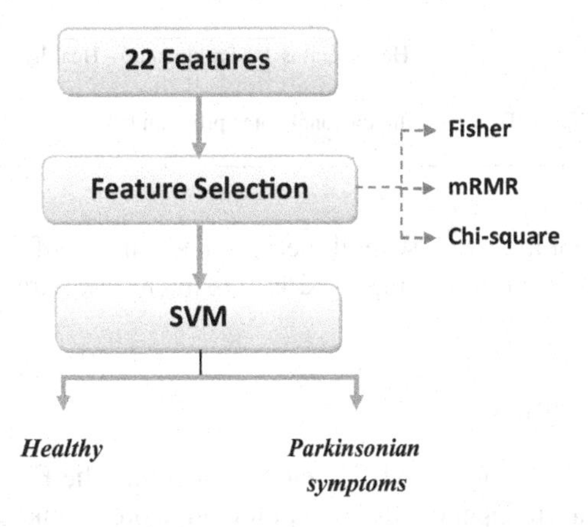

FIGURE 12.1 PD detection scheme.

TABLE 12.1
Parkinson's Dataset

Feature No.	Feature Name	Description
1	MDVP: Fo (Hz)	Mean pitch value
2	MDVP FHI (Hz)	Maximum of pitch
3	MDVP: Flo (Hz)	Minimum of pitch
4	MDVP: Jitter (%)	Jitter as a percentage (Pitch variations)
5	MDVP: Jitter (Abs)	Absolute Jitter in microsecond (Pitch variation)
6	MDVP: RAP	Relative amplitude perturbation (Amplitude variations)
7	MDVP: PPQ	Five point period perturbation quotient (pitch variations)
8	Jitter: DDP	Average absolute difference of differences between cycles, divided by the averaged period (pitch variations)
9	MDVP: Shimmer	local Shimmer (Amplitude variations)
10	MDVP: Shimmer (dB)	local Shimmer in decibels (Amplitude variations)
11	Shimmer: APQ3	Three-point amplitude perturbation quotient (Amplitude variations)
12	Shimmer: APQ5	Five-point amplitude perturbation quotient (Amplitude variations)
13	MDVP: APQ	11-point amplitude perturbation quotient (Amplitude variations)
14	Shimmer: DDA	Average absolute difference between consecutive differences between the amplitudes of consecutive periods (Amplitude variations)
15	NHR	Noise-to-harmonics ratio
16	HNR	Harmonics-to-noise ratio
17	RPDE	Recurrence period density entropy
18	DFA	Detrended fluctuation analysis
19	Spread1	Non Linear measure of pitch type 1
20	Spread2	Non Linear measure of pitch type 2
21	D2	Correlation dimension
22	PPE	Pitch Period Entropy
23	status	Health status 1 – Parkinson 0 - Healthy

*MDPV stands for KEYPENTAX multidimensional voice program [3].

performs feature ranking using several metrics. The number of features is chosen as needed. The methods we have employed in this manuscript are described as in the following text.

12.5 FISHER SCORE

This technique performs a supervised feature selection. The Fisher score algorithm chooses each feature independently based on their scores, ranked by their contribution to the classification problem at hand. The key idea of the Fisher score consists of determining a subset of features, such that in the data space spanned by the chosen features, data points in different classes are far apart, whereas data points in the same class are close to each other in term of a distance metric. The inter-class variance should be maximized, while the intra-class variance should be minimized [12].

12.6 MRMR (MINIMUM REDUNDANCY MAXIMUM RELEVANCE)

Minimal-optimal methods consist of determining a small set of features that, as a group, provides the maximum possible predictive power. On the other hand, individual powerful features that exhibit any predictive power are also selected to be part of the small set. So, mRMR not only values the notion of a group of features that can work together, but also appraises individual features that are powerful, even if they bring more or less the same information (or redundant) [11, 12].

12.7 CHI-SQUARE

This method is based on the computation of the Chi-square value (which is a sum of squared standard normal variables) between two variables: the predictor (the independent variable) and the response (the dependent variable). If this Chi-square value is high, then the feature predictor variable is more dependent on the response. Therefore, this feature can be selected for model training [11, 12].

12.8 CLASSIFICATION

Support Vector Machines (SVMs) are supervised learning models that produce a map of the sorted data with the margin between the two classes as far apart as possible. It was proposed by V. Vapnik [14]. The goal of SVMs to compute a hyperplane in a k-dimensional space (k represents the number of features) that unambiguously classifies the data points.

12.9 ASSESSMENT OF FEATURE SELECTION METHODS

A simplified diagram of the assessment methodology is depicted in Figure 12.2. We have divided the Parkinson dataset into two parts:

- o Training stage, which contains 75 percent of the data.
- o Testing stage, which contains the remaining 25 percent of the data.

The training stage aims to:

- Build the prediction model;
- Select the number of features to be used;
- Choose feature selection methods (Fisher, Chi-Square or mRMR);
- Select the hyperparameters of the SVM (c, gamma) via grid search and 10-fold cross validation method.

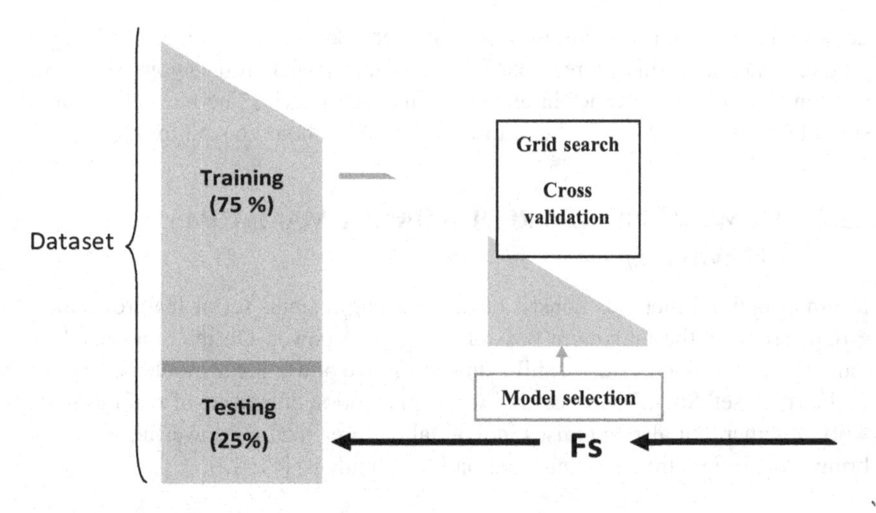

FIGURE 12.2 Assessment methodology*.

The testing stage aims to evaluate the performance of the global scheme. To improve the generalization power of the prediction model, the SVM model is re-estimated using the entire set of training samples (75 percent of the dataset). The testing phase is performed using the remaining 25 percent of data. The data used during testing have not been used to estimate the hyperparameters of the SVM. The number of features (NF) to be selected was set from 2 to 22.

For each value of NF, the optimization of the model parameters is performed using three different feature selection techniques: (i) Fisher; (ii) MRMR; and (iii) Chi-square.

12.10 RESULTS AND INTERPRETATION

The Feature Selection Algorithms toolbox developed by the University of Arizona has been employed during the experiments [15]. LibSVM has been used for classification [16]. We relied on the following metrics to assess the performance of feature selection methods: (i) classification accuracy (Ca); (ii) the ROC curves; and (iii) the AUC's criteria. Figure 12.3 shows the evolution of the classification accuracy as a function of the number of features (NF) using three different techniques of selection: Fisher, Chi-square and mRMR. We have found that:

i. The minimum number of features leading to the best classification accuracy depends on selection technique:
   - Fisher: 13 features are enough to get an accuracy of 95.8 percent,
   - mRMR: 13 features are necessary to get an accuracy of 93.7 percent,
   - Chi-Square: 14 features are necessary to get an accuracy of 93.7 percent.
ii. The increase of the number of features does not improve the classification accuracy,

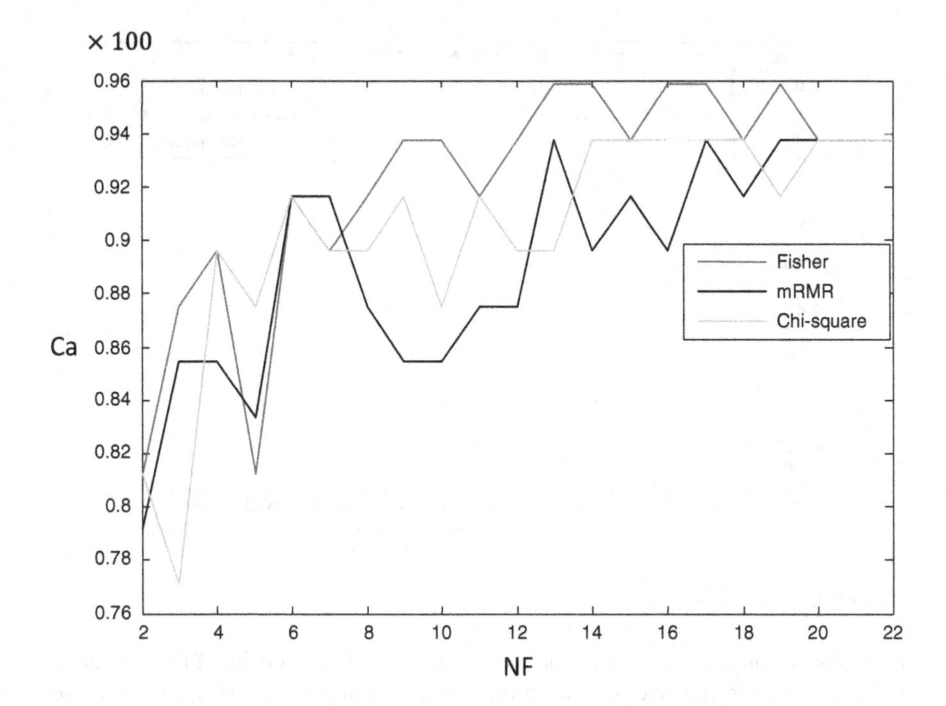

FIGURE 12.3 Classification accuracy using feature selection techniques.

iii. The decrease of the number of features (from 22 to 13 or 14) facilitates the learning process and reduces the computational cost,
iv. The combination of features depends on the technique of feature selection.

We have plotted the ROC curves using the three selection techniques. The number of features that maximizes the classification accuracy has been considered for each technique.

The results are shown in Figure 12.4, where TPR and FPR respectively stand for True Positives Rate and False Positive Rate. The AUCs are given as follows:

— AUC_F= 0.967 « Fisher »,
— AUC_mRMR= 0.953 « mRMR »,
— AUC_Ch= 0.921 « Chi-square ».

The AUC obtained using the Fisher selection technique is the most significant among the other ones.

12.11 CONCLUSION AND PERSPECTIVES

In this chapter, we have compared the performance of three feature selection methods, namely: Fisher, Chi-Square and mRMR. Subsequently, we relied on the SVM classifier due to its efficiency on small size datasets. The system has been evaluated

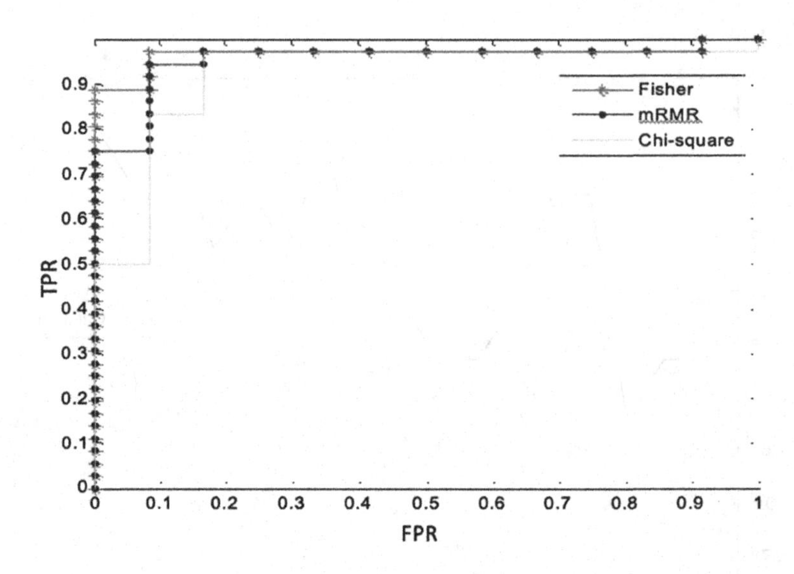

FIGURE 12.4 ROC curve.

on the Parkinson dataset which is publicly available on the website of the University of California at Irvine. We have used the classification accuracy (Ca), the Receiver Operating Characteristics (ROC), and the area under the curve (AUC) criteria to measure the performance of the proposed system. Experimental results demonstrate the outperformance of the Fisher technique compared to the other methods. The accuracy of the diagnosis system is equal to 95.8 percent using only 13 features among a set of 22 biomedical measurements. These results have been demonstrated on the ROC curves and the AUC's criteria. The proposed diagnosis system could be implemented on a smartphone device to detect early signs of Parkinson's disease using only the speaker's voice.

REFERENCES

1. Albano, L., Agosta, F., Basaia, S., Cividini, C., Stojkovic, T., Sarasso, E., … & Filippi, M. (2022). Functional connectivity in Parkinson's disease candidates for deep brain stimulation. *NPJ Parkinson's Disease*, 8(1), 1–12.
2. Gotardi, G. C., Barbieri, F. A., Simão, R. O., Pereira, V. A., Baptista, A. M., Imaizumi, L. F., … & Rodrigues, S. T. (2022). Parkinson's disease affects gaze behavior and performance of drivers. *Ergonomics*, (just-accepted), 1–30.
3. Little M.A., McSharry P.E., Hunter E.J., and Spielman J. (2009). Suitability of dysphonia measurements for telemonitoring of Parkinson's disease. *IEEE Trans. Biomed. Eng.* 56, 4, 1015–1022.
4. Gullapalli, A. S. and Mittal, V. K. (2022). Early detection of Parkinson's disease through speech features and machine learning: A review. *ICT with Intelligent Applications*, 203–212.
5. Massachusetts Eye and Ear Infirmary, Voice Disorders Database, Version. 1.03 [CD-ROM], Kay Elemetrics Corp., Lincoln Park, NJ, 1994.

6. Barry W.J. and Putzer M., Saarbrucken Voice Database, Institute of Phonetics, Univ. of Saarland, webpage: http://www.stimmdatenbank.coli.uni-saarland.de/.
7. Ykhlef, F., Benzaba, W., Boutaleb, R., Alonso, J. B., & Ykhlef, F. (2015, December). Yet another Approach for the Measurement of the Degree of Voice Normality: A Simple Scheme Based on Feature Reduction and Single Gaussian Distributions. In 2015 IEEE International Symposium on Multimedia (ISM) (pp. 335–338). IEEE.
8. Bouchaffra, D. and Tan, J. (2006). Structural Hidden Markov Models using a Relation of Equivalence: Application to Automotive Designs, in: Data Mining and Knowledge Discovery Journal, Volume 12: 1, Springer-V.
9. Nagasubramanian, G. and Sankayya, M. (2021). Multi-variate vocal data analysis for detection of Parkinson disease using deep learning. *Neural Computing and Applications*, 33(10), 4849–4864.
10. Pasupa, K. and Sunhem, W. (2016, October). A comparison between shallow and deep architecture classifiers on small dataset. In *2016 8th International Conference on Information Technology and Electrical Engineering (ICITEE)* (pp. 1–6). IEEE.
11. Hashemi, A., Dowlatshahi, M. B. and Nezamabadi-Pour, H. (2022). Ensemble of feature selection algorithms: A multi- criteria decision-making approach. *International Journal of Machine Learning and Cybernetics*, 13(1), 49–69.
12. Huan L. and Motoda H. (2007). *Computational Methods of Feature Selection*. Chapman and Hall/CRC.
13. https://archive.ics.uci.edu/ml/datasets/parkinsons
14. Vapnik V. and Cortes C. (1995). Support-vector networks. *J. Mach. Learn.* 20, 273–297.
15. Feature Selection Algorithms. Website: http://featureselection.asu.edu/software.php (last access 8/8/2021).
16. C.-C. Chang and C.-J. Lin. (2011). LIBSVM: A library for support vector machines. *ACM Transactions on Intelligent Systems and Technology*, 2:1–27.
17. Geetha R. and Sivagami G. (2011). Parkinson disease classification using Data Mining algorithms. *Int. J. Comp. App.* 32, 0975–8887. DOI 10.5120/3932-5571
18. Resul Das. (2010). A comparison of multiple classification methods for diagnosis of Parkinson disease 2010. *Expert Syst. Appl.* 37, 2, 1568–1572.

13 Enhancing Leaf Disease Identification with GAN for a Limited Training Dataset

Priyanka Sahu, Anuradha Chug, and Amit Prakash Singh
Guru Gobind Singh Indraprastha University, New Delhi, India

Dinesh Singh
Indian Agricultural Research Institute, New Delhi, India

CONTENTS

13.1 Introduction 195
13.2 Materials and Methods 197
13.2.1 Dataset 197
13.2.2 Method 197
13.2.2.1 DCGAN 198
13.2.2.2 StyleGAN 2 198
13.2.2.3 The Fine-Tuning of CNN for Classification 200
13.3 Experimental Setup 201
13.3.1 GAN Training 201
13.3.2 Generating Images 201
13.3.3 Results and Discussions 202
13.4 Conclusion 204
Acknowledgments 205
References 205

13.1 INTRODUCTION

Deep learning-based techniques [1] have been applied to constantly improve the state-of-the-art performance for many computer vision tasks, until they have even outperformed humans [2]. Numerous deep learning applications are currently facing a new challenge: learning from limited and unbalanced datasets. Techniques such as transfer learning [3], domain adaptability [4], and data augmentation [3] have been used to avoid these costs and learn from smaller datasets.

DOI: 10.1201/9781003267782-13

Although transfer learning and domain adaptation are common, they are still less well suited to applications where significant public dataset samples or pre-trained network parameters from a surrounding domain are not readily available, such as the recognition and classification of crops. Various researchers have reported their experimentations using basic data augmentation techniques for expanding the training set and balancing the classes [5–7].

However, the diversity and variability obtained using these augmented techniques, e.g., flip, brightness, rotation, scaling, translation, etc., is limited. This stimulates the adoption of synthetic data, which can have synthesized the samples and add more heterogeneity to the dataset, and can also enhance it further, in order to improve the accuracy and classification results.

Generative adversarial networks (GAN) have been extensively studied for various applications due to their power to generate synthetic data. GANs are used to generate synthetic images whenever the training samples are inadequate and image augmentation approaches have been incapable to improve the results. In [8], the authors have implemented a deep learning-based approach for the detection of tomato leaf disease using a Conditional GAN to produce synthesized images of tomato crop leaves. Subsequently, DenseNet121 was deployed for the training of the combined (original + synthetic) dataset. This model gave a classification accuracy of 97.11 percent, 98.65 percent, and 99.51 percent for tomato plant leaf image classification into 10-labeled classes, 7-labeled classes, and 5-labeled classes, respectively. In a similar work [9], the authors have deployed the DoubleGAN (a combination of Super-resolution GAN and Deep convolution GAN) for the generation of synthetic images of healthy and diseased plant leaves. This model gave a disease classification accuracy of 99.53 percent. In [10], Outlier Removal Auxiliary Classifier GAN has been used for the early identification of spotted wilt virus in tomato plants. Hyperspectral data were used for training and testing purposes. In [11], the authors have used a GAN to generate the image and convolutional neural network (CNN) for the classification of plant leaves deployed on an Android-based mobile application.

In this study, the authors have investigated plant leaf images to construct a deep learning- based system for the identification of crop leaf diseases. Rather than paying for expensive expert analysis, agronomists might use this technology to classify diseases impacting cultivation by merely capturing diseased leaves. In the proposed approach, the authors have used Deep Convolutional Generative Adversarial Networks (DCGAN) [12] and StyleGAN2 [13] to synthesize new images and merge these in the original image dataset to process the training. Following this, a ResNet50 architecture has been trained on original crop leaf images and the synthesized images produced by the deployed GANs.

The rest of the chapter is structured as follows: Section 13.2 describes the dataset used and the followed methodology of image generation and classification. This section also contains the network architecture details for image generation and also includes the details of CNN deployed for leaf disease classification. Subsequently, Section 13.3 entails the experimental setup and metrics used. Next, Section 13.4 elaborates the result findings and discussion. Finally, Section 13.5 concludes the study along with the future scope.

13.2 MATERIALS AND METHODS

13.2.1 Dataset

PlantVillage is a public image data repository for plant leaf diagnosis that comprises both diseased and healthy leaf images of various plants. This dataset contains 38 different classes and 54,305 colored leaf images. In this study, 1500 diseased and healthy leaf images were extracted from the same dataset (15-labeled classes with 100 images for each). Pepper, potato, and tomato are the main plant species whose classes are considered. Figure 13.1 shows the samples of diseased and healthy leaves.

13.2.2 Method

The proposed technique aims to observe how plant disease identification systems, given the class imbalance and sample deficiency in the training data, improve in accuracy. The system requires an additional data augmentation approach to expand the limited training dataset while maintaining classes balanced. The authors have proposed to synthetically produce additional training data using GAN and train the identification network using that data along with the original image data.

GAN has been designed and implemented to enhance the classification accuracy of CNN for the recognition of crop leaf diseases deploying over a limited training image dataset. Real images were used to train the GAN and, thus, the trained GAN was utilized to create supplementary labeled images. The generated images were merged with original input images. Finally, the dataset has used to train the CNN. Therefore, the proposed method consists of two components. The first component is GAN. It was used to generate additional images. The second component is CNN, which was used to classify plant disease types and plant species. Figure 13.2 shows a proposed pipeline that consists of two constituents: a synthetic data generation module with DCGAN and StyleGAN2. Subsequently, a classification system with a CNN, namely ResNet50, was used for conducting supervised learning. Once the

FIGURE 13.1 Instances of images from PlantVillage dataset.

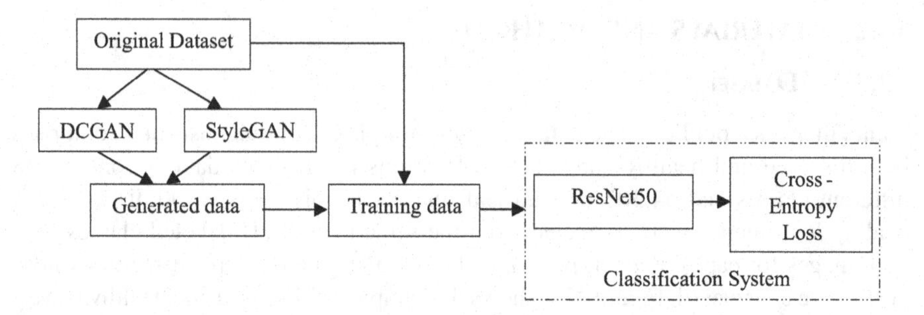

FIGURE 13.2 Proposed methodology for synthetic data generation and disease.

estimated probability of ground-truth categories is at its highest value, then the cross-entropy loss is minimized.

13.2.2.1 DCGAN

In this study, the authors have used DCGAN [12]. It is an extension of the basic GAN, with the exception that the discriminator uses a convolutional layer and the generator has convolutional-transpose (CT) layers. A strided convolution layer with LeakyReLU activations constitutes the discriminator. DCGAN is input with a 64 × 64 × 3 image and produces a scalar probability outcome. The generator is formed of CT layers with ReLU activations. The result is a 64 × 64 × 3 RGB image, with the input being a latent vector, *z*, taken from a typical normal distribution. The latent vector can be turned into a volume with an identical shape as an input image using the strided CT layers.

A basic DCGAN generator model is shown in Figure 13.3. The features of DCGAN are as follows:

- Strided convolutions (discriminator) and fractional-strided convolutions (fractional-strided convolutions) are used to replace any pooling layers (generator).
- Both the generator and the discriminator use batch norm.
- For deeper networks, fully connected (FC) hidden layers are being removed.
- All layers except the output use the ReLU activation function in the generator.
- Every layer deploys the LeakyReLU activation function in the discriminator.

13.2.2.2 StyleGAN 2

StyleGAN2 [13] is a generative adversarial network that expands on the original StyleGAN2. First, the adaptive instance normalization (AdaIN) is modified and substituted with a weight demodulation-based normalization technique. Second, an advanced training method is developed that achieves the same purpose progressively. However, the training process begins with low-resolution images and then gradually transfers focus to increasingly higher resolutions – although not modifying the network topology over training. Furthermore, different forms of regularization are

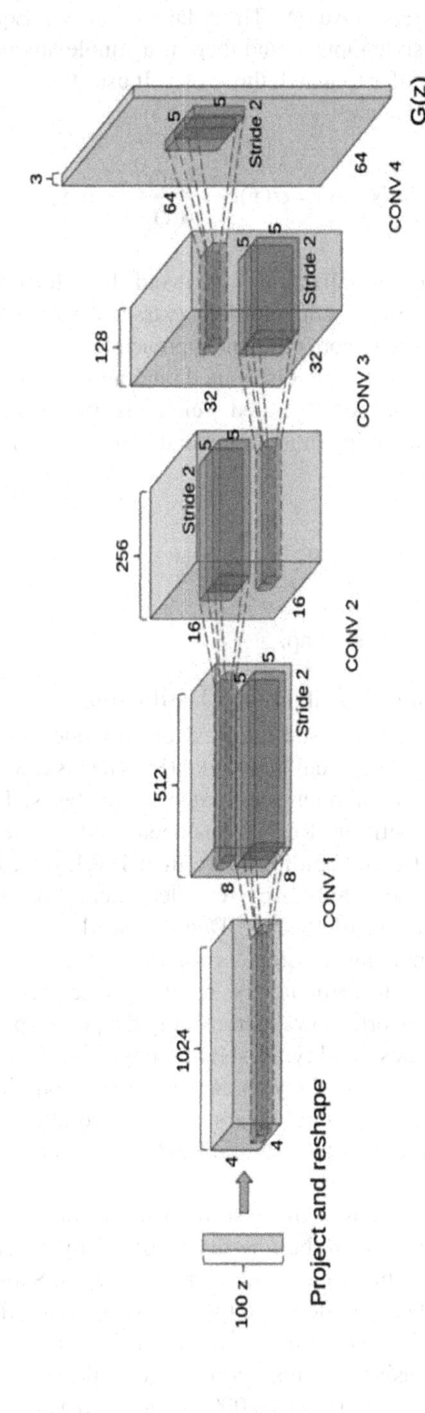

FIGURE 13.3 A DCGAN generator model. Source: [12]

introduced, such as lazy regularization and path length regularization. There are no learnable affine parameters in AdaIN. The AdaIN (shown in Equation 13.1) accepts a content input x and a style input y, and there is a simple alignment of the channel-wise mean and variance of x to match those of y. It uses the style input to adaptively compute the affine parameters:

$$\text{AdaIN}(x,y)=\sigma(y)\left(\frac{x-\mu(x)}{\sigma(x)}\right)+\mu(y) \tag{13.1}$$

The goal of instance normalization is to avoid the influence of s (feature map scales) from the metrics of the convolution's extracted output feature map vectors. Weight modulation is a much more efficient approach for achieving this goal. Assume that input activations are uniformly distributed random variables with a unit standard deviation. The L2 norm of the associated weights is used to scale the outputs. The output activations after applying modulation and convolution have a standard deviation of (equation 13.2):

$$\sigma_j=\sqrt{\sum_{i,k} w'^{2}_{ijk}} \tag{13.2}$$

Where j is the output feature map.

13.2.2.3 The Fine-Tuning of CNN for Classification

In this study, the authors have used the ResNet-50 model for multi-plant species and diseases recognition. Residual Networks (ResNet) is a well-known CNN that serves as the foundation for many computer vision tasks. In 2015, this model won the ImageNet competition. ResNet has been used for the successful training of very deep neural networks holding more than 150 layers. Due to the problem of vanishing gradients, the training of very deep neural networks was challenging before the development of ResNet. Deep networks are difficult to train due to the well-known vanishing gradient problem, which occurs when the gradient gets backpropagated to prior layers, resulting in extremely small gradients. Consequently, as the network grows increasingly deeper, its performance becomes saturated or even degrades quickly. ResNet 50 deployed the concept of skip connections between the convolution layers. ResNet-50 is subdivided into five phases, each containing three convolutional layers in the convolution block and three convolution layers in the identity block. In the ResNet-50, there are around 23 million trainable parameters.

The topmost FC layer, along with the softmax layer, was taken off from the architecture in order to proceed with the fine-tuning of the hyperparameters of the pre-trained ResNet50 architecture on the plant leaves' images. Subsequently, the ReLU activation function has been applied with two different convolutional layers, an average pooling layer, an FC layer, and a softmax layer were put in their place. The weights were modified using Adam's optimizer after the model was trained for 50 epochs. The applied learning rate was 0.0001 along with a batch size of 32 [8].

13.3 EXPERIMENTAL SETUP

In this chapter, two types of experimentations have been conducted. During the primary experiment, the GAN models were trained on the training samples for the generation of the synthetic images of plant leaves of 15 different varieties. For this purpose, DCGAN and StyleGAN2 models have been deployed for 250 epochs. After each epoch, the weights of the generator model and discriminator models have been modified to create synthetic images that were as identical as possible to real images. The authors have acquired 3,500 synthetic images of plant leaves from the DCGAN and StyleGAN2 models at the end of the network's training. The pre-trained ResNet50 model was deployed on the original training image dataset as well as on the mixture of the original training dataset (a subset of the PlantVillage repository) and synthesized plant leaves images. The required hardware and software specifications are also given as: (1) Graphics – 1xTesla K80, 2496 CUDA cores; (2) Memory – 12 GB GDDR5 VRAM; (3) Operating system – Windows 10 (64 bits); (4) Processor – AMD, Ryzen 5, 5000 series; and (5) Language- Python 3.7.

Table 13.1 shows several performance metrics, including classification accuracy, F1-score, precision, and recall that have been used for evaluation of the proposed model performance.

13.3.1 GAN Training

The authors have used '1' to denote the real label and '0' for the fake label. All the labels have been used to determine the discriminator (D) and generator (G) losses. The authors have generated a consistent batch of latent feature vectors derived from a Gaussian distribution or fixed noise to keep track of the generator's training progress. There is a need to input this fixed noise periodically into G during the training cycle, and images have been generated from this noise over defined iterations.

13.3.2 Generating Images

The qualitative performance of DCGAN and StyleGAN2 is visually inspected in Figure 13.4. It has been observed and seen in Figure 13.5 that generated images from

TABLE 13.1
Deployed Performance Metrics

Performance Metric	Equation	Naming Conventions
Accuracy	$TP + TN = TP + TN + FP + FN$	Notation meaning, TP = True Positive, FP = False Positive
Precision	$TP = TP + FP$	TN = True Negative, and
Recall	$TP = TP + FN$	FN = False Negative.
Fl − score	$2 * TP = 2 * TP + FP + FN$	

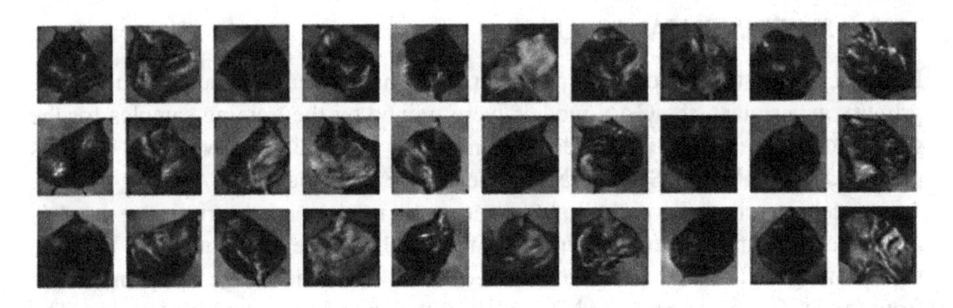

FIGURE 13.4 Instances of images generated by DCGAN.

FIGURE 13.5 Instances of images generated by StyleGAN2.

DCGAN are of average quality but generated images using StyleGAN2 are very comparable to the real ones.

13.3.3 Results and Discussions

Initially, the original dataset was applied to train models for plant leaf disease detection using the transfer learning technique ResNet50. The training, validation, and testing segments of the dataset were partitioned into 80 percent, 10 percent, and 10 percent, respectively. Figure 13.6 depicts the classification accuracy of the system by deploying the original dataset. Furthermore, a dataset based on simple image augmentation techniques was applied for the training of the pre-trained models with comparable hyperparameter ranges is also shown in Figure 13.6. The test findings depict that the augmented image dataset outperforms the original dataset in terms of classification accuracy.

In addition, the DCGAN and StyleGAN2 augmented dataset-based classification models were both trained using the same hyperparameter value. Once the training was completed, the models were put to test with unknown testing images. The results of the tests revealed that the classification models employing the DCGAN and StyleGAN2 augmented image dataset performed better than the previous non-augmented datasets. The class-wise performance of the DCGAN and StyleGAN2 deployed datasets is shown in Table 13.2. Finally, all of the augmentation approaches were used to construct the combined dataset, including simple augmentations, DCGAN and StyleGAN2. The accuracy of the pre-trained models employed on the

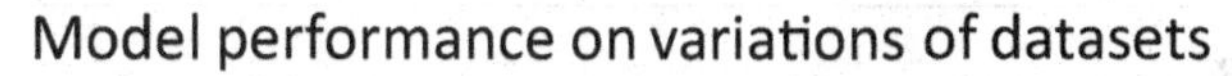

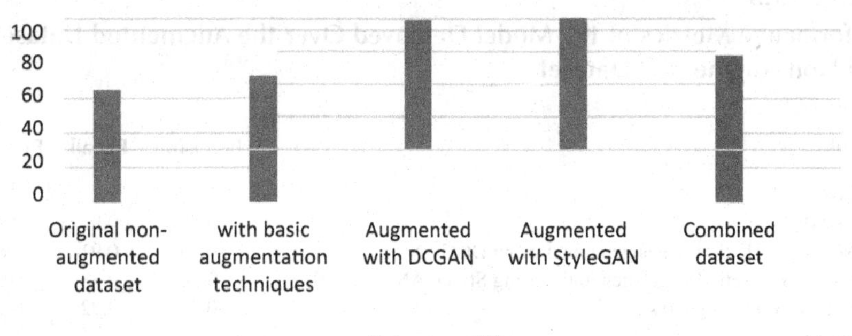

FIGURE 13.6 Performance of ResNet50 model deployed over original dataset and augmented dataset with different techniques.

TABLE 13.2
Precision, Recall, and F1-Score for Different Disease Classes of Pepper, Potato, and Tomato Plants

		DCGAN			StyleGAN2		
DL Model		**Precision**	**Recall**	**F1-score**	**Precision**	**Recall**	**F1-score**
Pepper Bell	Bacterial Spot	0.69	0.79	0.74	0.87	0.81	0.85
	Healthy	0.79	0.76	0.78	0.85	0.90	0.88
Potato	Early Blight	0.87	0.90	0.89	0.88	0.89	0.89
	Healthy	0.79	0.92	0.86	0.83	0.77	0.81
	Late Blight	0.64	0.65	0.65	0.83	0.86	0.85
Tomato	Bacterial Spot	0.81	0.69	0.76	0.78	0.80	0.79
	Early Blight	0.91	0.85	0.89	0.93	0.95	0.95
	Healthy	0.65	0.59	0.63	0.78	0.76	0.78
	Late Blight	0.89	0.84	0.88	0.92	0.87	0.89
	Leaf Mold	0.69	0.79	0.76	0.77	0.81	0.79
	Mosaic Virus	0.59	0.64	0.63	0.78	0.72	0.75
	Septoria Leaf Spot	0.89	0.87	0.89	0.89	0.85	0.88
	Two- Spotted Spider Mite	0.83	0.79	0.81	0.78	0.80	0.79
	Target Spot	0.78	0.80	0.79	0.79	0.78	0.78
	Yellow Leaf Curl Virus	0.92	0.86	0.90	0.92	0.94	0.93

combined dataset yielded a mixed set of results. It can be seen that StyleGAN2 generates real-like images when compared with the DCGAN model. Hence, on combining the dataset of both the GANs, somehow the quality and quantity of the merged dataset enhanced due to the images synthesized using GAN.

It has been observed that combined dataset performed best for classification as compared to all other mentioned augmentation techniques. The classification accuracy of the pre-trained model utilizing the combined image dataset is shown in

TABLE 13.3
Performance Metrics of the Model Deployed Over the Augmented Dataset and Non-Augmented Dataset

Model	Accuracy (%)	Precision	Recall	F1-score
5-classes				
ResNet50	90	0.88	0.91	0.90
ResNet50+Synthetic Images generated using DCGAN	86	0.79	0.91	0.86
ResNet50+Synthetic Images generated using StyleGAN2	91	0.92	0.88	0.91
ResNet50+Combined dataset	93	0.93	0.92	0.93
10-classes				
ResNet50	85	0.79	0.89	0.85
ResNet50+Synthetic Images generated using DCGAN	83	0.81	0.87	0.83
ResNet50+Synthetic Images generated using StyleGAN2	88	0.86	0.90	0.88
ResNet50+Combined dataset	89	0.88	0.89	0.89
15-classes				
ResNet50	82	0.80	0.83	0.82
ResNet50+Synthetic Images generated using DCGAN	79	0.89	0.69	0.79
ResNet50+Synthetic Images generated using StyleGAN2	84	0.87	0.81	0.84
ResNet50+Combined dataset	85	0.85	0.86	0.85

Figure 13.6 and Table 13.2. The experimental result shows that a composite dataset based on a mix of several data augmentation approaches outperformed the original dataset and individual augmentation techniques-based datasets. Furthermore, deep learning-based datasets complement existing image manipulation techniques. Table 13.2 shows the category-wise evaluation of the implemented method for a 15-labeled class classification target on an enhanced image dataset containing synthesized images.

The authors have analyzed the performance of the implemented model for 5-labeled class classification, 10-labeled class classification, and 15-labeled class classification. A results evaluation of the proposed approach for the PlantVillage dataset and the augmented leaf image dataset (original PlantVillage + Synthesized images) is shown in Tables 13.2 and 13.3. The presented method observed a classification accuracy of 93 percent, 89 percent, and 85 percent for 5-labeled classes classification, 10-labeled classes, and 15-labeled classes classification tasks, respectively. As demonstrated in Table 13.3, the ResNet50 model with generated images outperformed the original dataset in terms of accuracy, F1-score, precision, and recall for all classes. This gain in classification performance shows that using the Style-GAN model for data augmentation has helped the network to avoid over-fitting and become more generic.

13.4 CONCLUSION

To overcome classification difficulties in the limited size dataset for plant leaf disease detection, three different images augmented datasets were proposed in the study. The datasets were constructed using basic image augmentation techniques, DCGAN and

StyleGAN2, and, finally, a merged expanded dataset was generated. To identify plant leaf diseases, the datasets have been used for the training of the pre-trained learning algorithms. The vast simulation results provide two key insights for developing a plant leaf disease detection model. Firstly, deep learning-based augmentation outperforms simple image manipulation techniques in terms of performance. Secondly, the dataset based on a mix of different augmentation strategies provides a better result than all other datasets. In the future, the capacity of GANs could be improved for better image resolution for high-dimensional image space. Furthermore, more image generation and augmentation techniques could be used to generate and enhance the size of a limited image dataset.

ACKNOWLEDGMENTS

Authors are thankful to the Department of Science & Technology, Government of India, Delhi, for funding a project on the "Application of IoT in Agriculture Sector" through the ICPS division. This work is a part of the project.

REFERENCES

1. Y. LeCun, Y. Bengio, and G. Hinton, "Deep learning," *Nature*, vol. 521, no. 7553, pp. 436–444, 2015.
2. K. He, X. Zhang, S. Ren, and J. Sun, "Delving deep into rectifiers: Surpassing human-level performance on imagenet classification," in *Proceedings of the IEEE international conference on computer vision*, 2015, pp. 1026–1034.
3. P. Sahu, A. Chug, A. P. Singh, D. Singh, and R. P. Singh, "Implementation of CNNs for crop diseases classification: A comparison of pre-trained model and training from scratch," *IJCSNS*, vol. 20, no. 10, p. 206, 2020.
4. I. Goodfellow, Y. Bengio, and A. Courville, *Deep learning*. MIT Press, 2016.
5. J. Wang, L. Perez, and others, "The effectiveness of data augmentation in image classification using deep learning," *Convolutional Neural Networks Vis. Recognit*, vol. 11, pp. 1–8, 2017.
6. S. C. Wong, A. Gatt, V. Stamatescu, and M. D. McDonnell, "Understanding data augmentation for classification: when to warp?," in *2016 international conference on digital image computing: techniques and applications (DICTA)*, 2016, pp. 1–6.
7. M. Gour, S. Jain, and T. Sunil Kumar, "Residual learning based CNN for breast cancer histopathological image classification," *Int. J. Imaging Syst. Technol.*, vol. 30, no. 3, pp. 621–635, 2020.
8. A. Abbas, S. Jain, M. Gour, and S. Vankudothu, "Tomato plant disease detection using transfer learning with C-GAN synthetic images," *Comput. Electron. Agric.*, vol. 187, p. 106279, 2021.
9. Y. Zhao et al., "Plant disease detection using generated leaves based on DoubleGAN," *IEEE/ACM Trans. Comput. Biol. Bioinforma.*, 2021.
10. D. Wang et al., "Early detection of tomato spotted wilt virus by hyperspectral imaging and outlier removal auxiliary classifier generative adversarial nets (OR-AC-GAN)," *Sci. Rep.*, vol. 9, no. 1, pp. 1–14, 2019.
11. R. Gandhi, S. Nimbalkar, N. Yelamanchili, and S. Ponkshe, "Plant disease detection using CNNs and GANs as an augmentative approach," in *2018 IEEE International Conference on Innovative Research and Development (ICIRD)*, 2018, pp. 1–5.

12. A. Radford, L. Metz, and S. Chintala, "Unsupervised representation learning with deep convolutional generative adversarial networks," *arXiv Prepr. arXiv1511.06434*, 2015.
13. T. Karras, S. Laine, M. Aittala, J. Hellsten, J. Lehtinen, and T. Aila, "Analyzing and improving the image quality of StyleGAN2," in *Proceedings of the IEEE/CVF Conference on Computer Vision and Pattern Recognition*, 2020, pp. 8110–8119.

14 A Vision-Based Segmentation Technique Using HSV and YCbCr Color Model

Shamama Anwar, Subham Kumar Sinha, Snehanshu Vivek, and Vishal Ashank
Birla Institute of Technology, Mesra, India

CONTENTS

14.1 Introduction......207
14.2 Existing State-of-the-Art Gesture Recognition Systems......208
14.3 Proposed System Overview......209
14.4 Results......211
14.5 Conclusion......212
References......213

14.1 INTRODUCTION

Communication, being a vital part of the progression of mankind, is the art of conveying or exchanging information through speaking, writing, signalling or other means. Signalling or gesture forms an important mode of communication. Involuntarily, even while speaking, people tend to make gestures to emphasize certain points. It also forms the basis for the sign language. In the era of advanced technology and evolving computer systems, the Human–Computer Interaction has also evolved since its inception. Computing devices initially accepted input by means of punch cards. Most of these cards featured 80 columns and several punches in each column were used to represent either a letter or a number. The letter or number was also printed at the column header so that the cards could also be read by humans [1]. With the inception of personal computing devices, the *QWERTY* keyboard became the major input device; this was regarded as a significant improvement on the traditional punch cards. The earlier keyboard devices were bulkier, and more like a typewriter. Later, the devices gained their more ergonomic shape, as can be seen today. The mouse as an input device also gained entry in the personal computing era as a small hand-held device with limited interaction with the computer. With the introduction of hand-held computing devices such as tablets and phones, touch became an appropriate way for input. This proved immensely popular as it proved more convenient to users.

DOI: 10.1201/9781003267782-14

We are currently living in the age of smart devices: smartphones, smart televisions, smart washing machines, and so on. These devices have gone beyond all the existing input methodology and require no physical interaction on the part of the user to accept input. Gestures and voice input are the most recent trends in accepting input. Gestures are made in the air to control the devices, which are "smart enough" to understand the gesture and react accordingly.

Since gestures for computing devices are generally made by hand, the current work proposes a segmentation method to isolate the hand from the input video. The method is based on vision-based systems, and hence the input is captured using a camera which is further broken into individual frames; from these frames the hand is segmented for further processing. The segmentation is accurate; it eliminates any background noises and also works very well for a cluttered background.

This chapter is structured as follows: Section 14.2 presents a review of the literature with particular attention to the prevalent techniques for detecting hand gestures, along with a summary of the datasets on which the methodology has been tested. The proposed algorithm is described in Section 14.3, and the results are documented and discussed in Section 14.4. A concluding section ends the chapter.

14.2 EXISTING STATE-OF-THE-ART GESTURE RECOGNITION SYSTEMS

Gesture recognition systems are developed to capture and recognize the gesture with an increased level of accuracy. These systems are divided into two stages: (i) Hand detection; and (ii) Gesture recognition. Hand detection refers to identifying the hand performing the gesture and segregating it from the background for further tracking [2]. The earlier method of taking input in gesture systems made use of specially engineered gloves with sensors [3]. The user had to wear these devices and perform the gesture. The devices have inbuilt sensors that can detect movement. Although this is a more precise way to take input, the devices are cumbersome and awkward for the user. It also limits the scope of usage as it can be used with a stationary device [4].

The other more common and user-friendly method is the vision-based method. In this technique a camera records the motion of the hand [5]. Since the camera is recording the entire scene an additional task is to isolate the hand from the background [6]. Various methods have been proposed for the same, with the common approaches being background subtraction, the 3D hand model approach and skin color detection [7, 8]. The skin color detection method perceives regions that possibly have human limbs in images. This is attained by disintegrating the image into discrete pixels and categorizing them into skin coloured and background [9]. RGB (Red, Green, Blue), HSV (Hue, Saturation, Value) and YCbCr (Luminance, Chrominance) color models are the three main parameters for recognizing a skin pixel [10]. The skin color detection model is more widely used as it does not depend on the orientation and size, and it is also processed quickly.

Background subtraction is a technique used to eliminate the background scene in an image and to isolate the region of interest. In relevance to hand gesture recognition, the background would remain almost static. The background image (without any hand gesture) is initially taken. When the tracking begins, the input image is considered as the foreground image and to isolate the gesture image the difference principle is applied [11]. Edge detection is another method used to identify and isolate the hand region and is also employed to obtain different direction edges [12, 13].

After the hand has been identified from the image, it needs to be tracked continuously to identify the gesture. The methods used for the same includes principal component analysis, active shape models, feature extraction and template matching. For efficient template matching, a database is maintained for all probable gestures and any new input is matched with the database to identify them. Here, features need to be extracted to go about with the match. The angle count of hand gesture images can also be matched using the threshold filter, which initially selects possible types of hand gestures. Next, skin color angle values and non-skin color angle values were matched through threshold selection in the same way which further narrowed the selection [14]. The final decision is the identified hand gesture that was obtained by matching the Hu invariant moments feature. Another method is to use linear regression to accurately find the actual number of templates to be used for each gesture [15].

Apart from template matching, feature extraction is also widely used for gesture recognition. Features extracted from images tend to reduce the volume of input data as it is a reduced set of information that are relevant and hence preserved. An implementation for gesture recognition in sign language has also been emphasized [16]. Here, the input data are preprocessed using color and 3D depth map to identify the hand. 3D combined features of location, orientation and velocity with respect to Cartesian and Polar systems are then used to identify the location of the limb. Additionally, k-means clustering was also employed for the Hidden Markov Model (HMM), which led to the identification of the hand gesture path using Left-Right Banded topology (LRB). Since the disparity in gesture signalling is variant, the transition between states is essential [12]. The system was first trained to recognize a single gesture and, subsequently, repeated experiments were able to recognize 20 different gestures. The feature extraction-based methods, however, achieve a greater accuracy, but they are found to be computationally expensive.

The active shape methodology applies the active statistical model for hand gesture extraction and recognition. The hand silhouettes are constructed by a real-time segmenting and tracking system. A set of feature points are established along the silhouette. Various shape contours can be generated to match the hand edges extracted from the original images. The gesture is finally recognized after rigorous matching [17]. Machine learning-based techniques have also attracted considerable attention in the domain [18, 19]. A Convolutional Neural Network does eliminate the need for explicit feature extraction, but it is computationally expensive as it requires a large amount of data for training and testing purposes [20]. Apart from the different methods used in the hand gesture system, an abundance of datasets are also available. Table 14.1 summarizes the datasets publicly available for gesture recognition.

14.3 PROPOSED SYSTEM OVERVIEW

The proposed hand gesture recognition system first acquires images by using a static camera, making it a vision-based system. The acquired video shot is divided into frames at regular intervals. Figure 14.1 summarizes the workflow of the proposed methodology.

TABLE 14.1
Hand Gesture Dataset Available

Datasets	Number of Images	Number of Individuals	Number of Gestures
HGR1 [18]	899	12	25
HGR2A [19]	85	3	13
HGR2B [20]	574	18	32
Massey Gesture dataset [21]	1500	5	6
Sebastien Marcel Static Hand [22]	4872	10	6

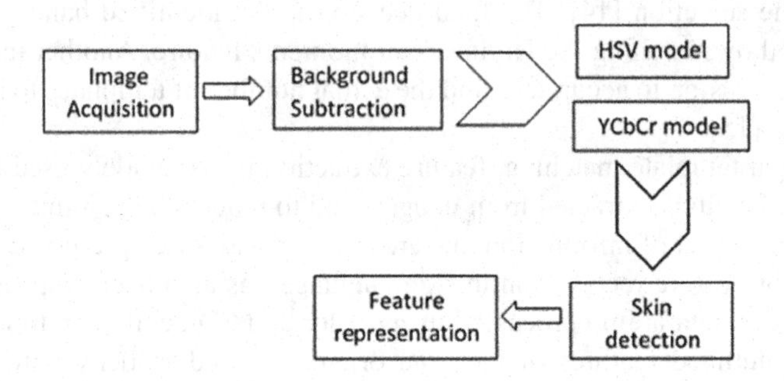

FIGURE 14.1 Proposed workflow.

The best hand segmentation results are achieved by using background subtraction and skin color detection. Each new frame in a captured shot is compared to a model against the scene background and hence helps to remove the recurrent body parts and also irrelevant background noise. The first input frame is treated as the background and subsequent frames then serve as the foreground. This method may work well for stationary devices as the background does not change drastically. In the case of moving devices, however, it is not generally appropriate to consider the first input frame as the background. To further solve this issue of any remaining noise, the skin color detection is done using the combination of RGB, HSV and YCbCr color space [23]. For the HSV calculation, the image after background subtraction is first normalized and H(Hue) S(Saturation) V(Value) is calculated as:

$$H = \begin{cases} 00^\circ, & \Delta = 0 \\ 60^\circ \times \left(\dfrac{G-B}{\Delta} \bmod 6 \right), & C_{\max} = R' \\ 60^\circ \times \left(\dfrac{B-R}{\Delta} + 2 \right), & C_{\max} = G' \\ 60^\circ \times \left(\dfrac{R-G}{\Delta} + 4 \right), & C_{\max} = B' \end{cases} \tag{14.1}$$

where R, G, B are the normalized Red, Green and Blue values, C_{max} represents the maximum of RGB, $\Delta = C_{max} - C_{min}$ and C_{min} is the minimum RGB value.

$$S = \begin{cases} 0, & C_{max} = 0 \\ \dfrac{\Delta}{C_{max}}, & C_{max} = 0 \end{cases} \tag{14.2}$$

and

$$V = C_{max} \tag{14.3}$$

YCbCr (Luminance and Chrominance) conversion is done as:

$$\begin{aligned} Y &= 16 + (65.481.R + 128.553.G + 24.966.B) \\ Cb &= 128 + (-37.797.R - 74.203 + 112.B) \\ Cr &= 128 + (112.R - 93.786.G - 18.214.B) \end{aligned} \tag{14.4}$$

The pixels in the image are classified as skin-coloured and non-skin-coloured if: (i) 140 Cr 165; (ii) Cb 195; (iii) 0.01 H 0.1. These values have been experimentally determined.

For feature extraction the regions that differ in properties, such as brightness or color, compared to surrounding regions, are detected. In these regions of an image the properties are constant or approximately constant; i.e. all the points in the region are considered to be similar. In order to detect such regions, the image is first binarized and a label image in initialized with no labels for any pixels. The image is then scanned to find the first non-zero pixel, which has no label. A label is created at that position in the labeled image. Next, all unmarked non-zero neighbors are marked as visited with the same label ID. This step is repeated for each neighbor and subsequently for the entire image. The labelled image shows the region of interest.

After the region is extracted, the next step is to extract features. The area of the region is the number of pixels of which it consists. This feature is often used to remove detected regions that are too small and that have been wrongly accepted since it may be skin coloured. A bounding box is constructed along the regions detected considering the maximum area covered and the others are discarded. The area feature is used to distinguish compact regions from non-compact ones. For example, a fist when compared with a hand with outstretched fingers.

14.4 RESULTS

The proposed method has been efficiently tested on the standard available datasets (Table 14.1). Since most of the images in these datasets are taken in an ideal background with almost no variations in lighting, illumination or background, the proposed method has also been tested on a few live images. These images are taken in a room with bright illumination so as to prove the efficacy of the proposed method. The

FIGURE 14.2 Dataset created for experiment.

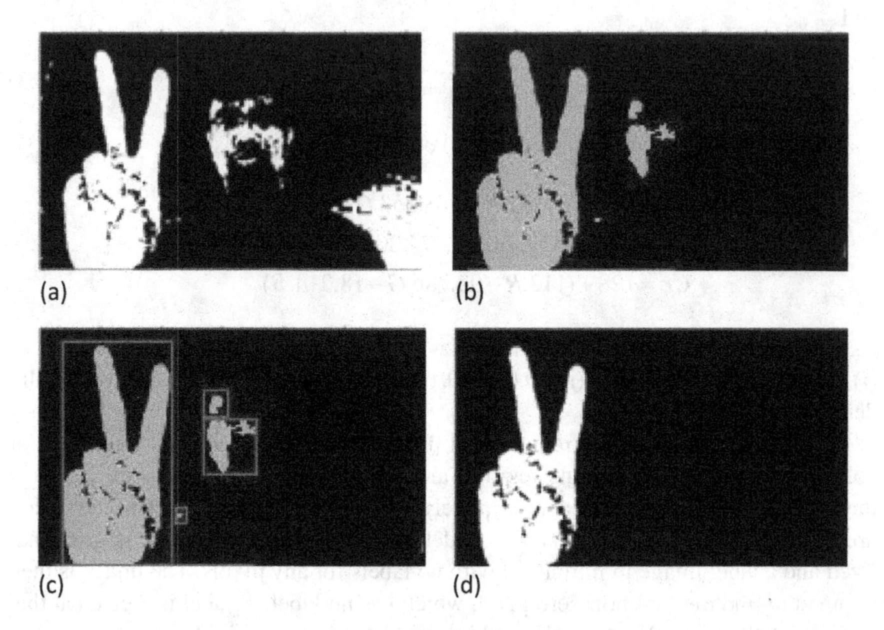

FIGURE 14.3 Segmentation Results; (a) Image after background subtraction; (b) Image after skin color detection; (c) Image with region of interest in bounding boxes; (d) Final segmented image.

results presented in this section show the step-by-step output of each stage in the segmentation process. The results included here are for one such image (Figure 14.2).

The visual outputs in Figure 14.3 depicts the different stages of the segmentation. Assuming the first captured frame in the input video as the background image, the first step is background subtraction (Figure 14.3(a)). The skin detection is then done (Figure 14.3(b)). The final segmentation result is then obtained (Figure 14.3(d)) after eliminating the smaller bounding boxes based on the area of the boxes around the region of interest.

14.5 CONCLUSION

As discussed above, a Human–Computer Interaction system can receive input using either a glove-based or a vision-based technique. The vision-based technique are generally easy to implement and portable and often there is no requirement for any specific or special hardware. A segmentation approach to identify hand gestures has been discussed in this work. The accurate segmentation achieved by the method is visually evident through the results provided. In addition, the simple implementation of the proposed work makes it an efficient choice for implementation in smaller hand-held devices.

REFERENCES

1. https://www.extremetech.com/computing/98287-from-punchcards-to-ipads-the-history-of-input-devices
2. Sturman, D.J. and Zeltzer, D., 1994. A survey of glove-based input. *IEEE Computer graphics and Applications*, 1, pp. 30–39.
3. Zhang, X., Chen, X., Li, Y., Lantz, V., Wang, K. and Yang, J., 2011. A framework for hand gesture recognition based on accelerometer and EMG sensors. *IEEE Transactions on Systems, Man, and Cybernetics-Part A: Systems and Humans*, 41(6), pp. 1064–1076.
4. Wang, R. Y. and Popovic, J., 2009. Real - time hand - tracking with a color glove. *ACM Transactions in Graphics*, 28(3).
5. Rautaray, S.S. and Agrawal, A., 2015. Vision based hand gesture recognition for human computer interaction: a survey. *Artificial Intelligence Review*, 43(1), pp. 1–54.
6. Wachs, J.P., Kolsch, M., Stern, H. and Edan, Y., 2011. Vision-based hand-gesture applications. *Communications of the ACM*, 54(2), pp. 60–71.
7. Horprasert, T., Harwood, D. and Davis, L.S., 1999, September. A statistical approach for real-time robust background subtraction and shadow detection. *IEEE ICCV*, 99(1999), pp. 1–19. Citeseer.
8. Kolkur, S., Kalbande, D., Shimpi, P., Bapat, C. and Jatakia, J., 2017. Human skin detection using RGB, HSV and YCbCr color models. arXiv preprint arXiv:1708.02694.
9. Bretzner, L., Laptev, I. and Lindeberg, T., 2002, May. Hand gesture recognition using multi-scale color features, hierarchical models and particle filtering. In Automatic Face and Gesture Recognition, 2002. Proceedings. Fifth IEEE International Conference on (pp. 423–428). IEEE.
10. Al-Tairi, Z.H., Rahmat, R.W.O., Saripan, M.I. and Sulaiman, P.S., 2014. Skin segmentation using YUV and RGB color spaces. *JIPS*, 10(2), pp.283–299.
11. Elgammal, A., Harwood, D. and Davis, L., 2000, June. Non-parametric model for background subtraction. In European conference on computer vision (pp. 751–767). Springer, Berlin, Heidelberg.
12. Chen, F.S., Fu, C.M. and Huang, C.L., 2003. Hand gesture recognition using a real-time tracking method and hidden Markov models. *Image and Vision Computing*, 21(8), pp.745–758.
13. Yang, M.H. and Ahuja, N., 2001. *Face detection and gesture recognition for human-computer interaction* (Vol. 1). Springer Science and Business Media.
14. Yun, L., Lifeng, Z. and Shujun, Z., 2012. A hand gesture recognition method based on multi-feature fusion and template matching. *Procedia Engineering*, 29, pp.1678–1684.
15. Carrera, K. C. P., Erise, A. P. R., Abrena, E. M. V., Colot, S. J. S. and Telentino, R. E., Application of template matching algorithm for dynamic gesture recognition of American sign language finger spelling and hand gesture. *Asia Pacific Journal of Multidisciplinary Research*, 2(4), pp.154–158, 2014.
16. Elmezain, M., Al-Hamadi, A., Pathan, S. S. and Michaelis, B., "Spatio-temporal feature extraction - based hand gesture recognition for isolated American Sign Language and Arabic numbers," in 6th International Symposium on Image and Signal Processing and Analysis, 2009.

17. Liu, N. and Lovell, B. C., Hand gesture extraction by active shape models. *Digital Image Computing: Techniques and Applications*, 2005.
18. Oudah, M., Al-Naji, A. and Chahl, J., 2020. Hand gesture recognition based on computer vision: a review of techniques. *Journal of Imaging*, 6(8), p. 73.
19. Parvathy, P., Subramaniam, K., Venkatesan, G.P., Karthikaikumar, P., Varghese, J. and Jayasankar, T., 2021. Development of hand gesture recognition system using machine learning. *Journal of Ambient Intelligence and Humanized Computing*, 12(6), pp.6793–6800.
20. Li, G., Tang, H., Sun, Y., Kong, J., Jiang, G., Jiang, D., Tao, B., Xu, S. and Liu, H., 2019. Hand gesture recognition based on convolution neural network. *Cluster Computing*, 22(2), pp.2719–2729.
21. Kawulok, M., Kawulok, J., Nalepa, J. and Smolka, B., 2014. Self-adaptive algorithm for segmenting skin regions. *EURASIP Journal on Advances in Signal Processing*, 2014(170).
22. Nalepa, J. and Kawulok, M., Fast and accurate hand shape classification, in *Beyond Databases, Architectures, and Structures*, S. Kozielski, D. Mrozek, P. Kasprowski, B. Malysiak-Mrozek, and D. Kostrzewa, Eds., vol. 424 of Communications in Computer and Information Science, pp. 364–373. Springer, 2014.
23. Grzejszczak, T., Kawulok, M. and Galuszka, A., 2016. Hand landmarks detection and localization in color images. *Multimedia Tools and Applications*, 75(23), pp.16363–16387.
24. Barczak, A.L.C., Reyes, N.H., Abastillas, M., Piccio, A. and Susnjak, T., 2011. A new 2D static hand gesture color image dataset for ASL gestures.
25. Marcel, S., Hand posture recognition in a body-face centered space. In Proceedings of the Conference on Human Factors in Computer Systems (CHI), 1999.
26. Shaik, K.B., Ganesan, P., Kalist, V., Sathish, B.S. and Jenitha, J.M.M., 2015. Comparative study of skin color detection and segmentation in HSV and YCbCr color space. *Procedia Computer Science*, 57, pp.41–48.

15 Medical Anomaly Detection Using Human Action Recognition

Mohammad Farukh Hashmi,
Praneeth Reddy Kunduru, Sameer Ahmed Mujavar,
Sai Shashank Nandigama, and Avinash G. Keskar
Department of Computer Science & Engineering,
NIT Warangal

CONTENTS

15.1 Introduction......215
15.2 Related Work......216
15.2.1 Keypoint Detection......216
15.2.2 Anomaly Detection......216
15.3 Technical Approach......217
15.3.1 Key Points Detection......217
15.3.2 Action Classification......218
15.3.3 Working of the Model......218
15.3.4 Optimizers and Training Process......219
15.4 Dataset and Experimentation......220
15.5 Conclusion......221
References......221

15.1 INTRODUCTION

Since the outbreak of Covid-19, it has become necessary to monitor medical anomalies in public places as these can prove beneficial in breaking the chain of spreading this disease. In general, automated processes are advantageous when compared with the existing primitive methods in terms of a number of factors: the amount of manual work required, their speed and their scalability [1].

In the existing primitive methods, a person has to manually call for medical assistance in the case of an emergency in public areas. When the large scale of such events are considered, there is a lot of manual work involved which can be reduced by using the automated process presented in this chapter. In our system we have used Human Action Recognition in order to detect anomalies in public areas.

Human Action Recognition is one of the challenging tasks in computer vision; it involves keypoint identification of the human body and the subsequent classification

DOI: 10.1201/9781003267782-15

of the action based on those keypoints. The aim of the project is to detect medical anomalies in public areas through the employment of advanced neural network architectures developed in the field of computer vision. Due to many critical real-life applications, the problem demands both high speed and accuracy, since the motion of a human can be described by the combined motion of his/her joints (or keypoints) [2]. The major challenges are broken into two parts: keypoints identification and anomaly detection through classification from the continuous stream of data. The existing methods for keypoint prediction employ a variety of architectures; a few of them have delivered state-of-the-art results in terms of both the high degree of accuracy and their speed in maintaining the resolution. Of all those models, HRNet [3] delivers state-of-the-art results, which is why we have adopted HR-Net in our project. The classification for a continuous stream of data is best solved by recurrent neural network-type architecture. The success of recently discovered transformer architecture is known for its attention mechanism, making it a suitable replacement for most of the standard recurrent neural network architectures.

Motivated by the success of transformers in other fields [4, 5] we decided to employ transformers for the anomaly classification.

15.2 RELATED WORK

15.2.1 Keypoint Detection

Highly accurate key point identification is generally achieved by many advanced neural network architectures, including CNN [6], R-CNN [7] Hour-glass [8], and Resnet-50 HR-Net architectures. Most of these architectures are held back by problems such as loss of information (Resolution), Occultation, substantial training time, and longer inference time.

Loss of information could lead to voids in keypoints, which cause errors in prediction. This can prove to be costly in the case of medical anomaly detection. HR-Net addresses this by maintaining high resolution throughout its process.

Occultation (that is, an event that occurs when one object is hidden by another object that passes between it and the camera) causes the loss of keypoints. This may result in null values in data, leading in turn to misclassifications. This can be addressed by using the HR-Net model.

HRNet-W32 [3, 1, 9, 10]: This is a state-of-the-art model, which achieves high performance by maintaining high resolution throughout the process with the help of parallel layers of high resolution and fusing the low resolution to maintain high resolution. It also uses a bottom-up approach which is very favorable for real-time detection as it is very fast.

15.2.2 Anomaly Detection

Traditional methods of anomaly detection for serial data involve using recurrent-type neural networks like RNN and LSTM [11] date back to 1997. RNN and LSTM have the capability of understanding the time-series data but they are both limited by their large computational cost, small memory windows and slow training speed. Transformers [20] are introduced as the models developed as better replacement for the RNNs to provide better performance driven by their attention mechanism which have delivered state-of-the-art results in sequential data classification applications like [4, 12, 13].

15.3 TECHNICAL APPROACH

The proposed system of our approach is as follows:

We use HR-Net to extract keypoints of the people present in the dataset and also the obtained key points are used to classify the action by using Transformer (Figure 15.1).

15.3.1 Key Points Detection

HR-Net helps to detect a person's keypoints. HR-Net does this through the adoption of a bottom-up approach. Since it is doing it in a bottom-up approach it can, in parallel, detect keypoints for up to 30 people in a frame. This makes it an exceptionally fast model for the detection of keypoints. HR-Net also overcomes an important problem that is encountered in most of the bottom-up approaches that we find in present-day models:, the problem with regard to the scale of the humans in a frame. HR-Net uses a scale-aware approach, which helps HR-Net to be able to detect keypoints accurately even if a person is far away from the observer or a person is very close to the observer. So even in a mix of both such cases HR-Net can predict accurate keypoints. HR-Net overcomes voids in keypoint detection by maintaining a high resolution throughout its pipeline. With regard to tasks such as Human Action Recognition, having voids in keypoints makes it highly difficult to predict the action; HR-Net helps to provide good data to the transformer model further down the pipeline. HR-Net also overcomes the problem of occlusion. When there are multiple people in a frame there is a high probability of people obstructing the view of the camera; this might hide some vital key points of other people, leading to gaps in the data. HR-Net was trained in such a way as to solve this problem. Thus we use HR-Net for keypoint detection.

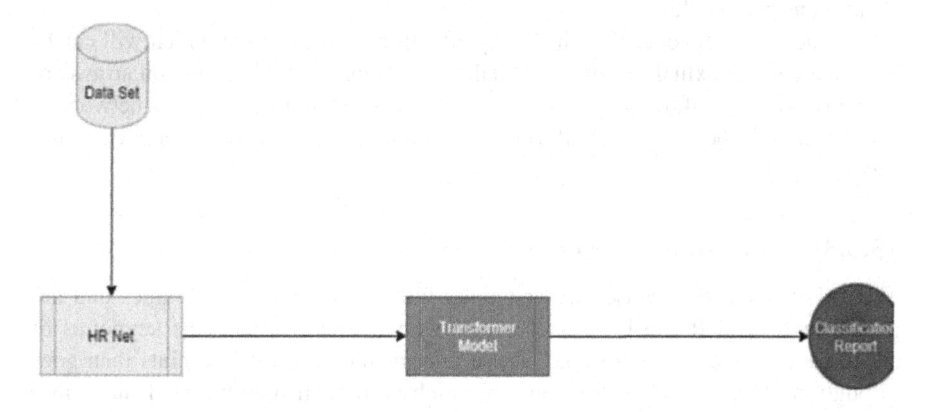

FIGURE 15.1. Architecture of the proposed model.

Here the input images are passed through HR-Net to obtain the keypoints of all the people present in the image. Once this has been done, the output of the HR-Net has a high-resolution representation of our keypoints and there are no voids in the data. This is then used as an input to the next phase, where we predict if any person is doing any anomalous action.

15.3.2 Action Classification

Sequence classification tasks are better handled by the recurrent-type neural networks. Many types of neural networks have evolved to handle the sequential data; the most famous of these include RNN-, LSTM- and GRU-based architectures. This also includes recent advancements in sequence-to-sequence translation in the field of Natural Language Processing using the novel transformer architecture. Because of the state-of-the-art results achieved and the possibility to deploy this system in real time, the transformer is considered as the replacement for many existing models in the handling of the sequential data. Due to the attention mechanism designed into Transformer the process can derive the contextual information and handle the sequential data. In this chapter we use the attention mechanism of the transformer to encode the sequential data and obtain the contextual information; this is then passed to a simple deep neural network for the classification of the data [14, 15].

Transformer: Transformer architecture is introduced as an efficient neural network for sequence translation in natural language processing. The introduced Transformer is basically divided into two blocks: an encoder stack and a decoder stack. In this project we only use the encoder stack to obtain the contextual information of the sequence. The encoder stack contains attention heads which are responsible for generating the contextual information.

Attention Mechanism: Attention mechanism is implemented using attention heads which compute the relevance of every vector (each point in the sequence is given as a vector of numbers) in the sequence with every other vector. At first three learnable vectors query(q), key(k) and value(v) vectors are computed by multiplying each vectors of the sequence with weight matrices the self-attention or relevance of each unit in sequence with other units will be calculated by computing scaled-dot-product attention as given below.

Encoder: each encoder contains many attention heads, each of which will calculate different contextual information finally all the contextual information arrays are merged and final information is obtained. A classification deep neural network is used to classify the contextual information obtained from the encoder into the action classes [16, 17].

15.3.3 Working of the Model

The videos from the dataset are converted into successive image frames and are passed onto the HR-Net [3] model which, as explained, extracts the keypoints of all humans present in each frame. These intermediate data of keypoints then goes through the Human Action Recognizer, which is implemented by the Transformer [4, 5], where the action is recognized based on the keypoints data. The actions are

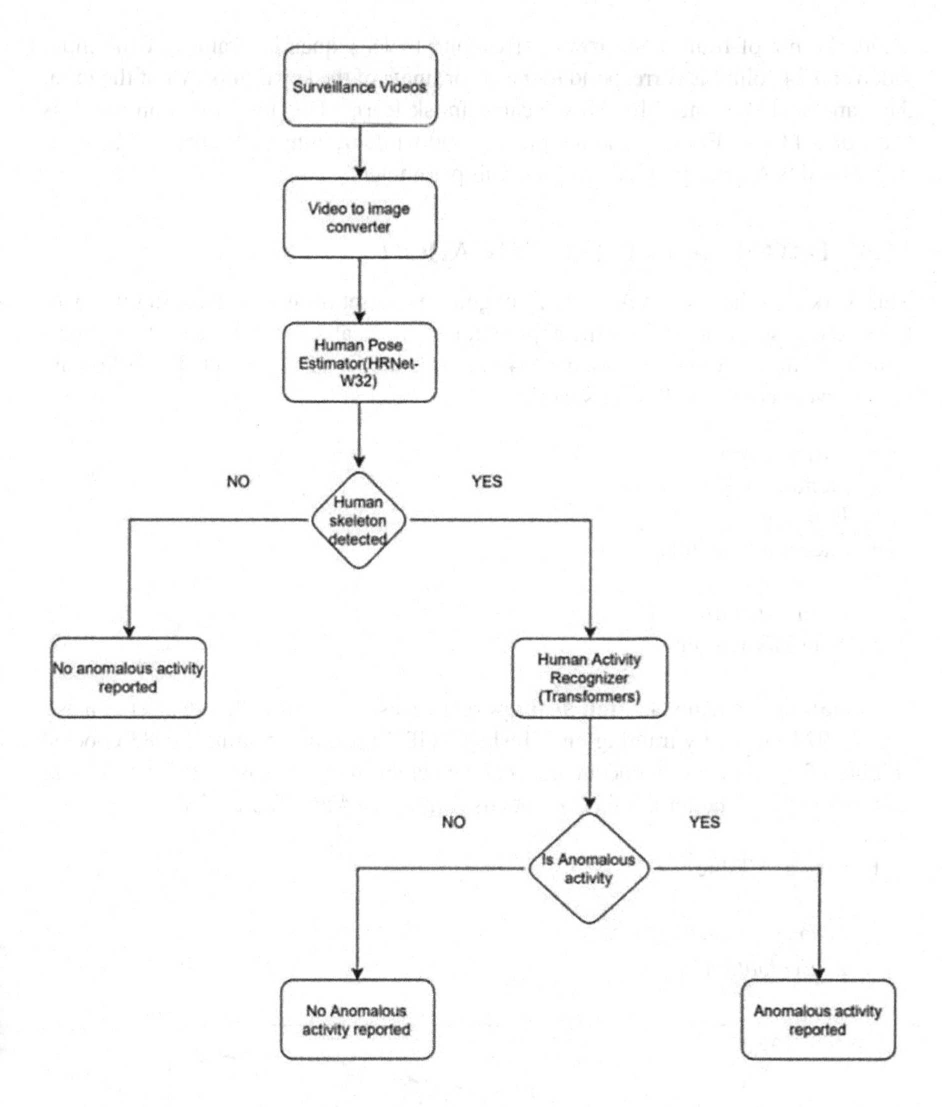

FIGURE 15.2 Flow chart of working of the model.

then classified based on the preset anomalous activities and the anomalous activities are reported whenever detected (Figure 15.2).

15.3.4 Optimizers and Training Process

The training process consists of two stages. The first stage is to obtain keypoints from the HR-Net. The next stage is training the Transformer model with the data acquired in the first stage.

The Transformer is trained to classify seven classes. In the input data corresponding to each output, the Transformer takes in a vector of shape [no. of frames × 34] where the no. of frames and rows correspond to 18 sequential frames of the input video and 34 columns correspond to the co-ordinate of the key points. All of the input data are scaled using Min-Max Scaler in sk-learn. The loss function used is Categorical Cross Entropy and the metric used while training is Accuracy. The optimizer used is Adam optimizer with default parameters.

15.4 DATASET AND EXPERIMENTATION

This work uses the Human Activity Recognition dataset from Rose labs. In this chapter we chose seven actions for the application of medical anomalous activity recognition. To train the system, we used 1974 videos from the same dataset. The following actions were considered for this work:

- Sitting down
- Standing up
- Jumping
- Sneezing/coughing
- Staggering
- Falling down
- Nausea/vomiting

A validation accuracy as high as 88 percent was achieved with a dataset consisting of 1974 videos by training on a Tesla T4 GPU for only 15 min. (for 83 epochs) (Figure 15.3). Even higher accuracies can be achieved by increasing the dataset size and running for a larger number of epochs (Figure 15.4 and Table 15.1).

Figures and Tables

1. *Performance metrics*
2. *Evaluation metrics*

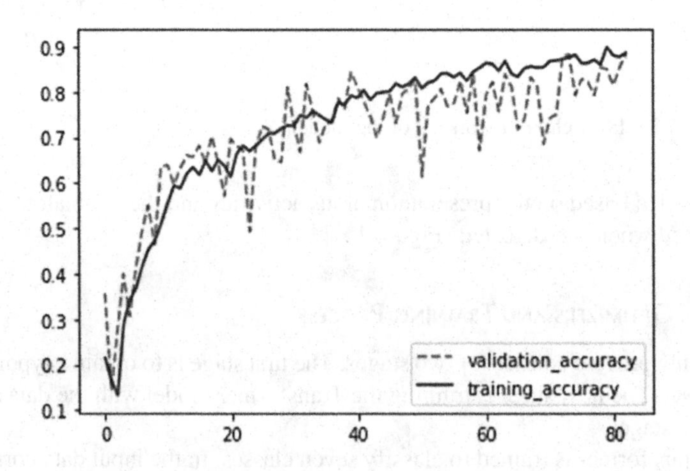

FIGURE 15.3 Training and validation accuracy.

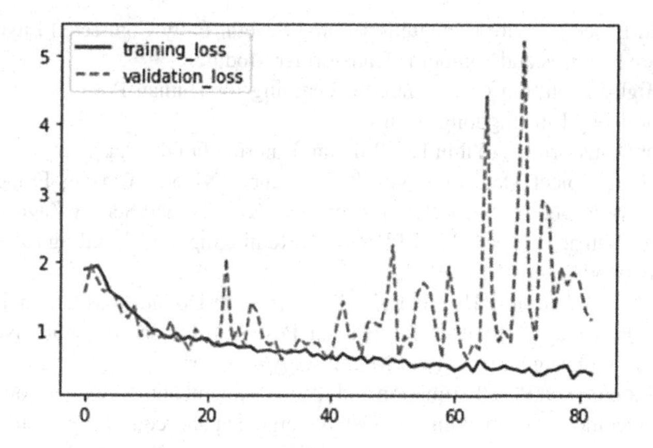

FIGURE 15.4 Training and validation losses.

TABLE 15.1
Precision, Recall, F1-score on the Rose Labs Validation Dataset

	Precision	Recall	Fl-score	Support
0	0.93	0.93	0.93	40
1	0.91	0.97	0.94	40
2	0.75	0.70	0.73	40
3	0.59	0.75	0.66	40
4	0. S3	0.72	0.77	40
5	1.00	1.00	1.00	40
6	0.76	0.65	0.70	40
accuracy			0.82	280
macro avg.	0.S2	0.82	0.82	280
weighted avg.	0.82	0.82	0.82	280

15.5 CONCLUSION

This work presents an automated solution for monitoring in public areas for any medical anomaly detection. The accuracy of the model is approximately 80 percent on the Rose Labs dataset. Our work adopts a novel approach to using Transformer architecture in the field of computer vision for sequence classification. This is a unique approach which no other work has advanced in the field of computer vision. This work is highly scalable in terms of expanding the set of action classes. The future scope for extending this work is to scale up this model for a set of a large number of actions and to implement the process in real-time applications.

REFERENCES

1. Distribution-Aware Coordinate Representation for HumanPoseEstimation (Feng Zhang, Xiatian Zhu, Hanbin Dai, Mao Ye, Ce Zhu).

2. Skeleton-Based Gesture Recognition Using Several Fully Connected Layers with Path Signature Features and Temporal Transformer Module.
3. Deep High-Resolution Representation Learning for Human Pose Estimation (Ke Sun Bin Xiao Dong Liu Jingdong Wang).
4. Epipolar Transformers (Yihui He, Rui Yan, Katerina Fragkiadak).
5. End-to-End Object Detection with Transformers Nicolas Carion, Francisco Massa, Gabriel Synnaeve, Nicolas Usunier, Alexander Kirillov, and Sergey Zagoruyko
6. Mingxing Tan and Quoc V Le. Efficientnet: Rethinking model scaling for convolutional neural networks. In ICML, 2019.
7. Mask R-CNN (Kaiming He, Georgia Gkioxari, Piotr Dollar, Ross Girshick).
8. Stacked Hourglass Networks for Human Pose Estimation Alejandro Newell, Kaiyu Yang, and Jia Deng University of Michigan, Ann Arbor.
9. PersonLab: Person Pose Estimation and Instance Segmentation with a Bottom-Up, Part-Based, Geometric Embedding Model (George Papandreou, Tyler Zhu, Liang-Chieh Chen, Spyros Gidaris, Jonathan Tompson, Kevin Murphy).
10. W. Yang, S. Li, W. Ouyang, H. Li, and X. Wang. Learning feature pyramids for human pose estimation. In The IEEE International Conference on Computer Vision (ICCV), volume 2, 2017.
11. Fundamentals of Recurrent Neural Network (RNN) and Long Short-Term Memory (LSTM) Network Alex Sherstinsky
12. Transformers in Vision: A Survey (Salman Khan, Muzammal Naseer, Munawar Hayat, Syed Waqas Zamir, Fahad Shahbaz Khan, and Mubarak Shah).
13. An Image is Worth 16x16 Words: Transformers for Image Recognition at Scale Alexey (Dosovitskiy, Lucas Beyer, Alexander Kolesnikov, Dirk Weissenborn, Xiaohua Zhai, Thomas Unterthiner, Mostafa Dehghani, Matthias Minderer, Georg Heigold, Sylvain Gelly, Jakob Uszkoreit, Neil Houlsby)
14. L. Ke, M.-C. Chang, H. Qi, and S. Lyu. Multi-scale structure-aware network for human pose estimation. arXiv preprint arXiv:1803.09894, 2018.
15. Y. Chen, C. Shen, X.-S. Wei, L. Liu, and J. Yang. Adversarial posenet: A structure-aware convolutional network for human pose estimation. In The IEEE International Conference on Computer Vision (ICCV), October 2017.
16. Human Pose Estimation via Improved ResNet-50 (Xiao Xiao, Wanggen Wand)
17. X. Chu, W. Yang, W. Ouyang, C. Ma, A. L. Yuille, and X. Wang. Multi-context attention for human pose estimation. CoRR, abs/1702.07432, 2017
18. Rethinking on Multi-Stage Networks for Human Pose Estimation (Wenbo Li, Zhicheng Wang, Binyi Yin, Qixiang Peng, Yuming Du, Tianzi Xiao, Gang Yu, Hongtao Lu, Yichen Wei, Jian Sun).

16 Architecture, Current Challenges, and Research Direction in Designing Optimized, IoT-Based Intelligent Healthcare Systems

B.S. Rajeshwari, M. Namratha, and A.N. Saritha
B.M.S College of Engineering, Bengaluru, India

CONTENTS

16.1 Introduction 223
16.1.1 IoT Integrated with a Cloud Computing-Based Healthcare System Basically Processes in Four Steps as Follows 224
16.2 Pros and Cons of IoT in Healthcare Intelligent System 225
16.2.1 Advantages of a Cloud IoT-Based Healthcare System 225
16.2.2 Limitations of an IoT-Based Intelligent Healthcare System 226
16.3 Applications of IoT in Intelligent Healthcare Systems 227
16.4 Current Challenges and Research Direction of IoT in an Intelligent Healthcare System 228
16.4.1 Current Challenges and the Research Direction of IoT in an Intelligent Healthcare System 229
16.4.2 The Research Background of IoT in an Intelligent Healthcare System 230
16.4.3 Hardware and Software Startups that Provide High-End Solutions for Current Healthcare Problems 231
16.5 Conclusion 232
References 232

16.1 INTRODUCTION

The rapid development in digital technological innovations has changed the integrated information management processes in all sectors. Digital technological innovation has generated a digital transformation, even in the healthcare sector, to optimize

DOI: 10.1201/9781003267782-16

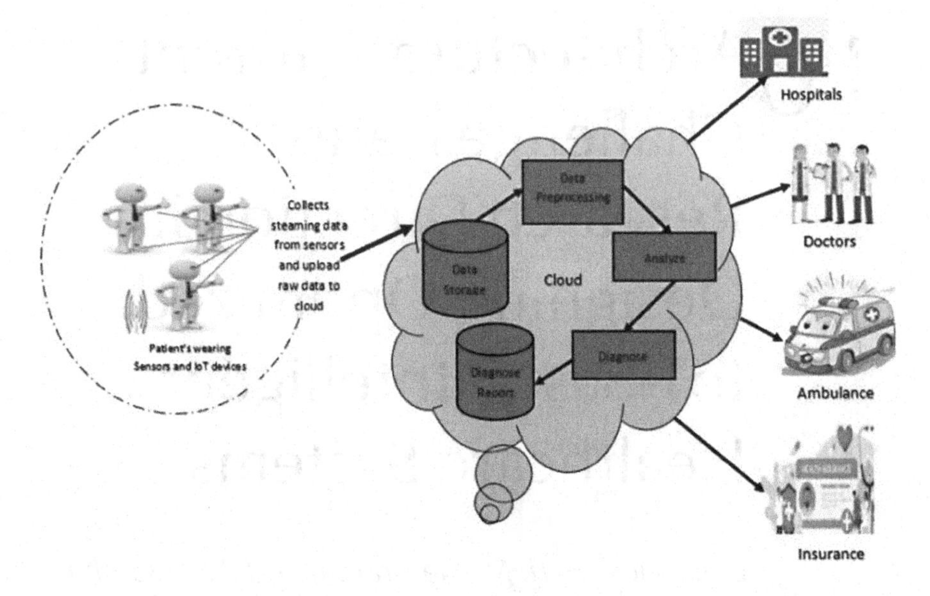

FIGURE 16.1 Architecture of IoT integrated with a cloud computing-based healthcare system.

healthcare management systems. Currently, IoT-based healthcare is a booming area in the healthcare system [1]. IoT technology has achieved great significance in the healthcare system, which enables systems to work in an integrated manner using sensors, integration methods, protocols, databases, cloud computing, and analytic algorithms [2]. The healthcare system, built upon the IoT, is very helpful for elderly in terms of monitoring their healthcare effectively [3]. For example, connectivity provided by the devices means that elderly people can stay safely in a single location. The combination of the IoT and cloud computing technology helps doctors deliberate with specialists all across the world, and to monitor patients' prolonged diseases. Thus, the integration of the IoT and cloud computing in the medical environment is a technological innovation in the health sector to optimize the healthcare management systems ensuring good care and quality treatment with reduced patients' costs [4].

Figure 16.1 shows the architecture of IoT integrated with a cloud computing-based healthcare system.

16.1.1 IoT Integrated with a Cloud Computing-Based Healthcare System Basically Processes in Four Steps as Follows

1. The interconnected devices, such as sensors, monitors, detectors, camera systems and other IoT devices, are deployed at the patient's bedside. These devices collect the data from a patient's body.
2. The raw data which are received from the sensors and other related IoT devices are accumulated and then moved to the cloud for further data processing.

3. These data are pre-processed, and analyzed using artificial intelligence (AI)-driven algorithms like machine learning (ML).
4. Finally, a report is generated based on the analysis which assists doctors and specialists by giving them the insights necessary for effective decision-making.

16.2 PROS AND CONS OF IOT IN HEALTHCARE INTELLIGENT SYSTEM

An IoT-based healthcare system works toward the rapid diagnosis of patients' disease, treatment of the disease at an early stage, recovery from the disease, and also monitoring the disease in place. It has both advantages and limitations.

16.2.1 Advantages of a Cloud IoT-Based Healthcare System

The key advantages of a cloud IoT-based healthcare system include [5, 6]:

➢ **Reduced Cost**: An IoT-based healthcare system empowers patient monitoring in real time, drastically reducing costly visits to doctors as well admissions to the hospitals. This makes testing more affordable. IoT-based homecare facilities allow patients to remain at home and to monitor their health status, which reduces hospital stays and re-admissions.

➢ **Remote Patient Monitoring**: Remote monitoring via connected IoT devices and smart alerts can diagnose illnesses, treat diseases and save lives of the patient in emergency cases such as heart attacks, variations in blood pressure and blood sugar levels, asthma attacks, etc. Using IoT healthcare solutions allows the remote tracking of medication and the patient's current status, meaning that the doctors are given better control over the treatment process. In the case of an emergency, patients may use a smartphone app to contact a doctor who may be many kilometers away.

➢ **Better Treatment**: IoT-based healthcare systems enable physicians and nurses to track the consumption of drugs, and response to the treatment, thereby reducing medical error. Cloud IoT-based healthcare systems support caretakers with the ability to access real-time data, which can be used to make well-informed decisions. IoT, combined with AI and ML, can offer multiple treatment options, since they have all the required patient medical data at hand.

➢ **Improved Healthcare Management**: Through the use of IoT devices, healthcare authorities can obtain information about the equipment in place and staff effectiveness. Sometimes even on normal days, hospitals have insufficient staff to look after all the patients under their care. The recent pandemic situation exacerbated this problem at times and highlighted weaknesses in the current healthcare systems regarding the lack of qualified specialists and the shortage of well-equipped rooms. Integrating IoT into the healthcare management system significantly accelerates and simplifies the process of monitoring patient admission, and can predict the arrival of patients during an epidemic period, meaning that patients need not be admitted into a healthcare setting since medical professionals can remotely access patient details, and diagnose and treat patients in their home setting.

- **Faster Disease Diagnosis**: Accessing real-time data through the sensors and continuous patient monitoring helps specialists to diagnose diseases at an early stage. On the basis of observed symptoms, specialists can identify any disease before it spreads and becomes more serious. Smart sensors examine person health conditions, and their lifestyle habits and recommend some preventative measures, which will reduce both the occurrence of diseases and also prevent and ailment from becoming serious.
- **Easy Maintenance of Drugs and Equipment**: The maintenance of drugs as well as the management of medical equipment is currently a challenging situation for the healthcare sector. The use of IoT devices can allow drug management to be done in an effective manner.
- **Medical Data Accessibility**: Data generated through IoT devices are moved to the cloud, where the received data can be analyzed through the application of ML algorithms and the generation of a report. This helps specialists to access data anywhere and at any time, leading to effective decision-making and ensuring smooth healthcare operations. With real-time data generated from the sensors and accessing them through the cloud, specialists and caretakers can carry out continuous monitoring of the patient's health condition.
- **Homecare**: A healthcare system built on the IoT supports the monitoring of patients who stay at homes. The patients' health is monitored using the sensors planted at the beds which send information. These data would be accumulated by the medical staff and a complete analysis is performed to check for the abnormalities of the patients and necessary action can be taken.
- **Better Medicine Adherence**: Under intelligent IoT-based healthcare management, caretakers have the remote access to find out whether or not the patient has consumed the necessary medicines/drugs; accordingly, they can also remind the patients.
- **Misconception due to the Human Factor**: Sometimes doctors and other medical specialists can reach incorrect conclusions. This may have serious consequences. With the IoT, this is largely avoided since AI and ML techniques, combined with human experience, enables specialists and doctors to get to the accurate diagnostic and make better decisions.

16.2.2 Limitations of an IoT-Based Intelligent Healthcare System

There are a number of limitations in an IoT-based Healthcare Intelligent System:

- The usage of the smartphones or wearable IoT devices can be overwhelming for older patients.
- During the tracking process, devices may not always be being by the end users.

- In those cases where the patient becomes unconscious, the devices may not be at all useful.
- Since the third party is involved in storing the patient's data, the patient may not be willing to provide their personal data.
- Battery life is not lengthy and hence they may run out of power during an emergency situation, meaning that any data stored or transmitted may be lost.
- Patients who reside in remote locations, where network coverage is a major problem, is an obstacle since the connectivity may not be properly established and thus messages may not be conveyed in a timely fashion.
- There has, as yet, been little testing of IoT devices in situations of extreme hot or cold

16.3 APPLICATIONS OF IOT IN INTELLIGENT HEALTHCARE SYSTEMS

IoT-based healthcare systems support hospitals and clinics to set up an environment where both patients and doctors can meet in a more portable way. The IoT also introduced a range of wearable devices which a person can wear more comfortably and safely. Some of the notable applications of IoT in the healthcare system include [7, 8]:

- Hearables
  Hearables are devices which provide aid for people who have difficulty in hearing and hence have limited interactions with the outside world. These devices are connected via Bluetooth and have to be synchronized with the smartphone which, in turn, amplify the sound and hence help people to suffer from hearing loss.
- Ingestible Sensors
  Ingestible sensors are used for irregularity detection within the human body and accordingly monitor the status of the body. These are used, in particular, by diabetic patients to check for symptoms and help in providing early warnings, hence these portable devices can be used to overcome critical issues.
- Moodables
  Moodables are wearable devices which send a low-intensity current to the brain of the person mounted on his head which, in turn, enhances their mood.
- Computer Vision Technology
  Computer vision technology helps in obtaining the patients' location data in the case of emergency and hence provides security to elderly people. Visually impaired people can navigate efficiently using such technology.
- Bedside Sensors
  These sensors monitor the patient's health while they are sleeping in bed and, when appropriate, send a warning to the medical staff. This is typically useful for patients who are unable to lift the phone and make a call during emergency.

➢ Insulin Pens
Patients suffering from diabetes can use these Insulin Pens to track the glucose level in their body. Any increase in the level can be reported over a dedicated mobile app. This report can be sent to their doctors and nurse for diagnose and get an improved treatment.

➢ Smart Video Pills
A smart pill travels through a patient's intestinal tract and takes a picture of their intestine. In the case of such pills, specialists and doctors can remotely view a patient's gastrointestinal tract and colon. It is in the form of a swallowable sensor that gives information about patients' stomach fluids.

➢ Temperature Sensors
Temperature sensors and disinfection systems ensure that food, blood, medications and other medical equipment are stored in a safer and secured manner.

➢ Occupancy Sensors
Occupancy sensors track the waiting areas in the hospitals and inform staffs to divert patients to other medical facilities as soon as the predefined capacity of the hospital is reached.

16.4 CURRENT CHALLENGES AND RESEARCH DIRECTION OF IOT IN AN INTELLIGENT HEALTHCARE SYSTEM

The existing healthcare management system is facing a number of problems, including the following:

➢ There is a requirement for specialized doctors at each local healthcare center.
➢ There is a requirement for digitized or latest medical equipment to record patient data and also generate necessary reports.
➢ There is limited time duration (a maximum of 48 hours) to receive any kind of report of the patient from the diagnostic laboratory or the hospital.
➢ There is a need to have Wi-Fi connection at local health centers and also to fund them properly.

A smart healthcare system based on recent technologies, such as the IoT, the Cloud, and based on AI and ML algorithms required to process a patient's data given the constraints of cost, security, and privacy. Over a remarkably short span of time, the IoT has brought about remarkable changes in the healthcare sector. Thanks to efficient data collection and management, IoT technology and Cloud technology in optimizing the healthcare system [9].

According to a recent business report, "The market for IoT healthcare technology will rise to $400 billion by 2022. Such growth will be due to the increasing demand, the improvement of 5G connectivity and IoT technology and the growing acceptance of healthcare IT software" [10]. According to another latest report from research and consulting firm Grand View Research, "The global healthcare sector will invest nearly $410 billion in IoT devices, software, and services in 2022, up from $58.9 billion in 2014" [11].

16.4.1 Current Challenges and the Research Direction of IoT in an Intelligent Healthcare System

Although the IoT has led to major advances in the healthcare system, there are still a few challenges that need to be addressed. The current challenges in the IoT-based healthcare system are as follows [12, 13]:

- ➢ **Security and Privacy**: Security and privacy is the major concern which prevent users from using IoT and cloud technology for healthcare and medical purposes, as the potential currently exists to breach or hack the person's sensitive information. It is still very challenging for companies in the healthcare sectors to secure large amounts of patient. In addition, there is significant ambiguity in data ownership regulation with the electronic devices. All these factors make the data highly susceptible to cybercriminals who can hack into the system and can misuse patient's health record, creating fake IDs and buying drugs and medical equipment which they can sell on at a later stage. Further, cybercriminals can hack patient data and file fraudulent insurance claims. Thus, providing security for the large repositories of patient data, and accessing patient data remotely in a more secure way is very much needed in order to secure complete success in cloud IoT-based healthcare sectors.
- ➢ **Risk of Failure**: Faulty sensors, anomalies, outliers and misconnected equipment results in risk for any healthcare operations. Apart from the above-mentioned issues, scheduling software updates must be taken care of, or else the situation is more dangerous compared to missing out on regular doctor's appointments. Thus, it is essential that future research that addresses issues such as identifying anomalies, outliers, and faults in the sensors and alerting the concerned parties, taking care of regular software updates, and achieving the optimized connection of equipment.
- ➢ **Integrating Multiple Devices**: Multiple IoT devices connected together to work in an integrated fashion may result in an interruption in the deployment of the IoT in the healthcare domain. This is because IoT devices are formed by different manufacturers and may therefore not work in cooperation with each other as there is no standard in IoT protocols. This causes variation in the behavior of the device and hence eventually reduced the scope of IoT in healthcare. In order to achieve the maximum impact of IoT in healthcare, it is very much required that we should seek to develop a framework which integrates the devices seamlessly, controlling and automating the whole process in the cloud IoT-based healthcare sector.
- ➢ **Time-consuming and Expensive to Implement**: An IoT-based healthcare system empowers real-time patient monitoring, drastically reducing the

need for costly visits to doctors as well as hospital admissions. However, the cost of implementation in hospitals is high and extra efforts will also need to be put into increased staff training. Thus, adopting a new cloud-based IoT architecture is, at present, one of the major challenges within the health domain.

Shifting the whole current facility in the hospital and clinic to an entirely new system consumes time, and the cost of investment is high, especially for smaller healthcare facilities and rural clinics. What is required, therefore, is a cost-effective framework for the deployment of IoT devices and equipment in hospitals and clinics.

- **Data Overload and Management**: One of the greatest challenges in the IoT healthcare system is with respect to the collection and management of the appropriate data. IoT devices produce a large amount of data and aggregating these data is tedious due to the use of dissimilar communication protocols and standards. However, the data generated are so tremendous that it is sometimes very difficult to derive insights from it, which ultimately comes down to the quality of the decision-making. This will eventually have an impact on patient health. Hence there is a need for the optimization of data collection and aggregation and for well-developed AL and ML algorithms to derive insights from it and generating informed reports.

In addition, research work needs to be done on:

- Innovations in sensor technology and integration;
- Establishing an intelligent IoT-based healthcare network;
- High-performance cloud computing;
- Advanced analytical software that can analyzes and generate reports in an optimized way.

16.4.2 The Research Background of IoT in an Intelligent Healthcare System

Kashif Hameed et al. [14] proposed an IoT- and ML-based intelligent healthcare system that was able to sense and process a patient's data through a medical decision support system. The proposed system is a low-cost solution for people in remote areas to provide health facilities through remote contact with local health facilities.

Uslu et al. [15] discussed optimization factors, challenging issues, currently available technologies, and also the infrastructure layers employed in IoT-based smart hospital environments.

Chaudhury et al. [16] proposed systems that offer continuous monitoring of patient health. In the event of an abnormality or any emergency and alert, such systems contacted staff, thereby ensuring the confidentiality and security of the patient. Temperature sensors, motion sensors, and pulse sensors are among the devices deployed in the proposed system.

Parthsarathy et al. [17] outlined a framework for monitoring a patient who is suffering from arthritis. The proposed framework works through the integration of three different processes: it collects the data from sensors; it then stores the collected data in the cloud; finally, the third level optimizes the collected information, detailing the levels of uric acid and C-reactive protein.

16.4.3 Hardware and Software Startups that Provide High-End Solutions for Current Healthcare Problems

The usage of IoT devices in healthcare is not limited to just smartwatches that detect heartrate and so on. There are also several other high-end solutions available which are handy to use and assist the patients in monitoring their health conditions.

➢ **Cardiomo**:
This is a wearable device which is used to monitor the health conditions of the body at regular intervals using built-in sensors. Tracking is carried out on biometric parameters, such as temperature, pulse, blood pressure, and so on, to give an accurate tracking of the person's overall health.

➢ **Elvie Pump**:
This is a device used by new mothers which proves useful in monitoring milk volumes, browsing the pumping history, and so on. This is a smart breast pump which simplifies our job since the breast pumping process is automated and no physical efforts are required.

➢ **Smart Hospital**:
This is a virtual hospital where the data of the patients from various wards are collected, gathered and monitored. Metrics such as heart rate, breathing rate, and so on is analyzed from the recorded data and the treatment process is adapted based on assessed changes or additional requirements.

➢ **Aira**:
This is solution designed specially for the visually impaired to assist them with movement around various places in the city. The glass sensors embedded have to be touched three times, at which point a member of the Aira support team will be connected. They will receive pictures of the current location as well as the patient history and accordingly guide them in the case of an emergency.

➢ **Amiko Smart Respirator**:
This device records details about the breathing rate of a person, analyzes these data and accordingly sends it to the doctor. This information can be useful for the treatment of the patient based on the ailments.

➢ **RapidSOS**:
This is a mobile application which includes the person's health profile data. This can be, in turn, connected to wearable devices or smart home security to alert the user in the case of an emergency or alternatively to call an ambulance. This is generally known as the life saver mode of operation, which

that contains all of the essential information, such as health profile, live incident data, and so on.

- **Thync**:

  Thync is a hardware startup that elevates a person's mood. It consists of two wireless devices: Calm, which helps to relieve person stress; and Energy, which helps a person to "recharge". Thync is capable of sending impulses to the brain. This affects the mood of the person and alters the mood in the case of either extreme excitement or extreme sadness.

16.5 CONCLUSION

The chapter elaborates on architecture of a cloud-based IoT intelligent healthcare system. There has been some discussion of the applications, and also the advantages as well as the limitations in IoT-based healthcare system. The current research challenges and direction toward designing an optimized IoT-based healthcare systems is also explained. This gives an insight for researchers, designers, and professionals in designing the best and most suitable IoT-based healthcare systems.

REFERENCES

1. https://www.intellectsoft.net/blog/iot-in-healthcare/
2. Uslu, Banu Çalıs, Okay, Ertug, Dursun, Erkan (2020), "Analysis of Factors Affecting IoT-Based Smart Hospital Design", *Journal of Cloud Computing: Advances, Systems and Applications*, 9 (1), (pp. 1–23), https://doi.org/10.1186/s13677-020-00215-5.
3. https://www.iotforall.com/5-challenges-facing-iot-healthcare-2019.
4. Liu, Yu, Beibei Dong, Benzhen Guo, Jingjing Yang, Wei Peng (2015), "Combination of Cloud Computing and Internet of Things in Medical Monitoring Systems", *International Journal of Hybrid Information Technology*, 8 (12), (pp. 367–376).
5. Pathan, Sana, Lad, Rashmi (2020), "Importance of Cloud Computing and Internet of Things in Healthcare Systems", *International Journal of Engineering Research & Technology*, 8 (5), ISSN: 2278-0181.
6. Yassein, Muneer Bani, Ismail Hmeidi, Marwa Al-Harbi, Lina Mrayan, Wail Mardini, Yaser Khamayseh (2019), "IoT-Based Healthcare Systems: A Survey." In *Proceedings of the Second International Conference on Data Science, E-Learning and Information Systems*, (pp. 1–9), DOI: https://doi.org/10.1145/3368691.3368721.
7. Kashani, Mostafa Haghi, Madanipour, Mona, Nikravan, Mohammad, Asghari, Parvaneh, Mahdipour, Ebrahim (2021), "A Systematic Review of IoT in Healthcare: Applications, Techniques, and Trends", *Journal of Network and Computer Applications*, (192), https://doi.org/10.1016/j.jnca.2021.103164.
8. Kulkarni, Alok, Sathe, Sampada (2014), "Healthcare Applications of the Internet of Things: A Review", *International Journal of Computer Science and Information Technologies*, 5 (5), (pp. 6229–6232).
9. https://www.iotforall.com/5-challenges-facing-iot-healthcare-2019.
10. https://www.iotforall.com/iot-healthcare-advantages-disadvantages.
11. https://www.businessinsider.com/the-global-market-for-iot-healthcare-tech-will-top-400-billion-in-2022-2016-5?IR=T.

12. Anmulwar, Sweta, Gupta, Anil Kumar, Derawi, Mohammad (2020), "Challenges of IoT in Healthcare", *IoT and ICT for Healthcare Applications, EAI/Springer Innovations in Communication and Computing book series*, Springer, (pp. 11–20), https://doi.org/10.1007/978-3-030-42934-8_2.
13. Selvaraj, Sureshkumar, Suresh Sundaravaradhan (2021), "Challenges and Opportunities in IoT Healthcare Systems: A Systematic Review", *SN Applied Sciences*, 2 (1), (pp. 1–8).
14. Hameed, Kashif, Imran Sarwar Bajwa, Shabana Ramzan, Waheed Anwar, Akmal Khan (2020), "An Intelligent IoT Based Healthcare System using Fuzzy Neural Networks", *Scientific Programming*, 2020, (pp. 1–15), Article ID 8836927, https://doi.org/10.1155/2020/8836927.
15. Uslu, Banu Çalış, Ertug Okay, Erkan Dursun (2020), "Analysis of Factors Affecting IoT-Based Smart Hospital Design", *Journal of Cloud Computing*, 9 (1), (pp. 1–23), https://doi.org/10.1186/s13677-020-00215-5.
16. Crowley, ST, Belcher, J, Choudhury, D, Griffin, C, Pichler, R, Robey, B, Rohatgi, R, Mielcarek, B (2017), "Targeting Access to Kidney Care via Telehealth: The VA Experience", *Adv Chronic Kidney Dis*, 24 (1), (pp. 22–30), https://doi.org/10.1053/j.ackd.2016.11.005.
17. Parthasarathy, P, Vivekanandan, S (2020), "A typical IoT architecture-based regular monitoring of arthritis disease using time wrapping algorithm", *International Journal of Computers and Applications*, 42 (3), (pp. 222–232), https://doi.org/10.1080/1206212X.2018.1457471.

17 Wireless Body Area Networks (WBANs) – Design Issues and Security Challenges

Jyoti Jangir and Khushboo Tripathi
Amity University Haryana Gurgaon, India

Deepshikha Agarwal
Indian Institute of Information Technology Lucknow, India

Abhishek Jain
Amity University Haryana Gurgaon, India

CONTENTS

17.1 Wireless Body Area Network Introduction 235
17.2 WBAN Architecture 237
17.3 WBAN Security and Privacy Requirements 238
17.4 Security Threats in Wireless Body Area Networks 239
 17.4.1 WBAN Current Measures for Data Security Which Are Important and Not to Be Ignored 239
17.5 Future Implementation for an Efficient Wireless Body Area Network 240
 17.5.1 Types of Attacks 242
17.6 Conclusion 242
References 243

17.1 WIRELESS BODY AREA NETWORK INTRODUCTION

The concept of body area sensors was firstly developed in the 1990s at the Massachusetts Institute of Technology. In the early stages of its development, the hypothesis was essentially based on attaching electrical devices to the human body for the purposes of medical rehabilitation and patient monitoring

The existing resources in the medical field will be insufficient to meet the demands of future patients, since affording a long-term stay in hospital is an impossibility for most patients due to their work culture, economic restrictions & other factors. At the same time, however, the health status of the patient must be monitored either in short-periodic time or in real time [1, 2]. Given this context, wireless monitoring will become an increasingly important part of healthcare in the near future.

DOI: 10.1201/9781003267782-17

A Wireless Body Area Network (WBAN) is built with the three following intelligent elements: Sensors, nodes, and actuators [3].

The work of the sensor node is to sense acoustic factors. These include: heart rate, ECG, blood pressure, pulse rate, sound, pressure and the temperature of the human body [4]. These sensor nodes are placed either subcutaneously or on the patient's clothes.

The WBAN involves two possible methods for data communication:

- Sensors – PDA (Personal Device Assistants) communication
- PDAs – base station communication

The connection between wearable devices and sensors is facilitated through the gateway node from the human to the Internet. Accordingly, the doctor can access patient data through the use of an internet connection. The consumption of high amounts of energy is the salient issue in WBAN because of the smaller size of the node. The importance of data security has increased to keep the data safe from being hacked during its transmission. At the same time, however, if we employ an effective clustering methodology and routing protocol that might reduce the security threats and energy consumption (Figure 17.1).

The WBAN consists of a personal device assistant, a transmission factor, parameters related to biology, a control unit and user access.

As shown in Figure 17.1, in WBAN, with the help of sensor senses the human body factors and get the biological information continuously from control unit. The ECG sensor [5] archives the electric impulse of the patient when it passes through the muscles of the heart. This assists the monitoring of the Patient-Heartbeat to track various movements such as moving, exercising, sleeping and resting.

The body temperature sensor [6] is used to detect the body temperature, i.e. forehead, ear, skin etc. The heart rate sensor is used to detect the pulse-wave and blood pressure during the pumping of the blood through patient body within arteries. The pulse-oximeter records oxygen saturation levels. In order to measure the body's respiration rate, an airflow sensor is positioned near to the nasal cavity.

A number of further processes take place during the process of data gathering:

1. PDA stores and collect information
2. Transmission of data to the base station
3. The transmission of data to the end user

The topology used in WBAN is known as star topology. Cloud computing allows patient data to be accessed by doctors from the server via the internet.

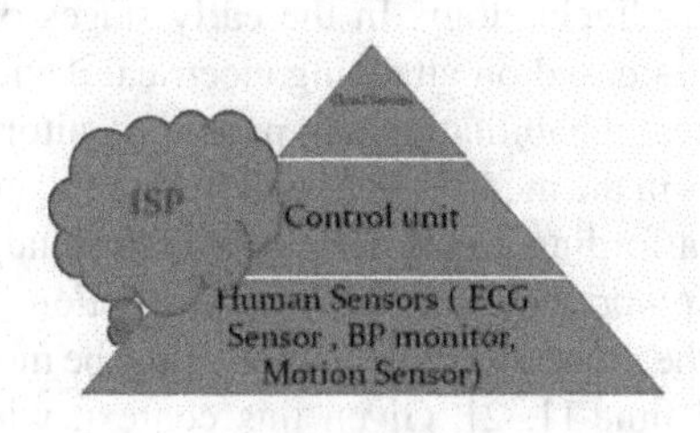

FIGURE 17.1 WBANs.

17.2 WBAN ARCHITECTURE

The wireless body area network is categorized into three different sections which are represented in Figure 17.2.

Intra-Wireless body area network communication:

- *Uses centralized design and star topology technology.*
- *Restrictions of coverage within a 2km range due to the connection between the sensors and a PDA*
- *ZigBee or Bluetooth is used for the transmission of data between the sensors and PDA.*
- *Here, the PDA is performing the role of coordinate node or centralized node to transmit and collect the data during transmission process (sensor to end users).*
- *The data communication is maintained by using external gateway with the help of Bluetooth.*

Inter-Wireless body area network communication:

- *Ad hoc architecture accompanying the connection between PDA and access points.*
- *Ad hoc architecture maintains a random topology, and is distributed to support the communication directly between nodes.*
- *Wireless devices and base station connected to each other, having a limited coverage range.*

Beyond Wireless body area network communication:

- *PDAs connects various networks by using gateway and are used either as coordinator nodes or as centralized nodes or coordinator nodes.*
- *The internet is used for the data communication and transmission between base stations and others such as cloud storage, doctor, ambulance, and the family members.*

By using a radio interface, the sensors in WBAN communicate and continuously monitor temperature, heart rate, blood pressure, respiration, motion sensor, EEG (Electroencephalogram), ECG (electrocardiogram), and blood oxygen level. As a

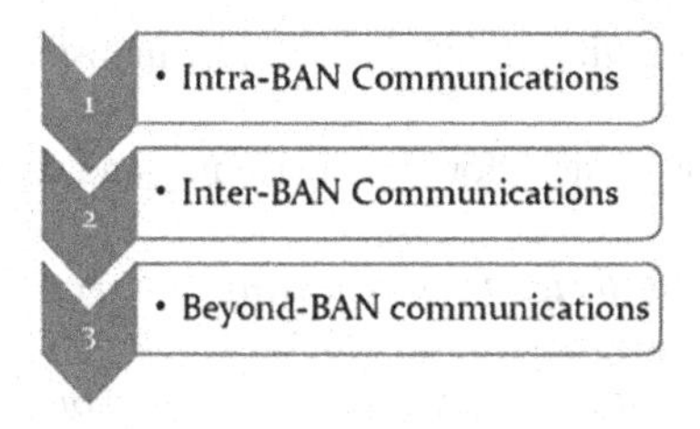

FIGURE 17.2 General Architecture of Wireless Body Area Networks (WBANs).

result of Covid-19, there have been considerable changes in human lifestyles and many aspects of healthcare have become increasingly digitalized.

Due to the social distancing guidelines introduced under Covid-19, humans are becoming accustomed to lockdown situations. Accordingly, WBANs are used for the communication of data to the doctor and patients can get an supervision and prescription through Cloud networks. In this manner WBANs have improved patients' experiences and reduced the costs of hospital stays.

The salient factors which we have to be concentrated during the implementation of WBAN are given by the following [7]:

- *Communication within short ranges*
- *Small-scale equipment*
- *Monitoring of patients on a 24/7 basis*
- *Confidential data storage*
- *Communication to stakeholders*
- *Due to real-time information, action can be taken quickly in emergency situations*

17.3 WBAN SECURITY AND PRIVACY REQUIREMENTS

In the context of patient information security, certain [8] measures have become matters of concern: these include data privacy, confidentiality and data integrity. During the implementation of WBAN, certain parameters should be taken care of which will guarantees all of the abovementioned features. Data security implies that the data should be protected from unauthorized users during the transmission of information. They should also be collected and stored safely in an unalterable state.

Data Exfiltration is the key concern, for illustration, if the patient confidential data shared among the insurance companies then they can use the leaked information to restrain themselves from coverage. The patient's data is extremely critical and sensitive; if leaked then it could lead to numerous negative consequences. On the other hand, if the intruder gets access to patient data then they can carry out alterations with the patient's confidential information. If passed on, the incorrect information can lead to the patient's demise.

Data privacy and security requirements that should kept in mind during the introduction of the WBAN system include the following:

1. **Data Confidentiality**
   - Patient Confidentiality of data [9] from being used by intruders.
2. **Data Integrity**
   - Protection of the content for its consistency and accuracy
3. **Data Freshness**
   - *Data are being protected from replaying and recording to maintain confidentiality and integrity.*
   - *It is important that the early provided data should not be recycled and that it should be provided in the correct format.*

4. **Availability of the network**
   - *The medical practitioner should have efficient access to patient information.*
   - *High availability of highly critical, sensitive and potentially lifesaving patient data should be ensured.*
5. **Data Authentication**
   - *A requirement for data authentication may help to confirm the authentication of the sender and the data.*
   - *The nodes used in WBAN should be capable of verifying the data sender i.e. it should be trusted and not an imposter.*
6. **Secure Management**
   - *Secure control is required during the decryption and the encryption of data*
7. **Dependability**
   - *The technique of the error-correcting code is used for the dependable and reliable retrieval of patient information during critical conditions.*
8. **Secure Localization**
   - *Securing the area of monitoring to find the location of the sensing devices.*

17.4 SECURITY THREATS IN WIRELESS BODY AREA NETWORKS

PII information of patients are critical issue as false information and alteration in data can lead to the patient coming to harm, or even dying.

It is also possible that a hacker can electronically damage and interfere in the data [10] during its transmission in order to acquire patient information. They can also exhaust the memory of the system by repeatedly sending information, which may overwhelm the system. In order to achieve this, they may use flooding techniques, such as DDoS.

17.4.1 WBAN Current Measures for Data Security Which Are Important and Not to Be Ignored

In the context of the privacy and security issues regarding data the following cryptographic techniques should be used to obtained efficient results such as high accuracy and reliability.

a. ***Bluetooth security protocols:***
   - *Various protocols included in it [11] : Logical Link Control and Adaptation, Link Manager Protocol and Baseband*
   - *The baseband is responsible for the connection between data exchanges in the form of packet and Bluetooth devices.*
   - *Encryption, decryption and authentication keys are taken care of by LMP.*
   - *The L2CAP support packets reassembly and multiplexing, which is responsible for service communication quality.*
   - *Nodes and MAC for the secure data packets during transmission.*

TABLE 17.1
Security Considerations in WBAN

Security Threats	Security Requirements	Possible Solutions
Unverified or unauthorized access	Verified or authorized access	Random key distribution Public key encryption
Information leakage	Confidentiality	Link layer or network layer encryption Access control
Tampering with message	Integrity	Type a secure hash function A digital signature
Deni a l-of-service attack (DoS)	Usability	Intrusion detection Redundant routing
Node capture, damaged nodes	The resilience of the damaged node	Consistency checking and node undo tamper-proof
Routing attacks	Secure Routing	Security routing protocol
Intrusions and advanced security attacks	Security group management, intrusion detection	Secure group communication, intrusion detection

b. ***Biometrics***
- *This method is used for communication using biometrics in the biomedical sensor. It uses the management of cryptographic keys of the sensors which are attached the human body.*

c. ***TinySec***
- *TinySec [12] is used for the authentication and encryption of patient data in the biomedical sensor-network.*
- *A group key is used for calculating the entire packet between sensors*

d. ***Wireless security protocols***
Wireless Equivalent Privacy (WEP) is the first protocol [13] used for Wi-Fi security based on a combination of user- and system-generated keys. This provides less security as hackers have found ways to breach the functions.

At present, Wi-fi protected Access (WPA) 2 and 3 are the most widely used systems because they can deal with several attacks, in contrast to WEP. This system involves a message integrity check, pre-shared key concept, cipher blockchaining and the simultaneous authentication of equals now makes it the most secure protocol.

e. ***Hardware Encryption***
The hardware equipment should be made secure so that access can be gained only by authorized personnel. Table 17.1 shows the security considerations in WBAN.

17.5 FUTURE IMPLEMENTATION FOR AN EFFICIENT WIRELESS BODY AREA NETWORK

The factors should be taken care in the near future during the implementation of an efficient wireless body area network.

Figure 17.3 shows the characteristics of a general WBAN network. These can be outlined as follows: interoperability, priorities, rapid communication, ease of use, simple design and constrained deployment.

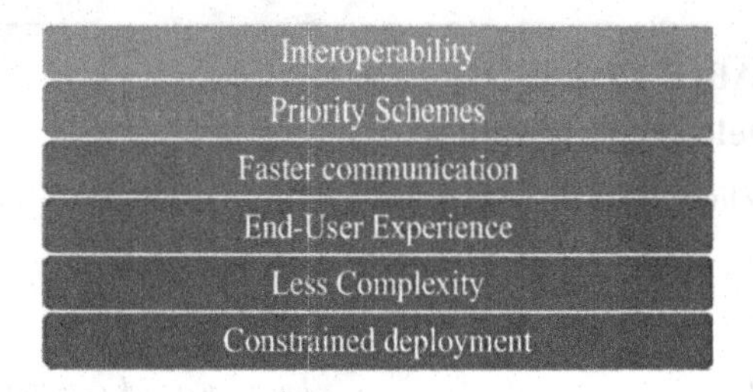

FIGURE 17.3 Characteristics of a WBAN network.

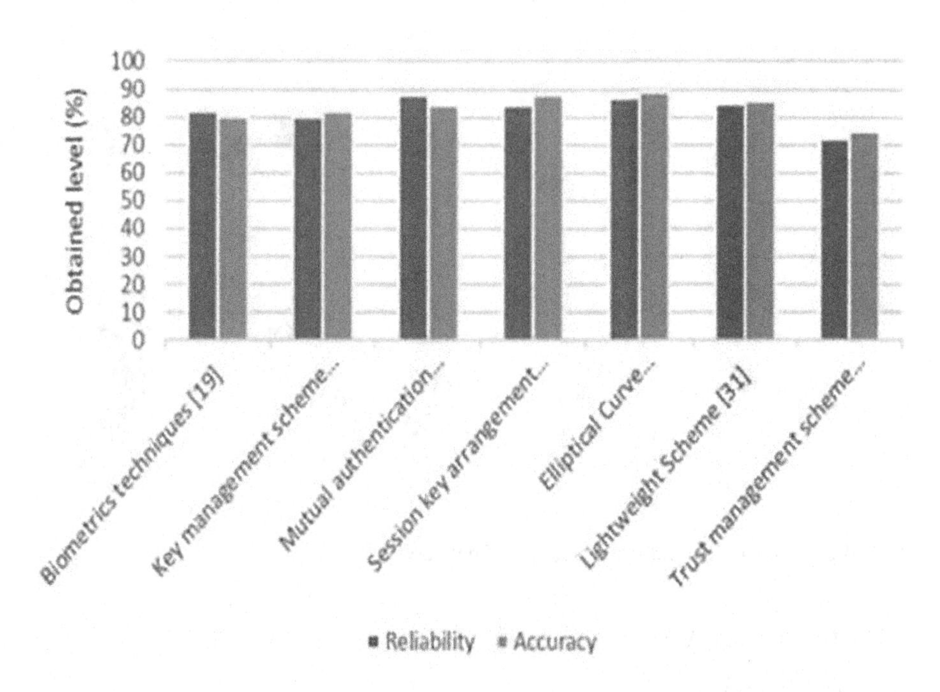

FIGURE 17.4 Cryptographic techniques – comparison.

The following factors should be taken care of in WBANs:

- *The scalability of the WBAN is improved with interoperability. The patient data need to be communicated with the doctors on a priority basis.*
- *The communication of information is based on the criticality of the patient.*
- *The communication should be more secure and faster.*
- *During network implementation, there should be less complexity.*
- *The data transmission should be successful and continuous.*

Detection techniques are being represented as follows. These are used to identify the attacker [14], which basically concentrates on the attacker's identification [10] and also a solution to overcome from the problem [15, 16], as shown in Figure 17.4.

TABLE 17.2
Delivery Rate Values

Delivery Rate (In Percentage)	Method
89	Low & high rate
72	DDoS attack
84	*Vulnerable attack*
72	*Jamming attack detection*
75	*Spoofing attacker detection*

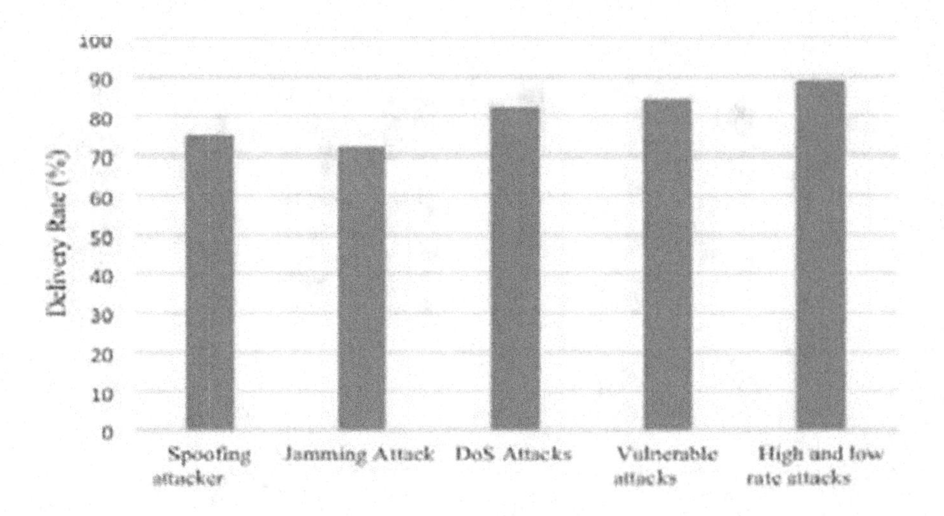

FIGURE 17.5 Type of attack versus rate of delivery.

17.5.1 Types of Attacks

- *DoS attacks* [17]
- *Vulnerable attacks* [18]
- *Spoofing attacker* [19]
- *High and low-rate attack* [20]
- *Jamming attack* [21, 22]

The following detection methods and their delivery rate is as follows in Table 17.2 as shown in Figure 17.5.

17.6 CONCLUSION

In WBAN, technology offers is a significant approach in the field of healthcare. The transmission of the patient information between the doctor and patient can be done without any disturbance by using this technology. The monitoring of e-health is an amicable scheme in biomedical applications. In WBAN, the system should be made efficient in terms of energy efficiency, privacy, network lifetime, security, rate of packet delivery and computational overhead. This discussion summarizes the WBAN challenges, issues, security limitations and developments. In this, many technologies are discussed, including DDoS attacks, cryptography, attacker detection, routing protocols, clustering schemes, and data privacy issues. The chapter has also outlined some of the issues regarding the future possibilities of such systems with regard to biological applications.

REFERENCES

1. D. Agarwal "Study of IoT and proposed accident detection system using IoT", *IOSR-JCE*, vol. 22, pp. 27–30, 2020.
2. S. Kumari, D. Agarwal, "Deployment of Machine Learning based internet of things networks for Tele-medical and remote healthcare", Springer Int. Conference on Evolutionary computing and mobile sustainable networks, Bengaluru, pp. 1–9, Sept. 2021.
3. CA Otto, E Jovanov, A Milenkovic., "A WBAN-based System for health monitoring at home", 3rd IEEE/EMBS international summer school and symposium on medical devices and biosensors, pp. 20–33, 2006.
4. D. Agarwal, K. Tripathi, A. Tyagi, "Combining the advantages of radiomic features based feature extraction and hyper parameters tuned RERNN using LOA for breast cancer classification", Biomedical Signal Processing and Control, pp. November 2021.
5. D. Agarwal and K. Tripathi, "Cyber and network security in IoT networks", Scholars book of world record, 2022.
6. D. H. Gawali and V. M. Wadhai, "Implementation of ECG sensor for real time signal processing applications," 2014 International Conference on Advances in Electronics Computers and Communications, pp. 1–3, 2014.
7. A Milenkovi, "Wireless sensors networks for personal health monitoring. Issues and an implementation", *Computational Communication*, pp. 2521–2533, 2006.
8. D Cavalcanti, "Performance analysis of 802.15.4 and 802.11e for body sensor network applications", 4th international workshop on wearable and implantable body sensor networks, Aachen, Germany. 2007
9. A. Jain and K. Tripathi, "Biometric Signature Authentication Scheme with RNN (BIOSIG_RNN) Machine Learning Approach", 3rd International Conference on Contemporary Computing and Informatics (IC3I), pp. 298–305,2018.
10. A. Jain and K. Tripathi, "Supervised AFRC (Ada boost fast regression) machine learning algorithm for enhancing performance of intrusion detection system", *International Journal of Engineering & Technology*, vol.7(4), pp. 5622–5628, 2018.
11. T. Sato, P. Moungnoul and M. Fukase, "Compatible WEP algorithm for improved cipher strength and high-speed processing," The 8th Electrical Engineering/ Electronics, Computer, Telecommunications and Information Technology (ECTI) Association of Thailand - Conference 2011, 2011, pp. 401–404.
12. S. Basu and M. Pushpalatha, "Analysis of energy efficient ECC and TinySec based security schemes in Wireless Sensor Networks," 2013 IEEE International Conference on Advanced Networks and Telecommunications Systems (ANTS), 2013, pp. 1–6.
13. T. Sato, P. Moungnoul and M. Fukase, "Compatible WEP algorithm for improved cipher strength and high-speed processing," The 8th Electrical Engineering/ Electronics, Computer, Telecommunications and Information Technology (ECTI) Association of Thailand - Conference 2011, 2011, pp. 401–404.

14. O. Jouini and K. Sethom, "Physical Layer Security Proposal for Wireless Body Area Networks," 2020 IEEE 5th Middle East and Africa Conference on Biomedical Engineering (MECBME), 2020, pp. 1–5, doi: 10.1109/MECBME47393.2020.9265157.
15. L. Mucchi, S. Jayousi, A. Martinelli, S. Caputo and P. Marcocci, "An Overview of Security Threats, Solutions and Challenges in WBANs for Healthcare," 2019 13th International Symposium on Medical Information and Communication Technology (ISMICT), 2019, pp. 1–6.
16. M. R. K. Naik and P. Samundiswary, "Wireless body area network security issues – Survey," 2016 International Conference on Control, Instrumentation, Communication and Computational Technologies (ICCICCT), 2016, pp. 190–194.
17. C. Jang, D. Lee and J. Han, "A Proposal of Security Framework for Wireless Body Area Network," 2008 International Conference on Security Technology, pp. 202–205, 2008.
18. K. Devisri, K. S. Indrani, A. L. Gayathri, K. Dedeepya, A. Roshini and M. Kommineni, "An Efficient hierarchical Routing Algorithm to Detect DoS in Wireless Body Area Networks," 2021 7th International Conference on Advanced Computing and Communication Systems (ICACCS), 2021, pp. 644–647, doi: 10.1109/ICACCS51430.2021.9442020.
19. S. K. Nagdeo and J. Mahapatro, "Wireless body area network sensor faults and anomalous data detection and classification using machine learning," 2019 IEEE Bombay Section Signature Conference (IBSSC), 2019, pp. 1–6.
20. S. Raguvaran, "Spoofing attack: Preventing in wireless networks," 2014 International Conference on Communication and Signal Processing, pp. 117–121, 2014.
21. N. Hoque, D. K. Bhattacharyya and J. K. Kalita, "A novel measure for low-rate and high-rate DDoS attack detection using multivariate data analysis," 2016 8th International Conference on Communication Systems and Networks (COMSNETS), pp. 1–2, 2016.
22. A. Bengag, O. Moussaoui and M. Moussaoui, "A new IDS for detecting jamming attacks in WBAN," 2019 Third International Conference on Intelligent Computing in Data Sciences (ICDS), pp. 1–5, 2019.

18 Cloud of Things
A Survey on Critical Research Issues

Adil Bashir and Saba Hilal

Islamic University of Science and Technology, Awantipora, India

CONTENTS

18.1 Introduction....245
 18.1.1 Delivery of Cloud Services....249
18.2 Integration Benefits of Cloud-IoT....249
 18.2.1 Benefits....250
 18.2.2 Applications of Cloud-IoT....251
18.3 Research Issues....255
18.4 Security Issues in Cloud-IoT....258
18.5 Conclusion....263
Acknowledgement....264
References....264

18.1 INTRODUCTION

Cloud Computing and the Internet of Things have both acknowledged an individualistic transformation. However, some mutual aspects have been identified in the literature as a consequence of their merger and further developments are anticipated in future. In particular, the Cloud provides a versatile tool for managing and designing IoT services, and even some applications that manipulate the information that they generate. From the other side, the Cloud will take advantage of the IoT by extending its purview to cope with issues in the actual environment in the most suitable and efficient manner, and to introduce new services in various real-life scenarios. In reality, the Cloud serves as an intermediary layer in between the program and the material, concealing all of the complexities and additional features that are required to execute the latter.

Internet of Things (IoT) functioning is focused on connected smart and self-configuring devices (things) in an evolving global network infrastructure. It is one of the largest discoveries allowing for inevitable and ubiquitous computing scenarios [1]. Typically, IoT is characterized as being composed of tiny items in the modern world, widely distributed with finite storage and processing capabilities. It generally focuses on issues such as efficiency, output, and privacy protection. On the other hand, Cloud

DOI: 10.1201/9781003267782-18

computing has substantially huge potential in terms of storing and processing power, and is a highly developed technology which helps the IoT to partially solve its problems. Consequently, the current as well as the future Internet should be transformed by a new IT paradigm that combines two complementary innovations. This model is called the Cloud-IoT. Evaluating the affluent and clear state of the art in the Cloud-IoT, both the concepts achieve reputation from some time, and only papers concerned with Cloud and IoT have seen an increasing pattern since 2008. Our key emphasis in this chapter is the convergence of Cloud and IoT and its stability, which is in reality a subject for both study and business, driven together by the growing attention toward Cloud and IoT [2].

The Internet of Things (IoT) has been an agent of the real world, because essentially each single system has sensors and actuators, being (uniquely) addressable and communicating together through the internet. Placing the IP stack together on embedded devices and adding IPv6 (which has incredibly broad addressing capabilities) enables the integration of both the digital and physical worlds, resulting in the rapid growth of the IoT. IoT systems enable users to conduct wide-ranging automation, review, and device integration. They boost the capabilities and productivity of these areas. The IoT takes advantage of current and advanced technology for sensing, networking, and robotics. Radio-frequency identification (RFID), Near Field Communication (NFC), Machine-to- Machine Communication (M2M) and Vehicle-to-Vehicle Communication (V2V) technologies in the market are used to implement the advanced concept of IoT [3].

The Internet of Things (IoT) involves the internet as the technological infrastructure that is used to collect sensed data from the physical world. In 2001 MIT Auto-IDC originally invented the IoT. The IoT describes a network of objects where each node (object) can be recognized and connected to the Internet through any means of communication and computing devices such as RFID, sensor, actuator, and mobile phone. In fact, artifacts can communicate with and connect to provide automated resources. Because of these embedded sensors, the standard and protection of life has increased in many fields where IoT technology includes omnipresent applications, including education, military monitoring, defense, transport, and logistics [4].

- *Total perception:* In fact, the Internet of Things encompasses many sensing technologies. Wherever and wherever feasible, the adoption of an RFID sensor, a two-dimensional code method for obtaining information on the subject, retrieves data in real time and continually updates data according to the received environmental information.
- *Reliable transmission:* The Internet of Things is an inevitable network collection, and the internet is the basic technology on which the Internet of Things relies. The Internet of Things transmits the knowledge reliably and effectively to the data centers in real time.
- *Intelligent processing:* The Internet of Things performs information retrieval, intelligent processing, and intelligent handling of information in addition to providing connectivity to the devices. The Internet of Things integrates sensors with intelligent processing to provide directions and

support for the implementation of smart decision-making and controlling of the facts by obtaining the useful information from the huge data retrieved for analysis and processing using Cloud computing, intelligent computing technology and fuzzy recognition.

Among envisioned IoT application areas are the following:

- Sustainable communities (i.e. sustainable parking, environmental safety, ambient noise charts, identification of devices, waste management etc.).
- Product fabrication.
- The automation of agriculture.
- Service management.
- Safety, monitoring, and supervision.
- Smart vehicles.
- Installation of safe ecological and technology homes (energy and water usage, remote control equipment, intrusion protection systems, art and products). Application of renewable and energy-effective homes (energy and water usage, remote control equipment, intrusion detection systems, art and preservation of products, etc.).
- Telemedicine and rehabilitation (fall prevention, medical freezer, sportsmen's treatment, patient monitoring, assessment of UV radiation).
- Product supervision and retail.
- Atmospheric protection (forest fire detection, air quality, avalanche and landslide avoidance, early notice of earthquakes).
- Smart agriculture and smart animal farming [5, 15].

The other important computing technology is known to be Cloud computing. This is a blend of emerging technologies that have merged their individual advantages to create a technological environment for Cloud computing. Cloud computing is often commonly used to explain how web-based software, networks, and storage services are delivered. Cloud consumers can benefit from certain organizations providing resources relevant to their files, applications and other computer needs on their behalf without having or operating on their own standard technical hardware (like servers) as well as applications (like email) [6]. Cloud computing marks the very next leap toward internet development, as it introduces a virtual network of elastic resources that incorporates on-demand provision of computing resources: Servers– Storage–Software–Services. Cloud computing allows users to perform the processing of complex data using virtual concepts. Virtualization means the creation of a virtual computer version or computer resource. This could be a device, a storage machine, a network or even an operating system. Devices, programs and human users may communicate with a machine object as though it were a completely special interactive device for them. Virtualization helps you "trick" the operating systems into believing that a server community is a single computing pool. All of this lets you operate simultaneous multiple operating systems on one device [7]. Virtualization is just the finest way to optimize the IT approach, because all permits, anti- viruses and system upgrades are carried out on one machine and not on another computer on the network. This not only ends up saving time and energy, but also saves money. The virtualization of the IT system is always the first step in a Cloud infrastructure approach. The following are the main reasons to use Cloud computing:

- Rising expenses to make recurring running expenditures more manageable.
- Allowing your staff to work from anywhere.
- Access data at any time, without any physical storage threats, because Cloud services manage this.
- Avoid complex disaster management plans, and enable Cloud service services to take control of things.
- Reach the broader class of technology, more developed competitors.
- Enable providers of Cloud services to operate your server for you, and free up your energy for more critical tasks.
- Boost the record management, enabling all the archives to operate from one main copy in one single location.

Cloud Computing architecture consists of five core functions, including resource pooling, immense network access, metered operation, on-demand self-service, and rapid elasticity. The Cloud model includes business models such as Infrastructure as a Service, Software as a Service, and Service Application. Cloud computing has also implemented frameworks such as Public Cloud, Private Cloud, Group Cloud, and Hybrid Cloud.

- *Software as a Service (SaaS):* This is defined as a situation where the service provider keeps the program and we do not need to install, maintain or purchase hardware. All you need to do is sign in, and use it. The service provider operates the device with SaaS, so you do not need to mount, support, or buy equipment to do so. Everything you have to do is sign in and use it. Among the examples of SaaS are the following: including managing customer relations (CRM) as a service; email; logistics software; order processing software; payroll software; and all other programs not physically mounted on the device, but accessible digitally. SaaS became and the way in which businesses continue their path into Cloud computing, typically beginning with the remote distribution of emails and the electronic storage of company details. This initially developed with Application Service Providers (ASP), which hosted and operated specialized business applications in the 1960s. Through a process of central administration they were minimizing costs [8–10].
- *Platform as a Service (PaaS):* The Software Platform is where the operating system (such as Windows, Android, BSD, iOS, Linux, Mac OS X and IBMz/OS) is held in the Cloud instead of having your own equipment installed physically. The PaaS layer provides simple mobile tools that programmers may use to create apps, in addition to their system infrastructure. These may involve development resources that are offered to create applications, data access and storage facilities, or payment systems as a company [8–10].

- *Infrastructure as a Service (IaaS):* Infrastructure as a service is a situation where space is leased and stored for physical servers in a vendor's data warehouse. Mostly as customer, users could indeed configure any legal software on your server, and as you see fit, allow it to be accessed as your staff and clients. The IaaS layer provides space and computer resources which the developers and IT organizations can use to deliver business solutions. The most basic form of Cloud services is Infrastructure as a Service. Users rent storage space, firewalls, and other hardware and software of any kind. As a user, you are responsible for any hardware element, from the OS to the installed and running applications. Applications developed either by the client, or by another vendor. Cloud service providers offer different products that deliver varying levels of functionality. Hardware, such as disk room, is the most common option, although more complex products provide repair services [8–10].

18.1.1 Delivery of Cloud services

The following frameworks should be considered for delivering Cloud infrastructure services [1, 9, 11]:

- *Private Cloud:* In this service the user and the business, which functions as a single entity, owned and managed the property on-site. The on-site infrastructure and the information behind the firewall are controlled by the user and the business.
- *Public Cloud:* Services may even be shared with other Cloud provider-provided data security organizations. Third-party vendors provide these Public Cloud services. It may be multi-tenant, or dedicated as a single entity. Multi-tenant ensures that the solution is shared by the company with other companies that keep data separate and secure.
- *Hybrid Cloud:* Single-agency services established by a combination of private and public Clouds. Hybrid Cloud is where a combination of public Clouds and private Clouds is utilized and combined by the same organization.
- *Community Cloud:* Public or private Cloud accessed by even more than one organization, with data secured and broken down by Cloud service providers.

18.2 INTEGRATION BENEFITS OF CLOUD-IoT

The Internet of Things model is characterized by the real world (physical world) and small things with little space for storage and computing. The Cloud, by contrast, has nearly infinite storage capacity and its computational power is much more technically advanced. IoT is a widespread model (things put everywhere) and Cloud is universal (resources available from all over the world) and the IoT uses the internet as an integration point and the Cloud uses the internet for the delivery of services. It speaks of the crucial need for the convergence of the two growing technologies.

In general, IoT will benefit from the Cloud's nearly infinite capacities and resources to compensate for its technical limitations (store, process, and connect, for example). The Cloud can provide an effective solution, to name a few examples for Managing and composing the IoT services, as well as implementing applications and services that leverage the products or data generated by them. From the other side, by extending its reach to tackle real-world problems in a more inclusive and diverse way and through offering creative solutions in a broad variety of various ways, the Cloud will benefit from the IoT. In certain instances, the Cloud can have an intermediary layer between the artifacts and the applications, disguising all the complexities and functionality required to execute them. This will affect the future development of applications where the collection, processing, and transmission of information will create additional challenges, particularly in a multi-Cloud environment. This part describes key Cloud IoT drivers i.e. the reason that drive the integration of the Cloud and the IoT. In fact, most literature papers see the Cloud as the lacking component in the integrated circumstance—in other words, they presume that the Cloud tries to fill several of the holes in the IoT (for example, restrictions on storage). By contrast, a few others see the reverse situation: the IoT having to fill the gaps in the Cloud [1, 2, 12].

18.2.1 Benefits

1. *Enabling the Cloud to Handle Data:* The Cloud can act as a means to resolve problems concerning IoT data storage and access. Because users now know how the Cloud operates and stores data, they are more likely to opt to split their data as they were or to have their own information services for personal records, while the company and perhaps also the government will handle time-honored large data. The Internet of Things' impact on infrastructure is double in terms of the data forms to be stored: large data (enterprise-focused) and personal data (consumer-focused) [13, 14]. In the case of users who use applications and smartphones, increased knowledge about the user can generate monotonous results. The IoT connects isolated assets and connects data between the assets and the centralized management systems. These resources may then be incorporated into current and new organizational processes that involve venue, availability, ranking, etc.

   IoT Cloud storage can deliver the benefits of increased accessibility and reliability, ease of deployment, high data backup, archiving and recovery from disasters and lower overall costs. By its nature, IoT includes a vast range of sources of knowledge (i.e. things) generating significant volumes of unstructured or semi-structured data capable of having the three common features of such Big Data, quantity (i.e. data size), diversity (i.e. data types actually) and speed (i.e. information production frequency) [2]. Massive and long-lived computing is a large Cloud-IoT platform, made possible by the effectively limitless, low-cost, and on-demand storage space offered by Cloud providers. Cloud is by far the most cost-effective and efficient approach for managing IoT-generated data and provides open possibilities

for data collection, collaboration and exchange with third parties in this regard.
2. *Computation:* IoT systems have limited computing and energy resources that do not require complex on-site processing of data. Collected data are typically transmitted to more efficient nodes where filtering and replication is feasible but scalability is difficult to accomplish without sufficient infrastructure. Cloud offers almost unlimited processing capability and a design of usage on demand. The processing needs of the IoT can be adequately addressed for real-time data collection, for dynamic, real-time, distributed, sensor-centered implementation, to facilitate the provision incidents and to promote energy-saving practices [9].
3. *Communication Resources:* IoT is the hardware generally identifying objects (things), which is IP-enabled so that it can be contacted by the most important products [5]. The Cloud fixes the problem, providing a reliable and simple solution for connecting, managing and monitoring it from anywhere utilizing personalized portals and technologies, and enabling access to the generated data in real time. Virtual things are monitored, or artifacts.
4. *New Models:* Integrating the Cloud IoT platform makes for new smart systems, smart objects, and system optimization implementation scenarios [2, 11]:
   - SaaS (Sensing as a service) gives inevitable exposure to sensor data.
   - EaaS (Ethernet as a Service) is dedicated to providing widely distributed access to remote devices at layer-2.
   - SAaaS (Sensing and Actuation as a Service) Integrated sensor logic applied to that service in the Cloud.
   - IPMaaS (Identity and Policy Management as a Service) offers broad accessibility to regulation and identification protection capabilities in that service.
   - VSaaS (Video Monitoring as a Service) offers Pervasive web exposure to recorded content and detailed analysis.
   - DBaaS (Database as a service) empowers the management of ubiquitous databases.
   - SEaaS (Sensor Event as a Service) sends Sensor-triggered networking services.
   - SenaaS (Sensor as a Service) empowers the readily available remote sensing control.
   - DaaS (Data as a Service) provides readily available exposure to information of every nature.

18.2.2 Applications of Cloud-IoT

For a vast number of characteristically specified applications, the convergence of two rapidly increasing technical fields makes sense. Some of the IoT applications in the Cloud are discussed below [2, 12, 15, 16, 18]:

- *Healthcare:* Adopting the healthcare Cloud-IoT model will offer many medical IT opportunities. Professionals agree it can greatly change healthcare and lead to its ongoing and systemic progress. Indeed, Cloud-IoT, which is used in this case, can lead to the improvement of healthcare processes and the provision of healthcare services can be increased by allowing collaboration between the various stakeholders [27]. In particular, the aim of Ambient Aided Living (AAL) is to lighten the daily lives of disabled people with serious medical conditions. By using Cloud-IoT in this area, a range of groundbreaking services can be provided, including: obtaining critical patient data through a range of sensors linked to medical equipment; transmitting information to the Cloud medical centers for collection; effectively handling sensing data; or ensuring open coverage or the sharing of health information, such as in the case of Electronic Healthcare Records (EHR). Cloud-IoT enables universal medical facilities to be cost-effective and of high quality, which are widespread in medical services. Deeply ingrained medical devices produce an immense volume of sensor data, which must be treated properly for further study and distribution. Cloud adoption is a positive path to handling healthcare sensor data effectively and allows for the extraction of technological details, removing the requirement for expertise or oversight in application infrastructure. In addition, this leads to a simple integration of the method of data analysis and of dissemination at a reduced cost. It also allows mobile apps ideal for storing, accessing, and sharing health information on the go. The Cloud enables this application scenario to address common challenges such as defense, anonymity, and confidentiality by increasing the protection of medical data and the quality and resilience of the service. In the field of health, specific problems related to the complete absence of consumer confidence in data privacy and security (susceptibility to intruder attacks, breach of medical records confidentiality, in-built security and lack of control, misuse of entitlement), erratic efficiency (depletion of resource base, data processing delays, impact on real-time infrastructure, Quality of Service (QoS) streaming), legal issues (contract law), and intellectual property. The absence of systematic work relevant to incorporating these advances in the context of critical implementations, intensified performance analysis as well as the limited number of case studies are also identified as the main obstacles.
- *Smart cities and human settlements inclusive:* The development due to CoT is of technologies which communicate with the natural world, creating new perceptions as well as area-awareness opportunities. Smart planning is a significant problem involving innovative, efficient, and user-friendly innovations with infrastructure. The goal is to maximize the shared potential of ICT networks (people's networks, information networks, sensors) to build a common and personal awareness of the diverse sustainability challenges confronting our community today on the financial, environmental and political levels. The resulting social influence will lead to more informed decision-making processes, and empower citizens, through participation and connection, to embrace more healthy person and group attitudes and

lifestyles [27]. Cloud-IoT could provide common software solutions for future-oriented major infrastructure systems, collect knowledge from different current managing services and infrastructure, access all sorts of geo-location and IoT technology (e.g., 3D representations through RFID sensors and geo-tagging), and display information effortlessly (e.g., via a dynamically annotated map). Mechanisms generally consist of a sensor platform (with sensing and actuating APIs) and a Cloud platform that delivers scalable, long-lived storage and processing services for automated, large-scale deployment management and real-world sensing device control. Because the IoT scenario is incredibly complex, sensor virtualization may be used to fill the void between existing heterogeneous technologies and their future consumers, enabling them to communicate on multiple levels with sensors. Furthermore, Cloud-based solutions enable third-party vendors to build and distribute Cloud- based IoT modules that connect any system. But that kind of fully advanced software system conceals the underpinning technology's difficulties and heterogeneity while meeting complex Cloud demands such as strong interaction and thoroughness, interoperability, reliability, ease of setup and flexibility. Popular health challenges relate to reliability, size, diversity and thoroughness. Nevertheless, providing the required resources, processing and computational capacities in a secure and transparent manner for large quantities of diverse and customized information (from multiple resources). The design and implementation of various skills possessed in such a fractured situation (where only different IoT natural systems cannot communicate with each other) are significant challenges. There are significant difficulties with real-time interactions in the presence of multiple sensing devices in the service production process, which enforces such a need to explore modifications to real-time operating systems for embedded applications and how they are enabled in a Cloud system. The integration of IoT services into the Cloud entails additional resource management requirements, but there is a need to improve not just to encoding and storing as well as I/O infrastructure, and also sensor reading periods, different sensors inquiries and mutual access to expensive IoT resources based on location. In the end, since urban areas share common interests there is the need to efficiently share data between and within city areas, and also through improved trans-border protocols. They have very little shared methodological framework for collaborative work, creating organizational and provincial fracturing that currently prevents creative opportunities for growth.

- *Smart Home and Smart Metering:* Domestic networks have been identified as the environment where consumers mainly operate. In home environments, Cloud-IoT has a wide-ranging implementation where the seamless integration of multiple digital devices and the Cloud allows typical in-house operations to be managed. Further, the integration of computing with physical entities makes it possible to transform ordinary items into Cloud-integrated knowledge instruments – they can reveal resources through a web interface. Many of the smart home solutions mentioned in the literary works provide (wireless) communication infrastructure linking smart devices to the Internet for the remote monitoring of their activity (e.g. power supply consumption analysis to improve power consumption patterns) and remote control (e.g. street light management, heating, and air ventilation). In recent years, smart lighting has, in addition, attracted increasing interest from the science community; lighting accounts for 19 percent of the world electrical resource consumption and accounts for around 6 percent of total greenhouse gas emissions. Smart lighting management systems have been shown to save up to 45 percent of lighting energy. The Cloud is the perfect choice in this scenario to build scalable applications with fewer source codes, rendering smart home technology an easy process, and providing the necessary infrastructure for operations beyond the reach of local stations (networks). The Cloud allows customers to interact directly with sensing devices/actuators (in other words, supporting event-based techniques) and therefore should satisfy many such essential aspects, including internal network interconnectivity (i.e. all wireless smart home hardware should be able to interconnect), mobile remote control (i.e. smart home appliances and facilities should be accessible smartly at any time). Cloud-based technologies allow an omnipresent environment to be generated in which each computer can be viewed independently in a structured manner and continuous, multi-access assistance can be assured through the Internet. Device management and control could be used through the deployment of quite powerful processing tools, as intermediaries between IoT devices and Cloud elements, adding advanced structures on top of them, reducing Cloud contact frequency to better address the potentially wide variety of devices and the density of their Cloud interaction. In this sense, when developing applications, many problems need to be addressed, which are primarily linked to the lack of consistency and dependability. Web-enabled domestic technologies and consistent interaction with those devices (which is to define a conventional web-based configuration for strategic level and interaction). Additionally, procedures for the identification of devices are needed to allow the simple exploration of equipment. Price issues are often associated with system failure, computer breakdown and not always functional QoS functionality.

- *Smart energy and power generation:* The IoT as well as the Cloud can indeed be adequately combined to deliver smart energy services and cost management across heterogeneous local and wide-ranging environments. In addition, IoT networks used with these processes have capacities with sensing, encoding, and networking but limited infrastructure. Computational activities can thus be properly requested from the Cloud, where they make quite comprehensive and innovative decisions. Implementation of the Cloud contributes to improved efficiency by providing self-healing processes and allowing user-shared operation and engagement, achieving distributed production, energy quality and responsiveness to demand. Cloud computing allows the collection and processing of large quantities of statistics and data from diverse sources spread through large networks in order to introduce intelligent object control. Many hurdles to understand the full value of such a framework should be properly addressed. Large-scale distributed sources raise concerns regarding complexity, data volume and production threshold, frequency fundamentals, and the costs of compliance with data protection. Legal issues arise from the dissemination of the collected data across different jurisdictions [20]. Lastly, customers should build up trust in data exchange in order to improve overall and optimize the services they offer.

- *Ecological Monitoring:* Joint use of the Cloud and the IoT will result in the delivery of high-speed information systems between wide-ranging monitoring organizations and the sensors/actuators that are efficiently installed in the region. Many schemes that contribute to the effective and long-term maintenance of water stages (now for rivers, reservoirs, sewage), environmental gas content (e.g., labs, deposits), soil moisture inclination of static structures (e.g. bridges, dams), location shifts (e.g. landslides), lighting conditions (e.g. identification of intrusions in dark places), electromagnetic radiation characteristics, static hit inclination (e.g. bridges, dams), changes in position (e.g. landslides), lighting conditions (e.g. detection of intrusions in dark places), and infrared radiation. Cloud and IoT hybrid usage will enable the introduction of high-speed information processing in between the wide-ranging control organization and the appropriately designed electronic sensors/actuators. Other potential uses of this kind include: agrarian data distribution and intelligent identification, smart irrigation monitoring, food standards tracking, effective irrigation, forest recognition, and tree tracking. The Cloud-based data exposure will address the acceleration-energy demands of low-power connectivity segments and the ever-present and accelerated cessation of service for us. It also enables dynamic activities to be coordinated and synchronized, generated by data transmitted from real-time devices. The key obstacles are the huge capacity of the infrastructure in this area. The innovation strategy specifically makes it extremely difficult to get sufficient computational strength to deal with evolving ecological circumstances. Furthermore, problems have something to do with data protection, since risks can be recognized in data leakage due to possible future problems associated by compromised customers or data transmission weaknesses. Eventually work remains to be done on implementing and supporting effective communications procedures (like IPv6 to discuss parts separately), establishing different IoT standards to encourage interoperability and reduce IoT facilities costs, and risk and uncertainty assessment.

18.3 RESEARCH ISSUES

- *The need for Standards:* Standard protocols, interfaces and APIs are needed in the Cloud IoT paradigm, while the scientific community has made a contribution to standardizing and Cloud and IoT framework execution. It is this interconnection with enhanced application creation and embedded computing artifacts which renders this as the embodiment of the Cloud-IoT framework. Mobile-To-Mobile (M2M) is the leading model, and has become something of a norm [14]. Existing solutions therefore use conventional internet, wireless communications, and internet technologies.

Few structures are also suggested from the Cloud around the center or from Wireless Sensor during the primary phase of IoT.

- *New Protocols:* Different protocols for connecting to the Internet will have to cohabit for different issues. Although there are relatively homogeneous structures such as an IOT sensing element or a wireless sensor network, various policies and procedures, including Wireless HART, ZigBee, IEEE 1451, Constrained Application Protocol (CoAP) can still be used by sensors and 6LOWPAN, for example [15]. Some of the procedures will encourage a safer alternative system, while others might not. With CoT this issue will increase, particularly due to mobile Cloud application development accessibility. The protocol support can play a key role for mobile phones and touchscreen computers when accessing various healthcare facilities and other sensor-based applications. Much of this depends on the interface and the detector used. Again from the user's point of view, a tendency would be for inexpensive or conveniently available sensors. Thus, it is not known whether or not a freshly appointed sensor will be efficiently activated. Surveying standard gateway protocols is one of the remedies to this category of challenge [2].
- *Energy-Efficient Sensing:* This would eventually result in a lot of transmitting data that consumes a great deal of energy, due to the omnipresence of sensor nodes and their availability in the Cloud. A typical wireless sensor node consists of four elements: detecting unit, control system, transmitter and receiver, and power unit. The power unit gets to play an essential role in decoding, video encoding and video sensing. Ordinarily video encoding seems to be more complicated, especially in comparison to decoding. Each interpretation behind it is that the encoder should first properly assess the redundancy in the video for suitable compression. It will not be sufficient to obtain a temporary power source, including batteries, and these will have to be replaced periodically. This is a monumental undertaking, incorporating hundreds of millions of sensing systems and low-power applications. It is important to have energy-efficient use and a permanent power supply. Sensors should have the atmospheric capacity to create energy through methods such as solar power, vibration, and air. In this way, even a productive sleep mode may be really beneficial. Another proposed solution here is to proceed with local Cloud services, known as Fog Computing. Being precise, Fog refers to a distributed Cloud, which can be used for process discharge purposes on the underlying IoT devices [14, 15].
- *Big Data:* With about 50 billion internet devices likely to be connected over the coming years, care must be taken to transmit, store, access and process the huge quantities of data which will be generated. The growing popularity of mobile devices and sensor perpetuation literally demand scalable computing systems (2.5 zillion bytes of data is being generated per day). The convenient management of such data is a crucial task, since the application's ultimate effectiveness is strongly essential for data organizational provider's assets. For example, Cloud-based techniques for summarizing Big Data are currently under investigation, based on the production of

semantic functionalities [21]. As a result, after the not only SQL (NoSQL) movement, commercial and open source systems implement appropriate database architectures for Big Data, hash values, data warehouse, huge column stores, and database schema. Sadly for the Cloud there is no ideal platform for the handling of big data [1]. In addition, data credibility is an essential consideration mostly for its effect on reliability, but also for its data protection aspects, notably with regard to outsourced data.

- *Advanced Data Mining:* Current systems are unable to completely address all the problems that are important to the complexities of big data. The amount of data an organization will receive is on the decline: due to the huge amount of big data sources and the increasing rate of data output, the gap between the data accessible to companies and the data that they can view is growing. Approximate testing typically involves commands of a magnitude faster than conventional query processing. Scientific research to tackle the big data hurdle is clearly needed. Modern technologies and query methodologies are highly required to much more effectively balance massive amounts of data with effective and cheaper asset and energy utilization [2]. Big data is highly valued data, combined to inexact and filthy info. A tough subject of machine intelligence science is obtaining valuable data at different geographic and contextual levels. Although state-of-the-art approaches employ subtle processing, deep learning is an emerging focus that aims to study several layers of abstraction that can be used to evaluate provided data. Diverse space-time-temporal (location-related and widely dispersed) IoT processing data are not often designed for the direct use of analytical techniques [12, 14].
- *Heterogeneity:* The main problem with Cloud-IoT is attributable to something like the broad variety of open computers, operating systems, networks, and services that are expected to be required by new or improved applications. The nature of the Cloud architectures is always an obvious problem. Cloud systems typically come with proprietary solutions, culminating in data aggregation and mash-up based on the various vendors being properly customized [19]. When users incorporate multi-Cloud alternatives, i.e. when services rely on multiple providers to improve the application features of vendors and durability or lock in, this problem can be compounded. Cloud brokering, which is mutually implemented by service vendors (in the context of a federation) or external parties, addresses these problems only partially. IoT applications and services were usually conceptualized as distinct vertical alternatives, under which all device elements are closely connected to the actual context of decision-making. Providers will assess goal scenarios, review requirements, pick combinations of hardware and software, implement heterogeneous systems, create and distribute computing infrastructure and manage devices for each potential application/function. Viewing differently, due to Cloud software distribution models, CoT can simplify the implementation of IoT services [20]. While platform as service-like models will going to be a standardized approach for promoting the delivery of IoT applications, their deployment implies addressing the

big problem of heterogeneity. For example, the communication of (managing) large quantities of quite diversified items (and the relevant information collected) in the Cloud at multiple stages must be handled properly [1]. This task encompasses many aspects where challenges to cohesive systems and interoperable programming interfaces are explored with middleware means to copy with data diversity etc.

- *Large Scale:* Cloud-IoT allows new technologies to be developed to incorporate and analyze information from (embedded) real-world equipment. Any of the instances seen are indirectly explicitly communicating with a vast variety of such programs, typically distributed through a broad range [22, 23]. The immense size of those same resulting devices has made this more difficult to tackle traditional challenges (as an example, prerequisites for storage space and computing power for more handling are hard to meet when dealing with long-lived high-rate gathering) [1]. Additionally, the delivery of the IoT system makes it more complicated to control activities because they face complexities of latency and communication problems [17].
- *Security and Privacy:* Problems occur as critical IoT apps move to the Cloud caused by a lack of faith on the part of providers, awareness of service-level agreements (SLAs) and awareness of the fact of where data are physically located [24]. Therefore, close consideration of potential problems is required. Such a centrally controlled platform is susceptible to various possible threats (e.g. client riding, SQL injection, cross-site scripting, and side channel) and significant vulnerabilities (e.g. user hijacking and virtual machine escape). Multi-tenancy may often undermine confidentiality and lead to the leakage of sensitive information. After all, public key cryptography cannot really be implemented at all levels due to the computational power limitations placed by the stuff [7, 8, 18].

Among these research issues, security is considered to be a critical issue for the integration of Cloud-IoT, which is discussed in detail in the subsequent section.

18.4 SECURITY ISSUES IN CLOUD-IoT

Security is among the main concerns that needs to be kept in mind when exchanging information in the Cloud-IoT environment. The various security attacks by insiders and outsiders to the IoT is because of its wireless nature. The on-going contact among the IoT devices or the IoT network and Cloud interface can be disrupted by an intruder. Infected Cloud-IoT connectivity adversely affects secure and effective Cloud data storage. Meanwhile, Cloud usage to enable IoT data storage poses privacy issues by requiring all users to access information globally. There is a requirement for secure communication between IoT gadgets and the Cloud framework, which has significant implications for the protection of personal privacy and security within the COT setting. Sadly, the integration of IoT with the Cloud has created a modern set of challenges. In this manner, an execution of a CoT requires advance change of Cloud innovations to properly oversee the information stream and the information source proprietorship inside CoT situations. Since information can often be accessed by third parties and IoT assets are virtualized, unused cross-examinations with respect to the source of information possession, irrefutable information origin and data reliability are made and ought to be addressed. Given the rapid rate of data expansion, assessing and monitoring without the assent and information of the individual being watched is another genuine issue. Within the CoT setting, gadgets can have ad hoc interconnects and can moreover communicate with a back-end Cloud. According to Vasic et al. [18], communication security requires an agreement protocol that will allow all parties to communicate to agree on a cryptographic algorithm and the keys used to protect communication in the messages exchanged. It is evident that the security problem has played a crucial role in preventing the acceptance of Cloud computing. Unquestionably, putting your outcomes, and indeed running your own programs on the hard disk of someone else's system, appears of great concern to many users. The information and computer programs of organizations are prone to well-known security issues such as data loss, phishing, and botnet (programs running remotely to gain control over other machines). In expansion, pooled computing assets and the multi-tenancy model in Cloud Computing have presented modern security challenges requiring new and unusual methods to address IoT security concerns within the field of security of interconnected gadgets and systems. The Web of Things incorporates the developing multiplication of objects and substances that have special identifiers and the capacity to transmit information over a network consequently [18, 19, 26, 28].

The following are specific security issues in a Cloud IOT environment:

i. *Data Confidentiality*

Confidentiality is the concept used to keep information from being exposed to unauthorized persons or systems. It is necessary to hide from observing devices on the internet while transmitting data seamlessly. Confidential transmission of data should be carried out in such a way as not to reveal unauthorized assets, e.g. a person's identity. In working with Cloud environments, confidentiality means that both the Cloud provider and all of their clients must keep the data and processing activities of a company confidential [10]. The confidentiality is among the most worrying concern in Cloud Computing. It is essentially due to consumers outsourcing their Cloud service data and computing exercises, which are controlled and taken care of by possibly untrustworthy Cloud suppliers.

Confidentiality of data is also a major problem in IoT. This requirement is considered a big challenge as almost every other sensing device collects personal information and, when combined, large quantities of such data become Personally Identifiable Information (PII), enough to identify a person. Encryption can be a successful way to protect privacy and confidentiality, but it does pose serious problems with time delay and efficiency [29]. To guarantee secrecy, all RFID Labels, IDs, and information ought to be encrypted on each computer before information is transmitted. However, powerful cryptographic encryption functions such as AES can be enforced in real-time data requirement as they have less latency time; that is, they utilize less time [13]. Also, Blowfish or RSA have lower power utilization and less processing control, and can be actualized successfully on gadgets with physical substrate (layer devices) [30].

Existing Defense Mechanisms
Elliptic Curve Cryptography (ECC) has been utilized to guarantee total security against security dangers. This design is unambiguous, it ensures security with much better productivity and makes a difference to realize a one smart card dream for all applications and exchanges 30]. In [34], Bai and Rabara suggested an integrated reliable and creative IoT and Cloud design. This creative architecture is reasonable for the universe, regardless of position, time, gadget, and network, to get to different keen applications within the Cloud. Elliptic Curve Cryptography (CC) has been utilized to guarantee total security against security dangers. In the absence of uncertainty, this architecture guarantees security with better efficiency and makes a difference realize the dream of "one brilliantly smart card' for any applications and affairs. Device to Device (Symmetric Encryption) in IOT prevents external gadgets from accompanying the sensor network (safeguarding privacy) or observing the information encompassed in the packets delegated as encrypted (safeguarding confidentiality) [31]. An IOT security algorithm based on cryptographic methods such as Triple DES (TDES), RSA, and AES. Cross-VM attack via side channels in the Cloud is addressed using Co-residency Detection to avoid co-residency [33]. Cloud users (particularly companies) need physical confinement, which may moreover be included in the Service Level Agreements (SLAs) [9]. A consumer should be allowed to check the exclusive use of a physical computer by his Virtual Management System (VMS) to ensure physical insulation. The Trusted Cloud Computing Platform (TCCP) ensures that guest virtual machines are run in trust. It also benefits service users to double-check the IaaS provider and to assess whether the service is safe prior its VMS are discharged into the Cloud [25]. Given the customer's dismay over the loss of data control in Cloud environments, Descher et al. [35] recommended that Cloud users enjoy data control by quietly accumulating encrypted VMS on Cloud servers.

The data cannot be used or altered in the Cloud unless an access key is available for encryption, ensuring both confidentiality and integrity. Fully Homomorphic Encryption (FHE) was proposed by the gentry to safeguard privacy in Cloud computing [25]. FHE allows encrypted data storage, whatever is stored on the Cloud provider's untrustworthy servers. Data can be handled without decoding. Cryptography is NOT always appropriate, i.e. Cryptography cannot single-handedly afford all the necessary explanations to all privacy and confidentiality problems in Cloud computing, despite having potent methods like FHE. A class of privacy issues can only be formally defined in terms of the different application schemes. There is no such Cryptographic algorithm that can be enforced to guarantee privacy when there is data sharing among the clients. Privacy Preservation Frameworks displayed a common information assurance system to address protection challenges within the provision of Cloud administrations.

ii. *Data Authentication*

Authentication is the process of identifying the legitimate entity of any web application in question. The process is a key to the successful incorporation of in-built devices and Cloud computing services. Authentication is generally referred to as a mechanism which sets out the validity of the individual's claimed identity. Authentication focuses primarily on two characteristics in Cloud-IoT: lightweight authentication clarification and identity/location privacy preservation. Distinguishing the authentication from the authorization is an essential process. A client is himself responsible for navigating and authorizing to his home provider during the authentication process. During this process only the user and the organization share information. After the successful authentication, the user is given access to resources based on the user credentials or parameters. After the successful authentication, the process of authorization is executed in which the user attributes are exchanged with the resource server, while there is no personal information leakage [9, 25, 36].

Authentication in IoT is one of the main challenges because of the number of devices available. Authenticating each and every device is no easy task. Many security mechanisms have been proposed because of the features of fast computation and energy conservation, based on private key cryptographic primitives. Information service providers shall apply any mechanism of access control to protect data from the misuse and damage by others of a user's private information. Data only need to be open to those it belongs. Identity management is a large administrative field in IOT that deals with object recognition through the use of different techniques in a system and managing its access by associating constraints and user privileges with the accepted identity. When a new device is added to the network, authentication and authorization processes are important to keep the malicious gadgets out of the system before the data are transmitted or received [18, 30, 37].

Cloud computing is a computing style in which robustly expandable and often-virtualized resources are made available over the Internet as a service. Users do not need to have knowledge of, or expertise in, the Cloud infrastructure that assists them. For Cloud protection, therefore, authentication becomes pretty necessary. A safe process of authentication and authorization is needed despite having a standard Cloud platform. It becomes difficult for users to authenticate on service providers repeatedly on their virtual offices and to retain multiple passwords. To eradicate this problem a secure authentication process must be used that enables users to access services through trusted parties.

Existing Defense Mechanisms

A certification authority is obligatory to approve digital entities in a Cloud domain, including the certification of physical framework servers, virtual servers, users of environments, and network gadgets. While registering within the Trust mesh, public key infrastructure (PKI) certification authority is responsible to generate the requisite certificates. In combination with

Single Sign On (SSO) and lightweight Directory Access Protocol (LDAP), digital signatures enforce the best authentication mechanism available in distributed environments, thus maintaining user usability and versatility. Shibboleth is open source middleware, a standard-based software that furnishes Web Single Sign On (SSO) beyond or inside organizational borders. It authorizes sites to create informed authorization choices in a privacy-conserving manner for specific access to secured online assets [36, 38]. An authentication scheme has been developed, based on biometrics for multi-Cloud-server environments. It employs biometric hashing as building blocks and the ECC. The productivity and achievements of the suggested theory is figure out to confirm its service [32].

The rapidly emerging IoT poses many Security and Privacy concerns. Considerable work has been carried out in the OAuth protocol regarding security. The OAuth protocol serves an authorization layer over the transport layer, i.e. http-over-TLS. The secure and safe Internet communication is presented by the TLS protocol [39]. It approves communication between client-server applications to inhibit the forgery of messages, eavesdropping or tampering. Authenticating large numbers of devices in real time is a challenge to solve these various handshaking schemes and algorithms and low-power pre-shared keys. RFID plays a major part in object identification. In identifying artifacts, it uses electromagnetic induction and the propagation of electromagnetic waves. From the point of view of security, RFID can also be adopted against replication, securing data on records and combined encryption, certificates, and other elements for the purposes of counterfeiting and control and management [4, 30]. A scheme used for system authentication in the IoT is one where user nodes are authenticated; however, it is not lightweight and it can affect both the lifetime and the efficiency of batteries. Another method used is the process of handshaking. A bit of time and definition of symmetric key cryptography is used for this method. To eradicate these issues a solution employing public key cryptography is used because of its high scalability, low memory requirements and no key pre-distribution framework requirement [36].

iii. *Data Integrity*

Integrity is the certainty and affirmation that the information is full and authentic. Data integrity confirms not only that the data are accurate but also that it is trustworthy and can be relied on. Integrity is a crucial feature of information security. Integrity assures that the data, software or hardware can be customized only by authorized users in an authorized manner. Integrity means the defense of data against unauthorized deletion, alteration, or manufacture. Given the limited ability of smart devices, the IoT and Cloud computing integration provides many advantages, principally in terms of computing and storage space. The large amount of data generated by the sensors stored on the Cloud servers and the Cloud instability are also a direct menace to the IoT`s protection and reliability [33, 37]. Users require checking for the remote integrity of the Cloud and IoT storage systems to safeguard its availability and data integrity. However, the latest schemes for checking remote data integrity mainly operate on the signature frameworks of RSA and BLS. An error detection procedure is implemented on every device to guarantee that the susceptible data are not manipulated. WH cryptographic hash function is generally used for strong error detection in spite of having many low-power utilization techniques such as Cyclic Redundancy Checks (CRC), Checksum and Parity Bit [30].

ACID (atomicity, consistency, isolation and durability) is used as a basis for ensuring data integrity in our database management systems, but these data integrity principles are not included by all the service providers in the case of data integrity problems. In addition, consumers often employ such a range of service providers that none of the providers are responsible for maintaining data integrity at the data entry and transaction processing stages. Over the period of time new standards have been developed for Cloud data management. These standards must be incorporated by cloud service providers to ensure the quality of Cloud data for their customers. He Internet is an essential medium for Cloud computing, and web apps offer an entry to this system. Data Integrity Field (DIF), SNIA Cloud Data Management Interface (CDMI), and XML-based solutions are among the emerging standards in the present-day Cloud world [10].

Existing Defense Mechanisms

The BLS signature-based PDP (Provable Data Possession) mechanism strengthens public verification and satisfies Cloud storage's lightweight design prerequisites [32]. Cryptographic hash functions are used for data integrity received from the IoT devices. In order to mitigate data tempering, time error correction techniques are employed. A Message Authentication code-based PDP system employing message authentication code as metadata authentication for confirming the validity of remote data is being suggested [41]. An RSA signing mechanism must be a foundation for the construction of a PDP mechanism to verify the remote data integrity, along with data transmission over the internet [40]. A third-party auditor must be employed to track the quality of the data outsourced in the Cloud domain along with the prevention of new susceptibilities and efficient auditing [25].

18.5 CONCLUSION

Throughout this chapter, we have systematically studied the research issues generally and specifically the security issues that arise due to integration of Cloud computing and IoT. Cloud computing provides different service benefits to IoT on the one hand and, on the other, IoT lets Cloud computing reach real-world objects. The Cloud system provides a practical situation to handle and analyze the large volumes of data generated by IoT devices. Similarly, the processing of complex data and its analytics can also be achieved using cloud computing services; however, there are certain research issues identified in this chapter which need to be addressed before the IoT can use Cloud computing services. Among the various security issues, the security of user data is pivotal. In this chapter, we have discussed the important security services, i.e. confidentiality, integrity, and authentication in the context of CoT. The chapter also discusses the existing defense strategies for these security services. The research carried out will serve as the important basis for further research, in addition to the issues put forth by the integration of cloud computing and the Internet of Things.

ACKNOWLEDGEMENT

This research work has been funded under the seed grant initiative of TEQIP-III project implemented at the Islamic University of Science and Technology, Awantipora, Jammu and Kashmir.

REFERENCES

1. Botta, Alessio, Walter De Donato, Valerio Persico, and Antonio Pescapé. "Integration of Cloud computing and internet of things: a survey." *Future Generation Computer Systems*, 56, pp. 684–700, 2016.
2. Botta, Alessio, Walter De Donato, Valerio Persico, and Antonio Pescapé. "On the integration of Cloud computing and internet of things." In *Proc. Future Internet of Things and Cloud (FiCloud)*, pp. 23–30, 2014.
3. Shah, Sajjad Hussain, and Ilyas Yaqoob. "A survey: Internet of Things (IOT) technologies, applications and challenges." In *2016 IEEE Smart Energy Grid Engineering (SEGE)*, pp. 381–385, 2016.
4. Xingmei, Xu, Zhou Jing, and Wang He. "Research on the basic characteristics, the key technologies, the network architecture and security problems of the internet of things." In *Proceedings of 3rd International Conference on Computer Science and Network Technology*, pp. 825–828. IEEE, 2013.
5. Kamilaris, Andreas, and Andreas Pitsillides. "Mobile phone computing and the internet of things: A survey." *IEEE Internet of Things Journal* 3, no. 6, pp. 885–898, 2016.
6. Shin, Seong Han, and Kazukuni Kobara, "Towards secure Cloud storage." *Demo for CloudCom* 2010.
7. Hashizume, Keiko, David G. Rosado, Eduardo Fernández-Medina, and Eduardo B. Fernandez. "An analysis of security issues for Cloud computing." *Journal of Internet Services and Applications* 4, no. 1, 2013.
8. Kuyoro, S. O., F. Ibikunle, and O. Awodele. "Cloud computing security issues and challenges." *International Journal of Computer Networks (IJCN)* 3, no. 5, pp. 247–255, 2011.
9. Padhy, Rabi Prasad, Manas Ranjan Patra, and Suresh Chandra Satapathy. "Cloud computing: Security issues and research challenges." *International Journal of Computer Science and Information Technology & Security (IJCSITS)* 1, no. 2, pp. 136–146, 2011.
10. Ajoudanian, Sh, and M. R. Ahmadi. "A novel data security model for Cloud computing." *International Journal of Engineering and Technology* 4, no. 3, 326, 2012.
11. Mahmood, Zaigham. "Data location and security issues in Cloud computing." In *2011 International Conference on Emerging Intelligent Data and Web Technologies*, pp. 49–54, IEEE, 2011.
12. Babu, Shaik Masthan, A. Jaya Lakshmi, and B. Thirumala Rao. "A study on Cloud based Internet of Things: CloudIoT." In *2015 global conference on communication technologies (GCCT)*, pp. 60–65. IEEE, 2015.
13. Srivastava, Pallavi, and Navish Garg. "Secure and optimized data storage for IoT through Cloud framework." In *International Conference on Computing, Communication & Automation*, pp. 720–723. IEEE, 2015.

14. Malik, A., & Om, H., Cloud computing and internet of things integration: Architecture, applications, issues, and challenges. In *Sustainable Cloud and Energy Services* (pp. 1–24). Springer, 2018.
15. Aazam, Mohammad, Eui-Nam Huh, Marc St-Hilaire, Chung-Horng Lung, and Ioannis Lambadaris. "Cloud of things: Integration of IoT with Cloud computing." In *Robots and Sensor Clouds*, pp. 77–94. Springer, Cham, 2016.
16. Díaz, Manuel, Cristian Martín, and Bartolomé Rubio. "State-of-the-art, challenges, and open issues in the integration of Internet of things and Cloud computing." *Journal of Network and Computer Applications* 67, pp. 99–117, 2016.
17. Stergiou, Christos, Kostas E. Psannis, Byung-Gyu Kim, and Brij Gupta. "Secure integration of IoT and Cloud computing." *Future Generation Computer Systems* 78, pp. 964–975, 2018.
18. Ari, Ado Adamou Abba, Olga Kengni Ngangmo, Chafiq Titouna, Ousmane Thiare, Alidou Mohamadou, and Abdelhak Mourad Gueroui. "Enabling privacy and security in Cloud of Things: Architecture, applications, security & privacy challenges." *Applied Computing and Informatics*, 2019.
19. Grozev, Nikolay, and Rajkumar Buyya. "Inter-Cloud architectures and application brokering: taxonomy and survey." *Software: Practice and Experience* 44, no. 3, pp. 369–390, 2014.
20. Li, Fei, Michael Vögler, Markus Claeßens, and Schahram Dustdar. "Efficient and scalable IoT service delivery on Cloud." In *2013 IEEE sixth international conference on Cloud computing*, pp. 740–747. IEEE, 2013.
21. Tan, Kian-Lee. "What's NExT? Sensor+ Cloud!?." In *Proceedings of the Seventh International Workshop on Data Management for Sensor Networks*, pp. 1–1. 2010.
22. Bo, Yifan, and Haiyan Wang. "The application of Cloud computing and the internet of things in agriculture and forestry." In *2011 International Joint Conference on Service Sciences*, pp. 168–172. IEEE, 2011.
23. Lazarescu, Mihai T. "Design of a WSN platform for long-term environmental monitoring for IoT applications." *IEEE Journal on Emerging and Selected Topics in Circuits and Systems* 3, no. 1, pp. 45–54, 2013.
24. Andrei, Traian, and Raj Jain. "Cloud computing challenges and related security issues." *A Survey Paper*. http://www.cse.wustl.edu/~jain/cse571-09/ftp/Cloud.pdf, 2009.
25. Xiao, Zhifeng, and Yang Xiao. "Security and privacy in Cloud computing." *IEEE Communications Surveys & Tutorials* 15, no. 2, pp. 843–859, 2012.
26. Mahmood, Zaigham. "Data location and security issues in Cloud computing." In *2011 International Conference on Emerging Intelligent Data and Web Technologies*, pp. 49–54. IEEE, 2011.
27. Benabdessalem, Raja, Mohamed Hamdi, and Tai-Hoon Kim. "A survey on security models, techniques, and tools for the internet of things." In *2014 7th International Conference on Advanced Software Engineering and Its Applications*, pp. 44–48. IEEE, 2014.
28. Zhou, Jun, Zhenfu Cao, Xiaolei Dong, and Athanasios V. Vasilakos. "Security and privacy for Cloud-based IoT: Challenges." *IEEE Communications Magazine* 55, no. 1, pp. 26–33, 2017.
29. Razzaq, Mirza Abdur, Sajid Habib Gill, Muhammad Ali Qureshi, and Saleem-Ullah. "Security issues in the Internet of Things (IoT): a comprehensive study." *International Journal of Advanced Computer Science and Applications* 8, no. 6, 2017.
30. Andrea, Ioannis, Chrysostomos Chrysostomou, and George Hadjichristofi. "Internet of Things: Security vulnerabilities and challenges." In *2015 IEEE Symposium on Computers and Communication (ISCC)*, pp. 180–187. IEEE, 2015.

31. Al-Turjman, Fadi, and Sinem Alturjman. "Confidential smart-sensing framework in the IoT era." *The Journal of Supercomputing* 74, no. 10, pp. 5187–5198, 2018.
32. Kumari, Saru, Marimuthu Karuppiah, Ashok Kumar Das, Xiong Li, Fan Wu, and Neeraj Kumar. "A secure authentication scheme based on elliptic curve cryptography for IoT and Cloud servers." *The Journal of Supercomputing* 74, no. 12, pp. 6428–6453, 2018.
33. Matsemela, Gift, Suvendi Rimer, Khmaies Ouahada, Richard Ndjiongue, and Zinhle Mngomezulu. "Internet of things data integrity." In *2017 IST-Africa Week Conference (IST-Africa)*, pp. 1–9. IEEE, 2017.
34. Bai, T. Daisy Premila, and S. Albert Rabara. "Design and development of integrated, secured and intelligent architecture for internet of things and Cloud computing." In *2015 3rd International Conference on Future Internet of Things and Cloud*, pp. 817–822. IEEE, 2015.
35. Descher, Marco, Philip Masser, Thomas Feilhauer, A. Min Tjoa, and David Huemer. "Retaining data control to the client in infrastructure Clouds." In *2009 International Conference on Availability, Reliability and Security*, pp. 9–16. IEEE, 2009.
36. Zissis, Dimitrios, and Dimitrios Lekkas. "Addressing Cloud computing security issues." *Future Generation Computer Systems* 28, no. 3, pp. 583–592, 2012.
37. Rehman, Sadiq Ur, Iqbal Uddin Khan, Muzaffar Moiz, and Sarmad Hasan. "Security and privacy issues in IoT." *International Journal of Communication Networks and Information Security* 8, no. 3, 2016.
38. Kalra, Sheetal, and Sandeep K. Sood. "Secure authentication scheme for IoT and Cloud servers." *Pervasive and Mobile Computing* 24, pp. 210–223, 2015.
39. Emerson, Shamini, Young-Kyu Choi, Dong-Yeop Hwang, Kang-Seok Kim, and Ki-Hyung Kim. "An OAuth based authentication mechanism for IoT networks." In *2015 International Conference on Information and Communication Technology Convergence (ICTC)*, pp. 1072–1074. IEEE, 2015.
40. Hashizume, Keiko, David G. Rosado, Eduardo Fernández-Medina, and Eduardo B. Fernandez. "An analysis of security issues for Cloud computing." *Journal of Internet Services and Applications* 4, no. 1, 2013.
41. Zhu, Hongliang, Ying Yuan, Yuling Chen, Yaxing Zha, Wanying Xi, Bin Jia, and Yang Xin. "A secure and efficient data integrity verification scheme for Cloud-IoT based on short signature." *IEEE Access* 7, pp. 90036–90044, 2019.

19 Evaluating Outdoor Environmental Impacts for Image Understanding and Preparation

Roopdeep Kaur, Gour Karmakar, and Feng Xia
Federation University Australia

CONTENTS

19.1 Introduction....268
19.2 Related Works....269
19.2.1 Applications that Do Not Consider the Impact of Rain, Shadow, Darkness, and Fog....269
19.2.2 Other Applications....271
19.3 Our Approach for Image Data Understanding and Preparation....271
19.3.1 Image Data Understanding....271
19.3.1.1 Image Data Gathering....273
19.3.1.2 Verifying Image Data Quality....273
19.3.2 Assessing the Consistency Among the Quality Values of the Images Captured Under a Particular Environmental Impact....273
19.3.3 Mapping Environmental Impact into JPEG Image Quality and Gaussian Noise Level....274
19.3.4 Applying Consistency and JPEG Image Quality and Gaussian Noise Level for Image Data Preparation....275
19.4 Experimental Method....275
19.4.1 Datasets....276
19.5 Results and Discussions....278
19.5.1 Analysis of Image Quality....278
19.5.2 Evaluating the Consistency Among the Quality Values for a Particular Impact Level....286
19.5.3 Assessing the Impacts in Terms of JPEG Image Quality and Gaussian Noise Levels....288
19.5.3.1 Mapping the Impact for PSNR....288
19.5.3.2 Mapping the Impact for ORB....290
19.5.3.3 Mapping the Impact for SSIM....290
19.6 Conclusions....294
References....294

DOI: 10.1201/9781003267782-19

19.1 INTRODUCTION

Digital image processing is widely used in many real-world applications and is presently driving the process of automation in industrial applications, especially using the Industrial Internet of Things (IIOT). Among examples of these applications are the following: object and event detections; robotic vision for automated assembling and manufacturing; environmental monitoring to detect hazardous conditions and chemical contaminations; and remote health monitoring. Many of these applications require the capturing of image data from outdoor environments through Internet of Things (IoT) devices. These sensed images are heavily impacted by dynamic and complex environmental changes.

The principal factors involved in capturing images in an outdoor environment are lightning, time (whether taken at day or night), camera orientation and position, and weather conditions (e.g. rain, wind, and fog). For instance, windy weather may lead to a decrease in the clarity of the objects. Consequently, it may be that the final captured pictures are blurry. The time, the camera setting, and the distance at which images are taken in uncontrolled conditions are other important factors. There are also differences in the images taken in the morning, the afternoon, and the night are also different because of the variations in the lighting conditions (Kapoor, Bhat, Shidnal, & Mehra, 2016), light sources, and the amount of shadows that affect the accuracy of image processing applications. Therefore, lighting effects need to be considered in the analysis of images captured in an outdoor environment (Fathi Kazerouni, Mohammed Saeed, & Kuhnert, 2019). Without a consideration of all possible outdoor environmental impacts, decisions derived from outdoor image analysis can be erroneous. Thus, research now exists, for example, to detect the amount of smoke in foggy images using a deep neural network (Khan, Muhammad, Mumtaz, Baik, & de Albuquerque, 2019). However, to our knowledge, no techniques are available that objectively assess the impact of environmental parameters such as rain, shadow, darkness, and fog, all of which can have an enormous impact on the image quality ability understanding images and apply that quality image preparation. However, such automatic image processing techniques require quantitative assessment of quality and mapping that quality into human perceptible terms that can be readily applied in image-filtering technique without any human interpretation. These types of image understanding and preparation not only reduce the cost and human time but also advanced industrial automation. Thus, this can create serious economic and other relevant consequences in the organization that uses image analysis applications. To reduce these consequences, in this project, we aim to measure the reliability of an image captured in an outdoor environment.

The major contributions of the chapter are as follows:

1. We are the first to solve this significant problem of assessing the impact of the outdoor environmental parameters such as rain, shadow, darkness and fog on the quality of images and mapped the objective quality into more human perceptual quality measure
2. We propose a new way for image data preparation by comparing the quality level of JPEG images and Gaussian noise to meet the application specification requirements

3. Extensive experiments are conducted using real-world image data and different popular image quality metrics. The results show that outdoor environmental changes has a huge influence on image quality, ranging from 1 to 100 percent of noise level

The structure of this chapter is as follows:

The literature related to the subject is summarized in Section 19.2. Section 19.3 explains our approach for image data understanding and preparation. An experimental method is mentioned in Section 19.4. The findings and discussions are described in Section 19.5. Section VI ends with conclusions along with future applications.

19.2 RELATED WORKS

Many studies have been done in various fields such as traffic control systems to calculate the traffic density, the identification of disease in the crops using image-processing techniques, automated plant recognition systems in unregulated outdoor environments, and smoke detection systems in outdoor environments. Existing applications in the image-processing field in outdoor environments can be divided into two categories:

(i) applications that do not consider the impact of rain, shadow, darkness and fog; and (ii) other applications.

19.2.1 Applications that Do Not Consider the Impact of Rain, Shadow, Darkness, and Fog

These applications consist of smart transportation systems and smart agriculture. In smart agriculture, to illustrate the influence of man-made or environmental factors on the plant's growth, Kapoor et al. (2016) first explained a method through a combination of IoT and image-processing techniques. They used the loT sensing network to capture the values of important environmental variables, as well as a picture of the leaf lattice. They presented the effect of sunlight, temperature, and excess fertilizers on the health of plants. However, this method will not work effectively in the case of leaves that are very dark in color because we will be unable to detect the disease in the dark leaves through this method. Also, they only considered the sunlight parameter in the outdoor environment; they did not include rain, night-time, or darkness and fog. To detect an early-disaster event in smart cities, smoke detection is a critical component. Earlier methods are less effective in detecting the smoke in the foggy and uncertain IoT environment.

After this, Thorat, Kumari, & Valakunde (2017) used sensor networks to test moisture, temperature, and humidity instead of using manual checks. Various sensors have been installed in various farm locations, and one controller, named Raspberry PI (RPI), is used to collect data from these sensors. The identification of disease in crops is achieved using image recognition. The camera is located next to the plant to take the picture of a leaf. The captured image is sent to the server and using image-processing techniques leaf disease is detected, the status of a leaf is sent back to the farmer on the webpage & app on the mobile phone. The key thing that determines the outcome is sunshine. Moreover, photos cannot be taken at night; similarly, if the illumination is too bright to capture the picture, or if the light is mirrored and the color of the leaf is not visible, the outcomes can also vary. There is another application area in Frank, Khamis Al Aamri, & Zayegh (2019), where real-time image and video-processing techniques are utilized to analyse the real traffic density on the lane. Images are taken while there is no traffic and saved on the network; these can be compared to the real-time image captured by the camera to determine the density. The theme is to control traffic by assessing the density of traffic on either side of the road and providing a traffic signal control option to the user through a software application. This greatly reduces average waiting time and improves traffic flow, which will reduce traffic congestion and pollution in the cities. It will greatly reduce the rate of on-road accidents and improve air quality. This system cannot work effectively in the night-time or darkness and it does not consider outdoor environments which have an impact on the image-capturing system. The efficacy of this method is limited, as it will be affected by the levels of traffic in the environment.

To address this issue, recently, an efficient image-processing-based system was presented in Pinto, Pais, Nisha, Gowri, & Puthi (2020), which tracks the traffic and changes the traffic signal state. After the counting of vehicles, the traffic density of each lane is calculated. If the traffic level is higher than the specific direction consisting of a high density, the traffic signal becomes green for the direction which has the highest traffic density, which leads to a reduction in traffic delays. This technique results in the reduction of traffic congestion which helps to reduce delays, fuel, transportation costs and improves resident's ability to access facilities and services on time. It also led to a sharp reduction in the levels of both air and noise pollution. Overall, it is saving money, time, and protecting the environment, which are some of the present-day's biggest challenges. However, this system may not work whenever there is an incident on the road because this method will give inappropriate density to such events and can result in more traffic congestion. In addition, this system does not take into account consider the effects of shadow, rain, fog, wind, and other environmental parameters while image capturing.

Later on, the Convolutional Neural Network (CNN) was introduced for automated plant recognition in unregulated outdoor settings in Fathi Kazerouni et al. (2019). Four different natural plant species were identified using deep learning. The suggested architecture allows for the use of deep neural networks and their transformative ability for plant identification in outdoor settings such as forests and fields, with 99.5 percent accuracy. In this system, the dataset is very complex even for the detection of four plants; if we increase the number, it will be difficult to handle such data which is expensive and they have just represented results visually; they have not quantitatively assessed the impact of the dynamic environment. In addition, they did not mention the number of images taken in a particular environment and did not take into account outdoor parameters such as fog, rain, and shadow.

The reliability of the Long Range (LoRa) network in the presence of temperature variations is demonstrated in (Boano, Cattani, & Römer, 2018) which shows that temperatures above 50 degrees Celsius has a considerable impact on data communications transmitted through the LoRa network.

Consequently, all of these applications and the associated studies have a common drawback that they did not consider outdoor parameters such as rain, shadow and darkness which plays a crucial role in assessing the quality of images.

19.2.2 Other Applications

There exists a study that considers the impact of fog while capturing images in the outdoor environment. Khan et al. (2019) made an energy-efficient system method for the detection of smoke at an early stage in the normal and foggy environment which is based on deep CNNs. When compared with other models, AlexNet has the lowest accuracy, and the highest false-positive and false-negative scores. In comparison to their proposed approach, GoogleNet achieves improved performance, but its accuracy is still poor, with a high false alarm rate. The proposed system outperforms the previous two methods, with a minimum false alarm rate of 2.30, a minimum false-negatives rate of 2.01, and a minimum false-positives rate of 2.01 and an accuracy rate of 97.72 percent, which is the highest recorded. This method represents both a quantitative and a qualitative comparison of images with and without fog but they have taken 20, 30, 50 percent training, validation and testing data, respectively.

The drawback of this work is that it can work only in an environment where a video surveillance system is available and the study considered only the impact of fog. However, the impact of influential outdoor parameters such as rain, shadow, and darkness has not been considered. In addition, they did not convert the impact into a human perceptible term that can be used in image filtering during data preparation.

19.3 Our Approach for Image Data Understanding and Preparation

In this section, for the first time, we introduce an approach for image data understanding and image data preparation, two main phases of the most popular data mining process model, namely the CRISP-DM model (Dinh, Karmakar, & Kamruzzaman, 2020). Image data understanding and preparation are achieved by verifying the quality of images captured in outdoor environments and filtering them that are not suitable for applications comparing the environmental impact equivalent to the impact of certain noise or JPEG image quality level, respectively. The architecture of our method is shown in Figure 19.1. This shows our proposed approach consists of different tasks such as image data understanding, checking the consistency among image qualities, mapping the quality into more perceptible impact level and image data preparation, which are explained below:

19.3.1 Image Data Understanding

In this project, image data understanding is represented in a twofold process: (i) gathering application-relevant image data; and (ii) verifying the quality of those images suitable for a particular application.

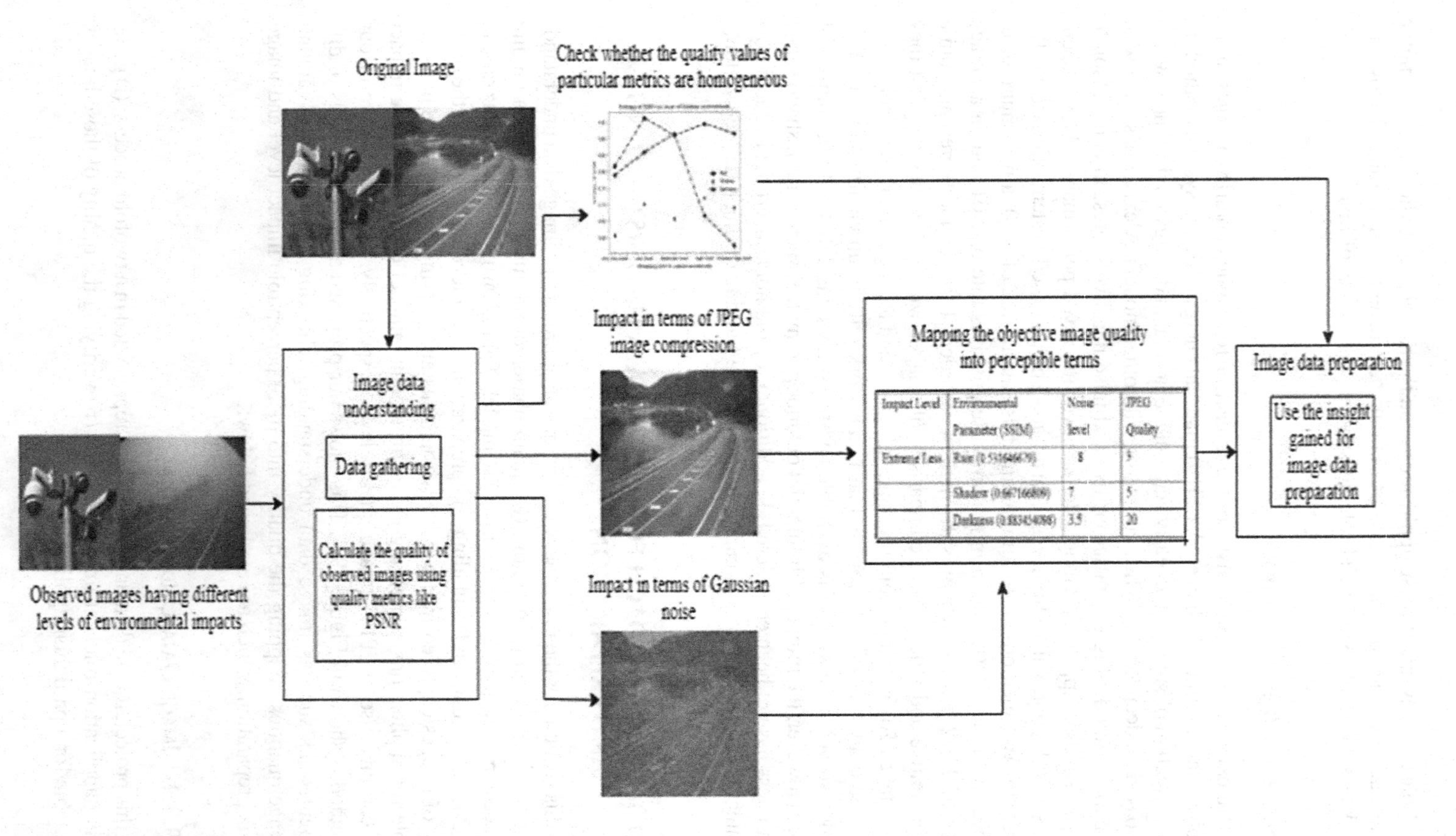

FIGURE 19.1 Block diagram for assessing the impact of outdoor environment on the quality of image data.

19.3.1.1 Image Data Gathering

Image data relevant to a particular application can be captured from outdoor environments in many different ways. For example, images can be collected through IoT vision sensors and IoT networks such as NB-IoT and Sigfox can be used to transfer those images to an edge device/server. Outdoor images can also be taken through traditional digital wired or wireless cameras, and those data can then be forwarded to the processing server through wired/wireless communication systems (e.g., cellular and Bluetooth-based communication systems).

19.3.1.2 Verifying Image Data Quality

In order to verify image data quality, we use various widely used image quality measurement metrics such as Mean Squared Error (MSE), Peak signal-to-noise ratio (PSNR), Structural Content (SC), Oriented FAST and rotated BRIEF (ORB) and Structural Similarity Index (SSIM). We can calculate these quality values of an image having different levels of environmental impact so that we can assess the impact on the quality with an increase in the levels of different outdoor environment parameters such as rain, shadow, darkness, and fog.

19.3.2 Assessing the Consistency Among the Quality Values of the Images Captured Under a Particular Environmental Impact

To assess how the values of a particular quality metric for an environmental impact conforms to an intuitively defined impact level, we measure the consistency among the quality values. There are various ways to assess consistency among the quality values for images. We consider their entropy as a consistency measure. Because entropy is widely used to measure the homogeneity of values of a variable. The higher the value of entropy, the lower the homogeneity is. The entropy of the quality values using a histogram (H) is given as below:

$$H = -\sum_{i=1}^{N} p(b_i)\log(p(b_i)) \tag{19.1}$$

$$p(b_i) = \sum_{i=1}^{N} Y_i / Y_i \tag{19.2}$$

Here, $p(b_i)$ is the probability of the quality values belong to i^{th} bin b_i, γ_i represents the frequency of quality values falling in bin b_i and N is the total number of bins (Hassan, Karmakar, & Kamruzzaman, 2013). If the values of image quality metrics such as MSE, PSNR, SC, and SSIM vary too much, H will be higher, representing more uncertainty and less consistency (homogeneity).

The maximum value of entropy H_m is,

$$H_m = \log N \tag{19.3}$$

So far, we assess the impact of dynamic outdoor environments and also analyze the uncertainty with entropy. This entropy will be highly useful for image data preparation in various applications to filter out low-quality images (Okafor, 2005).

19.3.3 Mapping Environmental Impact into JPEG Image Quality and Gaussian Noise Level

In this section, we introduce a mapping function to covert the objectively assessed image quality into more human perceptible terms such as JPEG image quality and Gaussian noise level. This conversion will help an application understand image quality in a more intuitive and easier way and enable it to decide image data preparation. The mapping function (f) for converting the image quality (θ_{ij}) estimated by applying a jth quality assessment metric over an image can be defined as:

$$f : \theta_{ij} \rightarrow I_s \tag{19.4}$$

where θ_{ij} is the ith image for the jth quality metric and I_j indicates the equivalent impact percentage of interference agents with jth factor level. Here, we consider two perceptual interfering agents, JPEG quality level and Gaussian noise level, respectively, to assess the quality of distorted images. $f(\theta_{ij}, s)$ can be defined in many ways. One of the feasible and practical approaches is to use a lookup table for mapping. Using the lookup table, $f(\theta_{ij}, s)$ can be presented as:

$$f(\theta_{ij}, s) = l(\theta_{ij}, s) \tag{19.5}$$

where $l(\theta_{ij}, s)$ returns the equivalent impact level of Gaussian noise and JPEG compression quality for $s = 1$ and $s = 2$, respectively. An example of a such lookup table is shown in Table 19.1, in which NL means noise level and JQ means JPEG quality. In Table 19.1, if we consider the value of SSIM (0.48) at an extremely high level, for $s = 1$ and $s = 2$, it will return us to a noise level equal to 45 percent, and JPEG compression quality is equal to less than zero, respectively. Similarly, with regard to rain, shadow, darkness and fog, we can find the equivalent to Gaussian noise level and JPEG compression quality for the values of other image quality metrics using their respective lookup tables. Even though we utilize these two perceptual interfering agents because Gaussian noise appears commonly on images from natural sources,

TABLE 19.1
Comparison of SSIM for Rain in terms of Gaussian Noise Level and JPEG Compression Quality

Level	SSIM of Rain	NL	JQ
Extremely less	0.53	3.5	20
Moderate	0.52	15	less than 0
High	0.51	25	less than 0
Extremely high	0.48	45	less than 0

which is independent at each pixel, and also does not depend on the intensity of the signal. Gaussian noise from various natural sources can influence most computer and communication systems. We use JPEG image quality level because it exploits the characteristics of human vision and is widely used by image compression research community and related applications.

19.3.4 Applying Consistency and JPEG Image Quality and Gaussian Noise Level for Image Data Preparation

Given the assessment of the impact in terms of JPEG image quality and Gaussian noise level presented in the previous section, and depending on the application's requirements, we can filter out low-quality images for image data preparation.

In addition, as mentioned before, in this section, we can also apply the entropy derived in section 19.3.2 to select application-specific outdoor image data. The use of entropy for such image data preparation purposes can be justified by the fact that entropy is widely used in data mining to retrieve application-specific required information (Okafor, 2005). Therefore, from these insights, we can filter out low-quality images which reduce the efficacy of image processing applications.

19.4 EXPERIMENTAL METHOD

Firstly, we took reference and the distorted images having different levels of outdoor environments. We completed the experiment using Python programming language and the total number of bins is set to $N = 10$.

Next, we calculated the values of image metrics MSE (Aziz, Tayarani-N, & Afsar, 2015), PSNR (Welstead, 1999), SC (Vora, Suthar, Makwana, & Davda, 2010), ORB (Rublee, Rabaud, Konolige, & Bradski, 2011) and SSIM (Wang, Bovik, Sheikh, & Simoncelli, 2004) using Python. The similarity of the two images is measured by calculating PSNR, which is a reciprocal of MSE (Aziz et al., 2015). Typical PSNR values in the lossy image and video compression are between 30 and 50 dB. PSNR values for 16-bit data are usually between 60 and 80 dB (Welstead, 1999; Hamzaoui, Saupe, & Barni, 2006). Acceptable levels of wireless communication quality loss are between 20 and 25 decibels (Thomos, Boulgouris, & Strintzis, 2005; Li & Cai, 2007). There are also several other metrics, such as Structural Content (SC), that are used to assess an image's quality. SC is defined as follows (Vora et al., 2010):

$$SC = \frac{\sum_{i=1}^{M} q(m,n)^2}{\sum_{i=1}^{M} q'(m,n)^2} \tag{19.6}$$

where, $M = m \times n$. ORB is another parameter that is used to compare the similarity between images.

ORB (Rublee et al., 2011), which is built on the FAST keypoint detector (Rosten & Drummond, 2006) and the BRIEF feature descriptor (Calonder, Lepetit, Strecha, & Fua, 2010), is a fast and reliable visual feature detector. In image matching, the

ORB descriptor is commonly used (Karami, Prasad, & Shehata, 2017). It produces consistent results and is a suitable substitute for the ratio test suggested by D. Lowe in a SIFT paper (H. Liu, Tan, & Kuo, 2019).

Structural similarity index (SSIM) is a popular image quality assessment approach for evaluating the similarity between two images and the value of SSIM ranges from 0 to 1 (Z Wang et al., 2004).

19.4.1 Datasets

For these experiments, we have taken a foggy real-world database (W. Liu, Zhou, Lu, Duan, & Qiu, 2020) and rainy, shadow and dark data from IEEE data source that is challenging real-environment traffic sign recognition (Temel, Kwon, Prabhushankar, and AlRegib, 2019). Both of these data are real-world databases. The MRFID (Multiple Real-World Foggy Camera Dataset) includes foggy and clear pictures of 200 outdoor scenes. From photographs obtained from these scenes over a calendar year, one clear image and four foggy images of various densities described as slightly foggy, moderately foggy, heavily foggy, and extremely foggy are selectively chosen for each scene.

The images shown in Figure 19.2 are taken from an IEEE data source that is challenging real-environment traffic sign recognition having five different levels of outdoor environmental impacts. The original and impacted rainy images are shown in Figures 19.2(a), 19.2(b), 19.2(c), 19.2(d), 19.2(e) and 19.2(f) respectively.

Similar to rain, the original images and their corresponding images impacted by shadow and darkness, and fog are shown in Figures 19.3–19.5, respectively. Note,

FIGURE 19.2 (a) Original rain image and different levels of impact: (b) Extreme less (c) Less (d) Moderate (e) High (f) Extreme high.

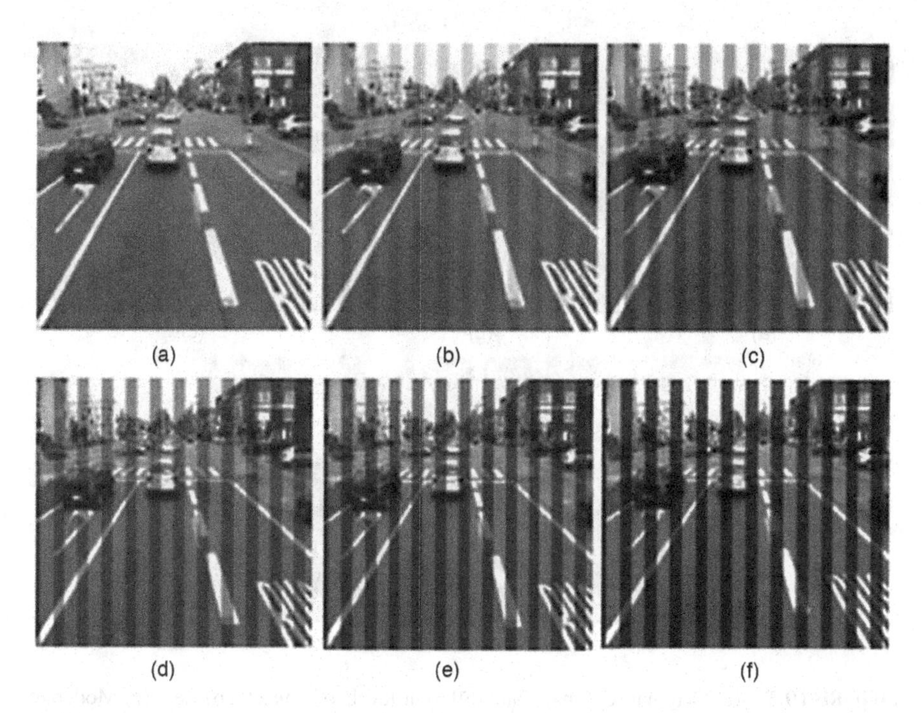

FIGURE 19.3 (a) Original shadow image and different levels of impact: (b) Extreme less (c) Less (d) Moderate (e) High (f) Extreme high.

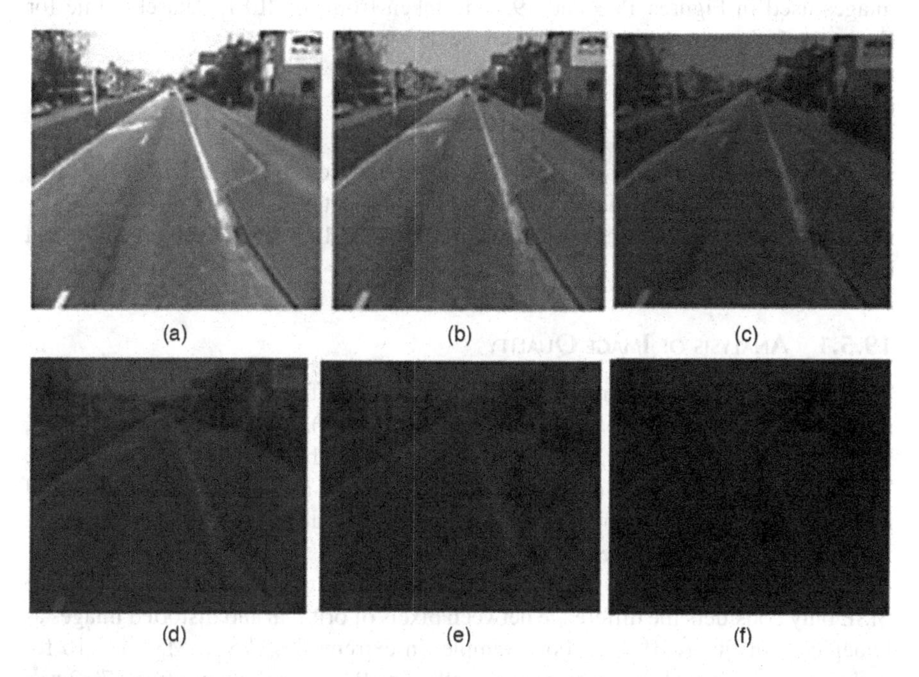

FIGURE 19.4 (a) Original darkness image and different levels of impact: (b) Extreme less (c) Less (d) Moderate (e) High (f) Extreme high.

FIGURE 19.5 (a) Original fog image and different levels of impact: (b) Less (c) Moderate (d) High (e) Extreme high.

images used in Figures 19.3 and 19.4 are taken from the IEEE dataset, while for Figure 19.5, they are from the foggy image dataset.

19.5 RESULTS AND DISCUSSIONS

We examine the image quality based on image metrics such as PSNR, MSE, SC, ORB and SSIM. Image quality consistency is inspected based on our experimental results and we also assess the image quality based on JPEG image quality and Gaussian noise levels.

19.5.1 Analysis of Image Quality

The image quality values in terms of MSE for the different impact levels of rain, shadow, darkness, and fog are shown in Figure 19.6(a), which shows as expected, MSE values are generally in an increasing trend with the rise in impact levels. However, there are some exceptional circumstances at an extremely high-level impact because MSE cannot determine the perceptual difference between the original and distorted image pixels at this impact level.

In the case of darkness, MSE is high as compared to rain and shadow because MSE only considers the difference between pixels of original and distorted images as is depicted in Figure 19.6(a). For example, in extreme less level, MSE is 110 for darkness which is high as compared to rain having 94.1 and shadow having 70.2 values of MSE, respectively, as shown in Figure 19.6(a).

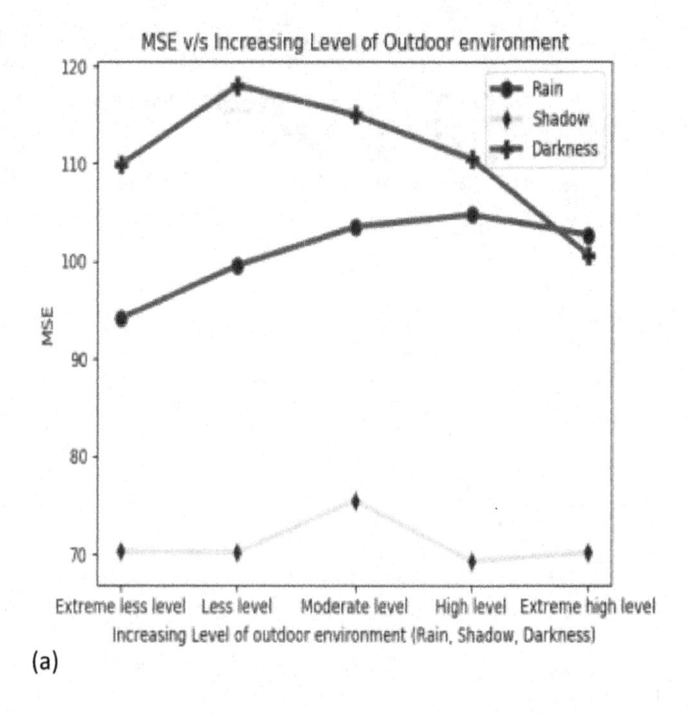

(a)

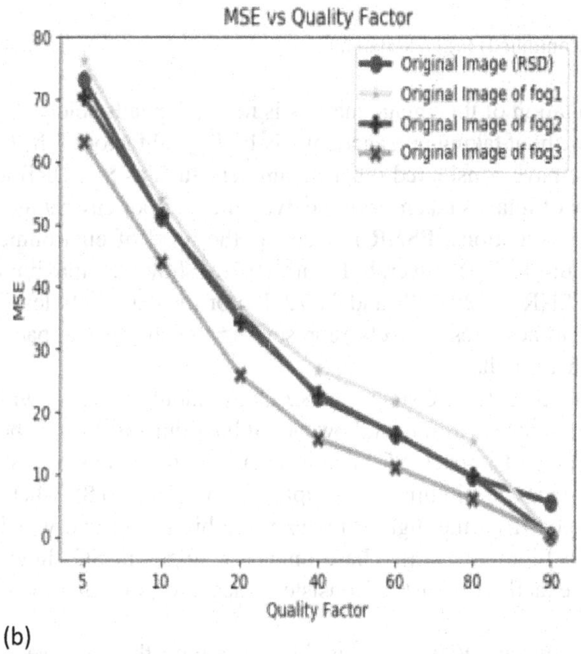

(b)

FIGURE 19.6 Comparison of MSE with various parameters at different levels of outdoor environment. (a) MSE for rain, shadow and darkness, (b) JPEG compression quality, and

(Continued)

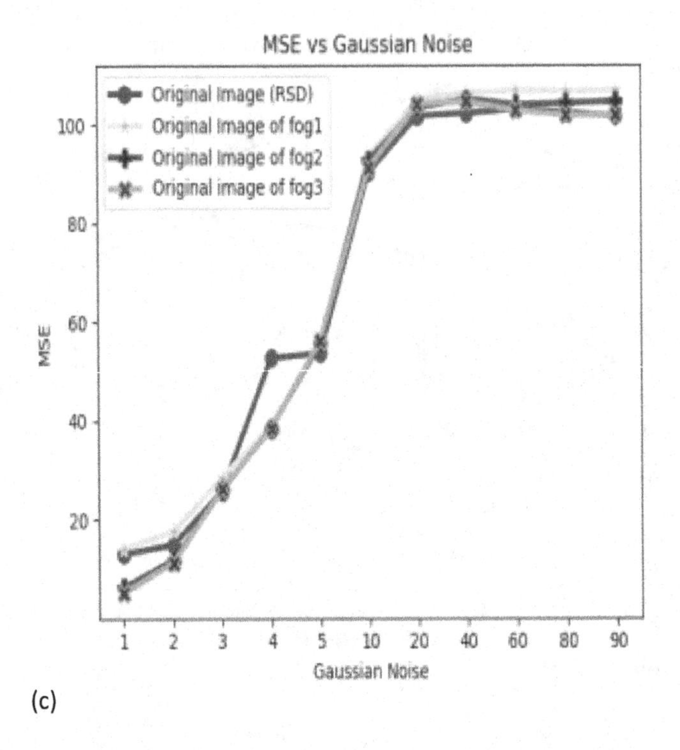

FIGURE 19.6 (Continued) (c) Gaussian noise.

The main limitation of this image metrics is that it depends solely on the numerical comparison without taking into account any biological factor of the human visual system. Thus, we have considered other parameters such as SC, ORB and SSIM to assess the quality of images taken from the dynamic outdoor environment. As usual, albeit with some exceptions, PSNR reduces as the level of environmental impact rises (refer to Figure 19.7(a)). Overall, Figure 19.7(a) shows the maximum and minimum values of PSNR are 29.7 dB and 27.72 dB for the extremely less impact level of shadow and darkness, respectively representing the stronger impact of darkness compared with that for shadow.

Concerning the Structural Content, generally, it should increase with the rise in the level of the outdoor environment, however, it has some exclusions because at the extremely high level of impact, SC cannot determine the pixels of distorted image perfectly. For instance, it plummeted abruptly from 1.55 to 0.85 when the level of darkness is changed from the high to the extreme high level and which was unexpected as seen in Figure 19.8(a). These uneven changes in SC show that the SC parameter is not exactly a robust or consistent measure to predict the quality of an image.

For the image metric ORB, its value decreases with the increase in the impact level. For example, in Figure 19.9(a), as the level of the outdoor environment rises from the extreme less level to the extreme high level, the value of ORB consistently alleviates from 0.45 to 0.19 for rain because with an increase in the level of the

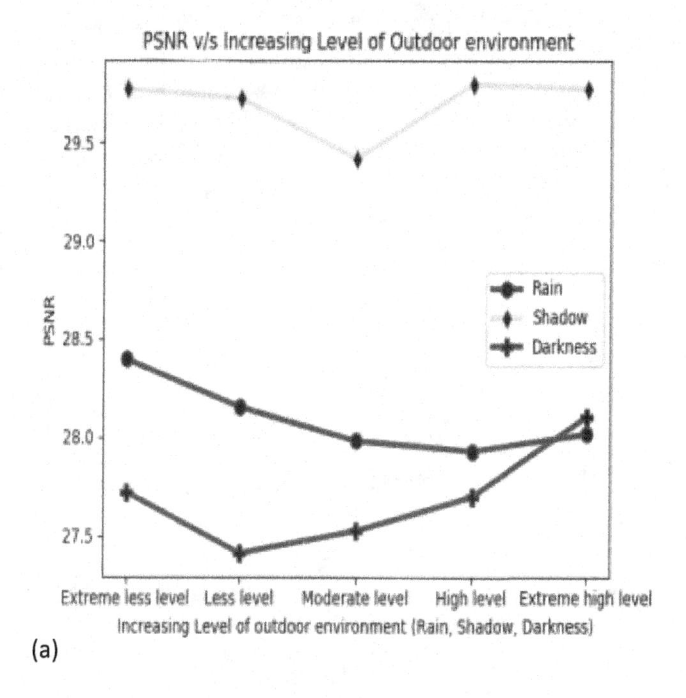

(a)

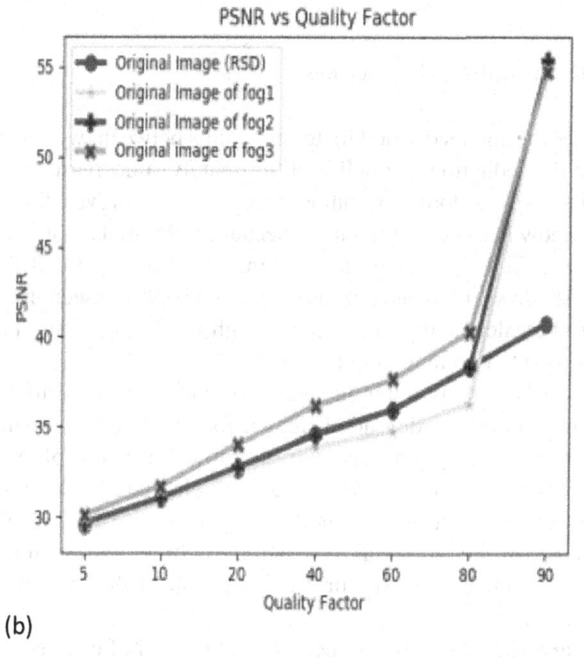

(b)

FIGURE 19.7 Comparison of PSNR with various parameters at different outdoor environment impact levels. (a) PSNR for rain, shadow and darkness, (b) JPEG compression quality, and

(Continued)

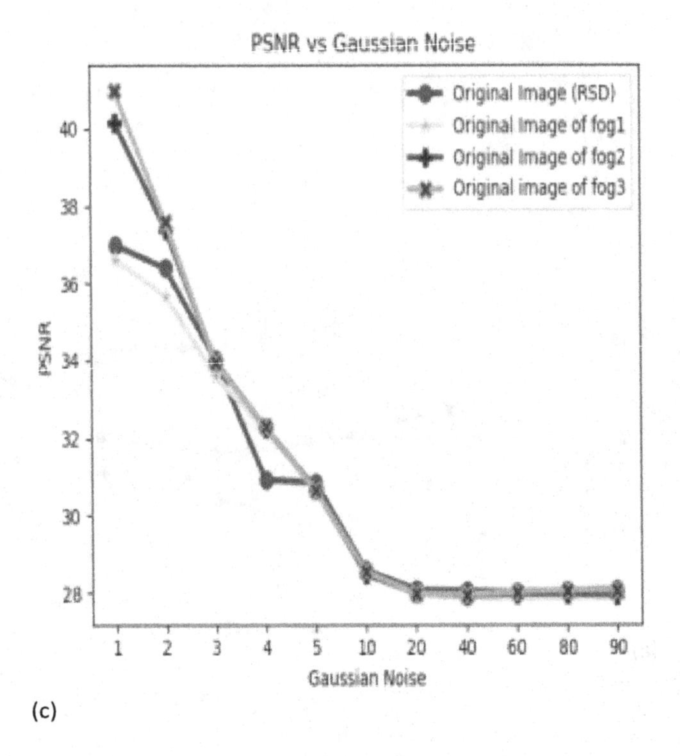

(c)

FIGURE 19.7 (Continued) (c) Gaussian noise.

outdoor parameter the counts of good matching points between two images decreases which results in the reduction of ORB and, ultimately, a decrease in the quality of images as the level of outdoor environment increases. However, there is only one exception for shadow at extreme high level because ORB finds more good matching points in this scenario. Overall, rain has a minimum value of ORB that is 0.45 as compared to the shadow (0.56) and darkness (which is 0.99) as seen in Figure 19.9(a) and ORB cannot be calculated at the extreme high level of darkness because of the incapability of finding good matching points.

SSIM, which is a very popular measure to calculate the similarity measure between the original and the degraded image, its value comes down as the level of environmental im pact plummets. For example, its value plummets sharply from 0.88 to 0.06 for darkness and because of changes in the luminance and structure of the images as it is presented in Figure 19.10(a). Overall, rain has a lower value of SSIM i.e., 0.45 as compared to other parameters because of changes in the luminance and structural information of the images, as is clearly shown in Figure 19.10(a).

Overall, it is analysed that with an increase in the level of different outdoor environments such as rain, fog, shadow and darkness, SSIM decreases consistently. So, the SSIM parameter is one of the most reliable and consistent parameters to assess the quality of an image because the human visual system is more sensitive to

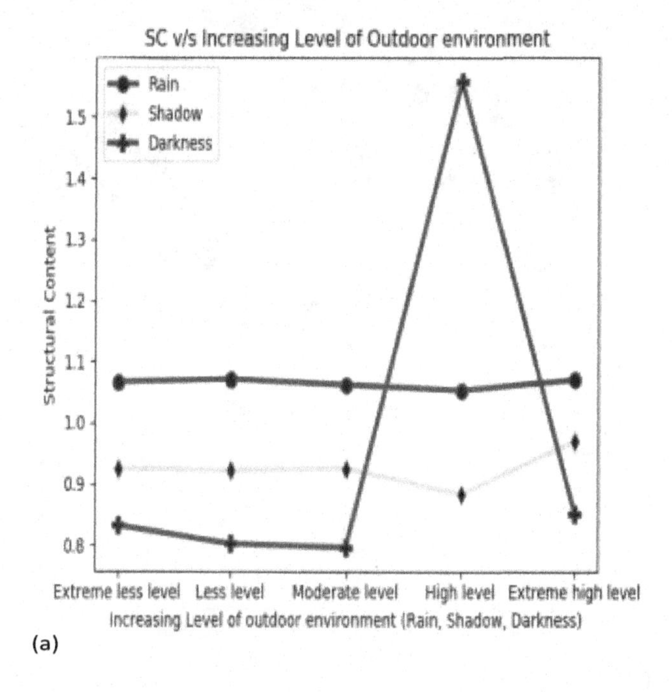

(a)

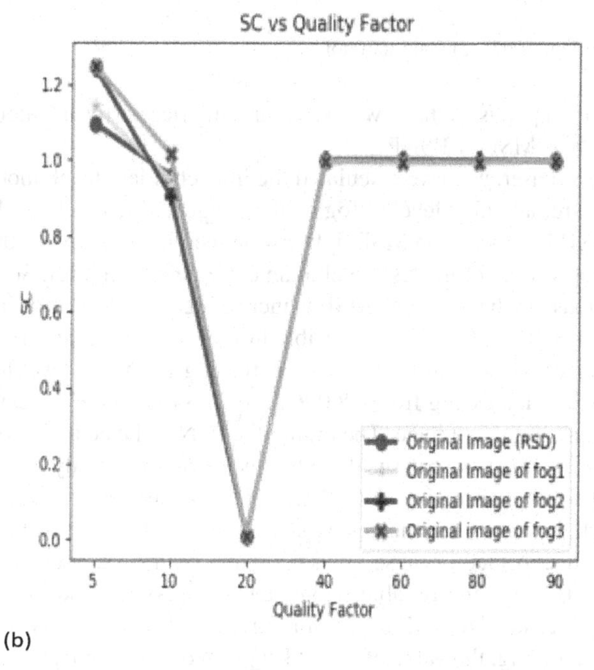

(b)

FIGURE 19.8 Comparison of SC with various parameters at different outdoor environment impact levels. (a) SC for rain, shadow and darkness, (b) JPEG compression quality, and

(Continued)

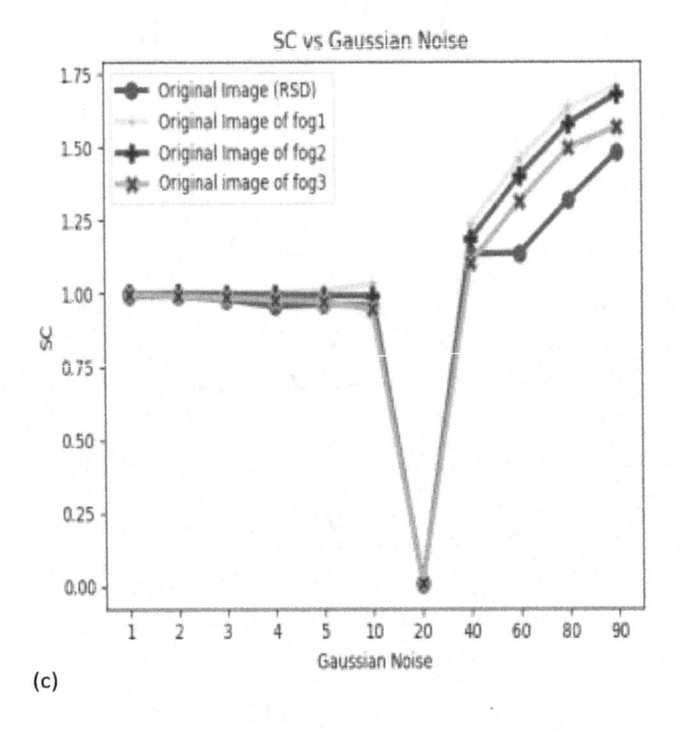

FIGURE 19.8 (Continued) (c) Gaussian noise.

structure rather than pixels so that's why structural metrics perform better than pixel-wise fidelity such as MSE or PSNR.

For the foggy dataset, we have calculated the impact of less level, moderate level, high level and extremely high level of fog on the image metrics such as MSE, PSNR, SC, ORB and SSIM. Concerning MSE, it keeps on incrementing from 100.8 to 106.7 as the level of fog varies from a less level to an extremely high level. So, we can say that with an increase in the fog level, MSE is increasing and PSNR is decreasing; this is exactly as expected generally as it is visible in Figure 19.11(a) and 19.11(b).

In the case of PSNR, as the fog level is increasing from less level to extremely high level, PSNR is decreasing from 28.09642 to 27.84648, which means there is a decrease in the quality of an image. The higher the PSNR, the better the quality of an image. In Figure 19.12(a), we can see that ORB is varying randomly with an increase in the fog level. It is plummeting to 0.02 from 0.04 as the level is increasing from moderate to high; however, it is increasing sharply from 0.02 to 0.06 at extreme high levels, which is unexpected because good matching points are not calculated accurately. Thus, ORB is not the reliable parameter to assess the quality of an image, particularly in the case of fog. Moreover, in Figure 19.12(b), it is visible that as the level of fog is increasing, the value of SSIM keeps on decreasing because of changes in the luminance and contrast of an image. This means that fog is affecting the image quality, which needs to be considered in the image-processing applications. Overall,

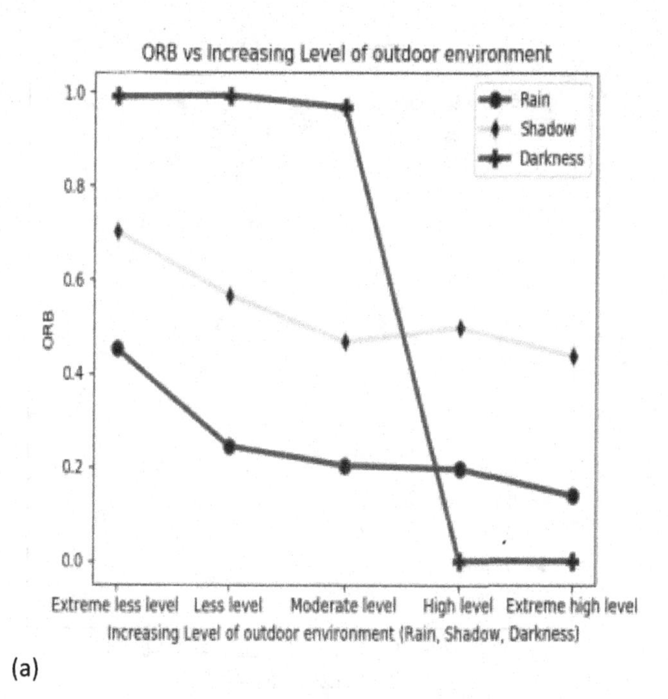

(a)

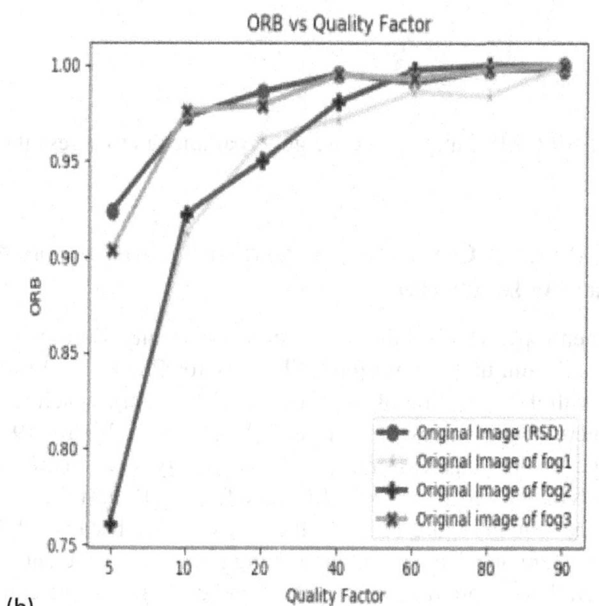

(b)

FIGURE 19.9 Comparison of ORB with various parameters at different outdoor environment impact levels. (a) ORB for rain, shadow and darkness (b) JPEG compression quality, and

(Continued)

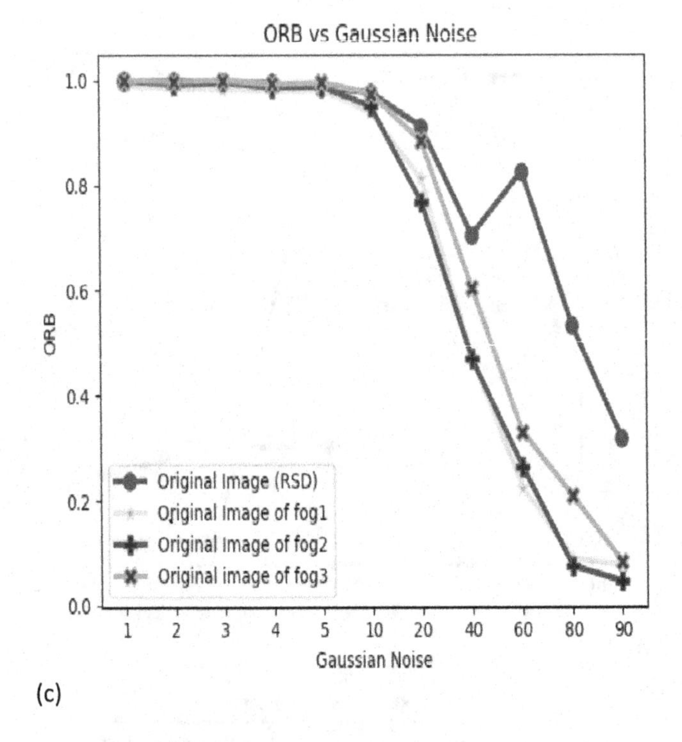

FIGURE 19.9 (Continued) (c) Gaussian noise.

we can say that SSIM, MSE and PSNR are good parameters to assess the quality of the foggy images.

19.5.2 Evaluating the Consistency Among the Quality Values for a Particular Impact Level

We next analyse entropy, which is the measure of uncertainty. Entropy is at a maximum in the case of rain, that is 0.90 for MSE, 0.91 for PSNR, 0.90 for ORB, and 0.88 for SSIM, with the exception of SC (that is, 0.73 in rain), which means higher uncertainty is there in the values of rain, as can be seen in Figure 19.13. On the opposite side, the shadow has less entropy or uncertainty as compared to rain and darkness, which is 0.74 for MSE, 0.63 for SC, 0.66 for PSNR, 0.81 for ORB and 0.60 for SSIM as can be seen in Figure 19.13. In the case of ORB, because of the extreme darkness level, we cannot calculate good matching points; thus, the entropy of ORB is not given. Overall, there are random increases and decreases in the entropy values of rain, shadow and darkness.

Note, the minimum value of entropy is 0. The maximum value of entropy for 10 bins histogram is 1. In Figure 19.13, entropy values for all image quality metrics for all impact levels are high. In particular, for the rain image under the moderate impact level, the entropy values are 0.79, 0.70, 0.80 and 0.90 for MSE, PSNR, SC and SSIM,

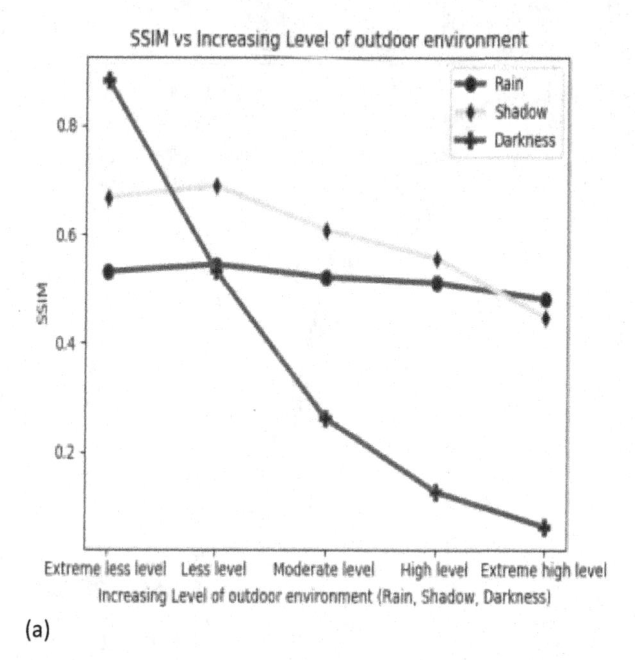

(a)

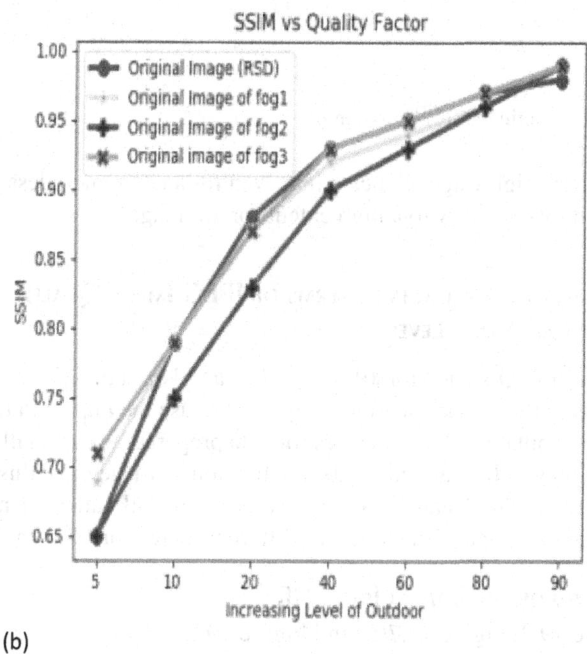

(b)

FIGURE 19.10 Comparison of SSIM with various parameters at different outdoor environment impact levels. (a) SSIM for rain, shadow and darkness, (b) JPEG compression quality, and *(Continued)*

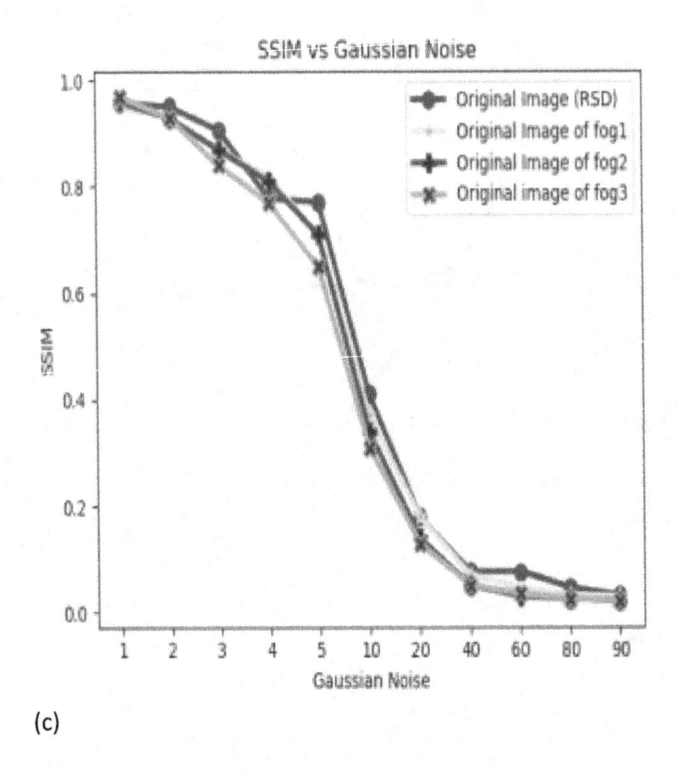

(c)

FIGURE 19.10 (Continued) (c) Gaussian noise.

respectively. These high values indicate that even for an extremely less level impact, the image quality values vary to a high extent for all images.

19.5.3 Assessing the Impacts in Terms of JPEG Image Quality and Gaussian Noise Levels

In Table 19.2, JPEG Q stands for JPEG Quality and NL stands for Noise level. As alluded to above, different applications may need to use the impact in terms of more familiar and perceptual impact indicators for data preparation, especially with regard to filtering the images that can reduce the application's efficacy. For this purpose, the equivalent Gaussian noise and JPEG quality levels for all values of metrics under many environmental impact scenarios for different images are shown in Table 19.2.

19.5.3.1 Mapping the Impact for PSNR

In PSNR, Table 19.2, Figure 19.7(a) and Figure 19.7(b) exhibit less than zero JPEG quality for rain and darkness. In contrast, their corresponding noise levels are very high for high levels of impact (e.g., the noise levels are 100 and 60 for rain and fog). However, in the case of shadow, the relevant JPEG quality levels are close to 6 or 7 percent and the noise level is at a minimum (7 percent).The quality level decreases with an increase in the level of shadow. Thus, we can conclude that in terms of

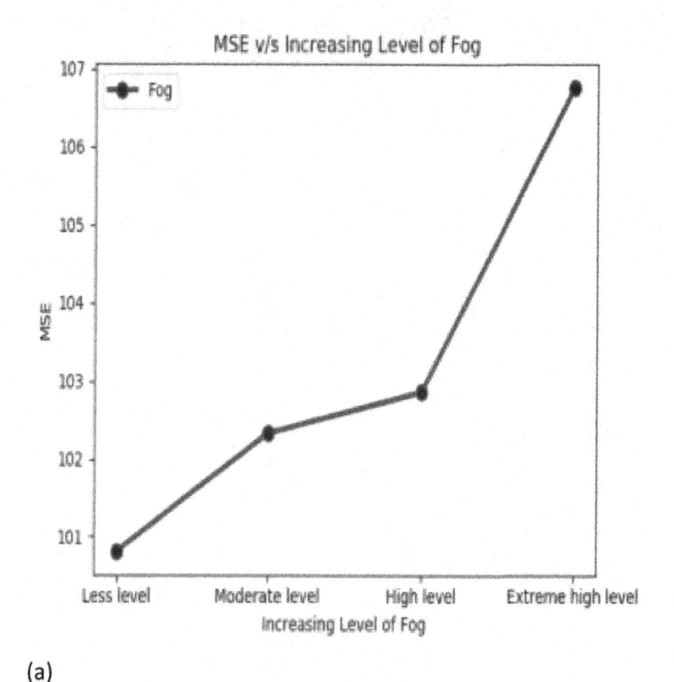

(a)

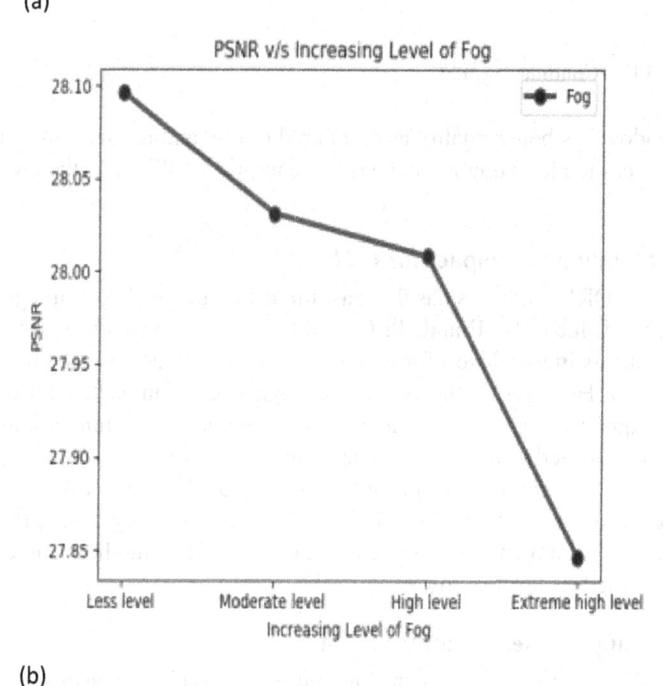

(b)

FIGURE 19.11 Comparison of fog with various parameters at different outdoor environment impact levels. (a) MSE, (b) PSNR, and *(Continued)*

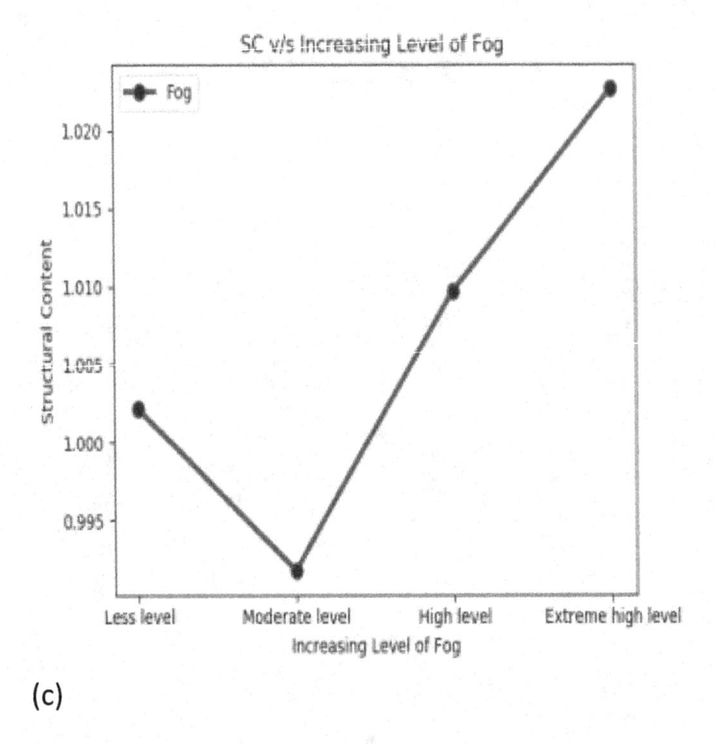

FIGURE 19.11 (Continued) (c) SC.

PSNR, a shadow has better quality as compared to rain and has less impact and also that darkness has the least quality and has more impact on PSNR as the level of darkness increases.

19.5.3.2 Mapping the Impact for ORB

With respect to ORB, darkness has the least impact on the quality of images as darkness has a high value of ORB and JPEG quality and low noise levels (e.g. 1 percent noise level for less impact level) for less and moderate impact, as seen in Table 19.2 and Figure 19.9. However, in the case of the high level of impact, we are unable to calculate the quality because of the inability to count good matching points between reference and distorted images. In contrast, the fog has very less JPEG quality (e.g. less than 0) and also the noise levels are increasing as the level of fog is increasing from less level to extremely high level. For the extreme less fog level, JPEG quality and noise levels are shown, since foggy data are taken from another source which do not have extreme less level of fog.

19.5.3.3 Mapping the Impact for SSIM

For SSIM, JPEG Quality is decreasing and the noise level is increasing as the level of the outdoor environment is increasing as expected. However, for less shadow impact level, JPEG quality contradicts with SSIM because for the decrease in SSIM, JPEG

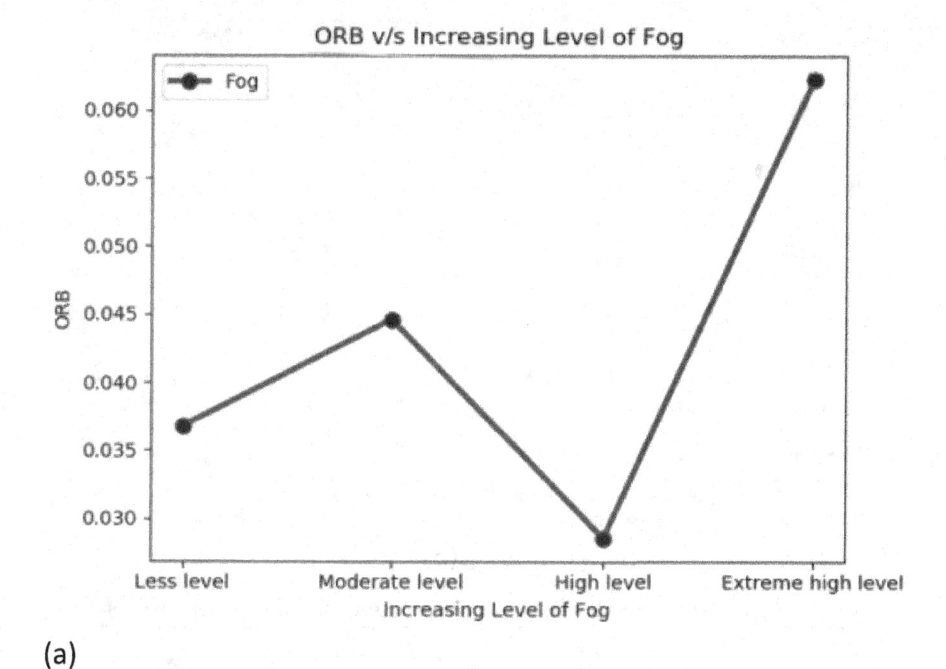

(a)

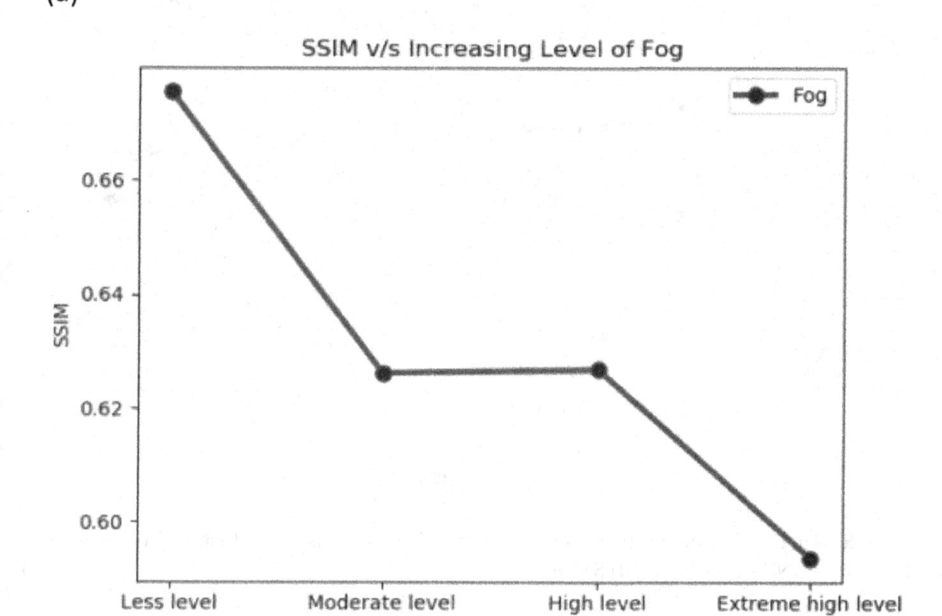

(b)

FIGURE 19.12 Comparison of fog with various image metrics at different levels of outdoor environment. (a) ORB and (b) SSIM.

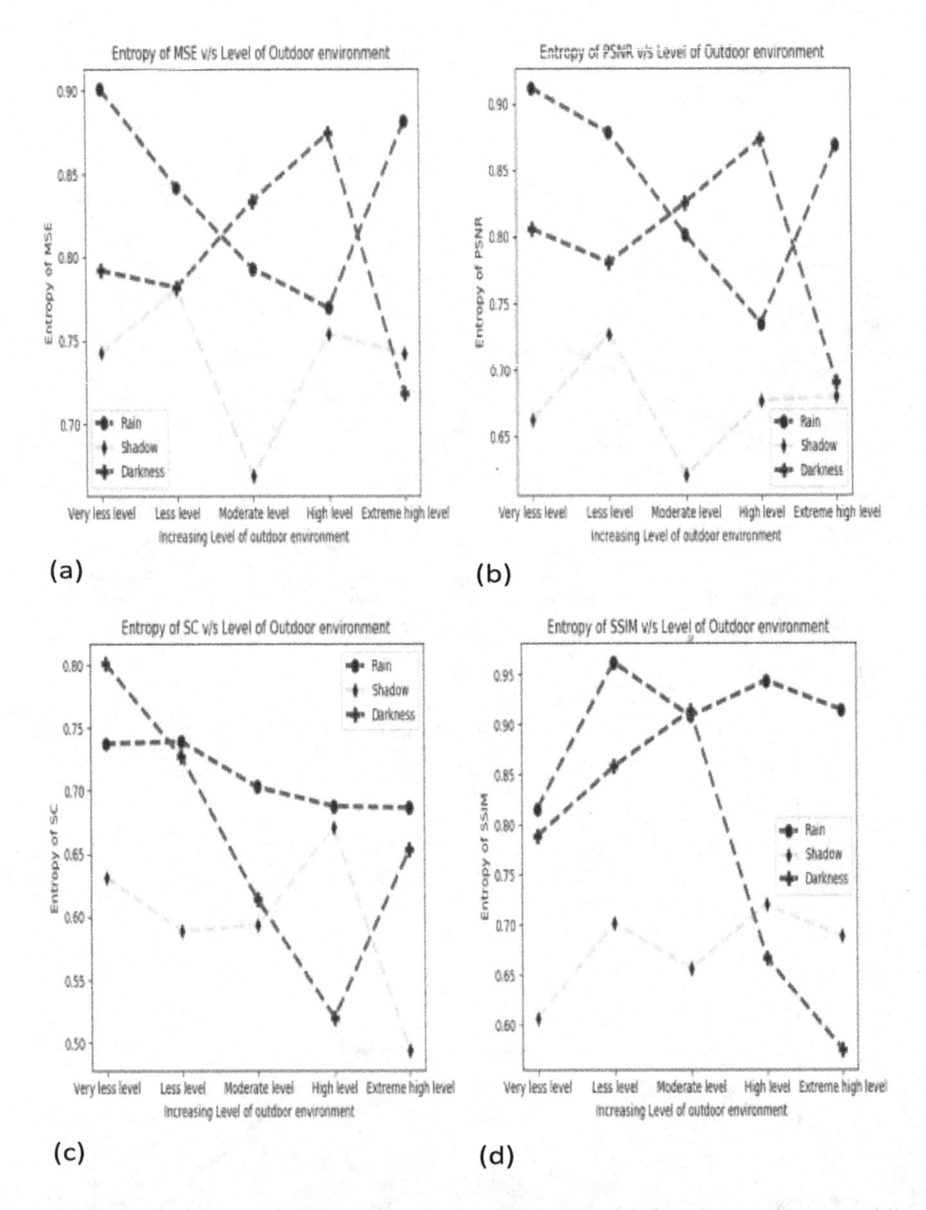

FIGURE 19.13 Entropy of image metrics with different levels of outdoor environment. (a) MSE, (b) PSNR, (c) SC, and (d) SSIM.

quality increases from 5 percent to 7 percent, as seen in Table 19.2 and Figure 19.10(b) and Figure 19.10(c). We can conclude that SSIM is the consistent parameter to assess the quality of images. Overall, darkness has the most impact and shadow has the least impact of dynamic outdoor environment on the quality of images.

TABLE 19.2
Mapping the Values of Different Image Metrics into Various Gaussian Noise and JPEG Quality Levels

	Rain	NL	JPEG Q	Shadow	NL	JPEG Q	Darkness	NL	JPEG Q	Fog	NL	JPEG Q
Extreme less	PSNR 28.39	15	less than 0	29.77	7	6	27.72	100	less than 0	–	–	–
	ORB 0.45	65	less than 0	0.70	40	4	0.99	3	60	–	–	–
	SSIM 0.53	8	3	0.66	7	5	0.88	3.5	20	–	–	–
Less	PSNR 28.15	15	less than 0	29.72	7	6	27.41	100	less than 0	28.03	18	less than 0
	ORB 0.24	85	less than 0	0.56	54	less than 0	0.99	1	63	0.04	94	less than 0
	SSIM 0.54	8	4	0.68	6	7	0.53	8	3	0.62	7	4
Moderate	PSNR 27.98	100	less than 0	29.41	7	5	27.52	100	less than 0	28	19	less than 0
	ORB 0.202662	96	less than 0	0.46	64	less than 0	0.96	9	9	0.02	96	less than 0
	SSIM 0.52	8	2	0.60	7	5	0.26	15	less than 0	0.62	7	4
High	PSNR 27.92	100	less than 0	29.79	7	6	27.69	100	less than 0	27.84	60	less than 0
	ORB 0.19	97	less than 0	0.49	64	less than 0	–	–	–	0.06	91	less than 0
	SSIM 0.51	8	3	0.55	8	4	0.12	25	less than 0	0.59	8	less than 0
Extreme high	PSNR 28.01	50	less than 0	29.77	7	6	28.10	15	less than 0	28.09	17	less than 0
	ORB 0.13	98	less than 0	0.43	65	less than 0	–	–	less than 0	0.036	95	less than 0
	SSIM 0.48	9	less than 0	0.44	9	less than 0	0.06	45	less than 0	0.67	6	4

19.6 CONCLUSIONS

We introduce a technique for assessing the impact of environmental parameters on the image quality in terms of more perceptual impacts created by the different levels of JPEG image quality and Gaussian noise. Based on our experimental observations, SSIM is one of the most consistent parameters and by comparing the values of image quality measurement metrics with JPEG image quality and noise levels, we can have image data preparation. The techniques for data understanding and preparation presented in this chapter can be used in any image analysis-based applications where the captured images are affected by the outdoor environmental conditions. Examples of these applications include smart agriculture (leaf disease detection, plant species recognition, soil analysis and crop yield prediction), smart transportation (traffic flow prediction, traffic accident hot spots and vehicle license plate recognition), disaster management (fire detection system, landslide recognition) and defense (target detection and tracking, missile guidance, vehicle navigation and automatic target recognition).

REFERENCES

Aziz, M., Tayarani-N, M. H., & Afsar, M. (2015). A cycling chaos-based cryptic-free algorithm for image steganography. *Nonlinear Dynamics*, *80* (3), 1271–1290.

Boano, C. A., Cattani, M., & Römer, K. (2018). Impact of temperature variations on the reliability of lora. In *Proceedings of the 7th International Conference on Sensor Networks* (pp. 39–50).

Calonder, M., Lepetit, V., Strecha, C., & Fua, P. (2010). Brief: Binary robust independent elementary features. In *European Conference on Computer Vision* (pp. 778–792).

Dinh, L. T. N., Karmakar, G., & Kamruzzaman, J. (2020). A survey on context awareness in big data analytics for business applications. *Knowledge and Information Systems*, *62* (9), 3387–3415.

Fathi Kazerouni, M., Mohammed Saeed, N. T., & Kuhnert, K.-D. (2019). Fully- automatic natural plant recognition system using deep neural network for dynamic out- door environments. *SN Applied Sciences*, *1* (7), 756. https://doi.org/10.1007/s42452-019-0785-9

Frank, A., Khamis Al Aamri, Y. S., & Zayegh, A. (2019). Iot based smart traffic density control using image processing. In *2019 4th MEC International Conference on Big Data and Smart City (ICBDSC)* (p. 1–4). doi:

Hamzaoui, R., Saupe, D., & Barni, M. (2006). Fractal image compression. *Document and Image Compression*, 145–177.

Hassan, M. R., Karmakar, G., & Kamruzzaman, J. (2013). Reputation and user require- ment based price modeling for dynamic spectrum access. *IEEE transactions on Mobile Computing*, *13* (9), 2128–2140.

Kapoor, A., Bhat, S. I., Shidnal, S., & Mehra, A. (2016). Implementation of iot (internet of things) and image processing in smart agriculture. In *2016 International Conference on Computation System and Information Technology for Sustainable Solutions (CSITSS)* (p. 21–26).

Karami, E., Prasad, S., & Shehata, M. (2017). Image matching using sift, surf, brief and orb: performance comparison for distorted images. *arXiv preprint arXiv:1710.02726*.

Khan, S., Muhammad, K., Mumtaz, S., Baik, S. W., & de Albuquerque, V. H. C. (2019). Energy-efficient deep cnn for smoke detection in foggy iot environment. *IEEE Internet of Things Journal*, *6* (6), 9237–9245.

Li, X., & Cai, J. (2007). Robust transmission of jpeg2000 encoded images over packet loss channels. In *2007 IEEE International Conference on Multimedia and Expo* (pp. 947–950).

Liu, H., Tan, T.-H., & Kuo, T.-Y. (2019). A novel shot detection approach based on orb fused with structural similarity. *IEEE Access, 8*, 2472–2481.

Liu, W., Zhou, F., Lu, T., Duan, J., & Qiu, G. (2020). Image defogging quality assessment: Real-world database and method. *IEEE Transactions on Image Processing, 30*, 176–190.

Okafor, A. (2005). *Entropy based techniques with applications in data mining*. University of Florida.

Pinto, M., Pais, S. L., Nisha, Gowri, S., & Puthi, V. (2020). An efficient approach for traffic monitoring system using image processing. In S. Smys, T. Senjyu, & P. Lafata (Eds.), *Second international conference on computer networks and communication technologies* (pp. 264–270). Cham: Springer International Publishing.

Rosten, E., & Drummond, T. (2006). Machine learning for high-speed corner detection. In *European Conference on Computer Vision* (pp. 430–443).

Rublee, E., Rabaud, V., Konolige, K., & Bradski, G. (2011). Orb: An efficient alternative to sift or surf. In *2011 International Conference on Computer Vision* (pp. 2564–2571).

Temel, Dogancan, Kwon, Gukyeong, Prabhushankar, Mohit, AlRegib, Ghassan (2019). *Cure-TSR: Challenging unreal and real environments for traffic sign recognition*. IEEE Dataport. Retrieved from https://dx.doi.org/10.21227/n4xw-cg56

Thomos, N., Boulgouris, N. V., & Strintzis, M. G. (2005). Optimized transmission of jpeg2000 streams over wireless channels. *IEEE Transactions on Image Processing, 15* (1), 54–67.

Thorat, A., Kumari, S., & Valakunde, N. D. (2017). An iot based smart solution for leaf disease detection. In *2017 International Conference on Big Data, IoT and Data Science (BID)* (pp. 193–198).

Vora, V., Suthar, A., Makwana, Y., & Davda, S. (2010). Analysis of compressed image quality assessments, m. *Tech Student in E &C Dept, CCET, Wadhwan-Gujarat.*

Wang, Z., Bovik, A. C., Sheikh, H. R., & Simoncelli, E. P. (2004). Image quality assessment: From error visibility to structural similarity. *IEEE Transactions on Image Processing, 13* (4), 600–612.

Welstead, S. T. (1999). *Fractal and wavelet image compression techniques* (Vol. 40). Spie Press.

20 Telemedicine
A New Opportunity for Transforming and Improving Rural India's Healthcare

Seema Maitrey

Department of Computer Science and Engineering, KIET Group of Institutions, Gzb, India

Deepti Seth

Department of Applied Science, KIET Group of Institutions, Gzb, India)

Kajal Kansal

Department of Computer Science and Engineering, KIET Group of Institutions, Gzb, India

Anil Kumar

Department of Mathematics and Statistics Swami Vivekanand Subharti, University Meerut, India

CONTENTS

20.1 Introduction 298
20.2 Rural Healthcare 298
20.3 Benefits of Telemedicine to Patients 300
20.4 ISRO'S Move with Telemedicine 301
20.5 Development Challenge 301
20.5.1 Awareness Building 301
20.5.2 Acceptance 302
20.5.3 Availability 302
20.5.4 Affordability 302
20.6 Conclusion 302
References 303

DOI: 10.1201/9781003267782-20

20.1 INTRODUCTION

The majority of the population in our country, India, lives in rural areas where healthcare facilities are both inefficient and insufficient. Steps toward the initiation of telemedicine are now working as a bridge between the requirements and their fulfilments. Telemedicine makes the use of electronic information and communication technologies to provide healthcare facilities for people living in remote areas. It takes care of those patients to whom the transfer of medical information, physicians, other healthcare providers and medical institutions is a very big challenge [1]. In a very short duration of time, all these facilities can be made available to patients living in the remote areas. It could be life-changing for those patients who need specialized care. They need not to travel long distances and can receive a diagnosis at reduced cost Figure 20.1. Due to lack of Telemedicine in the present scenario, death rate is high in rural areas. If they will get treatment at required time, definitely the death rate will be reduced in our country. There are several aspects of telemedicine, which are shown in Figure 20.1

Although there are a lot of barriers to the execution of telemedicine, the positive outcomes of its implementation are the opportunity it gives to provide medical services to those who would otherwise not have access to medical care [2]. At present, most people in rural areas possess mobile phones and computers, so, they can get easily connected to a provider with the improvements in accessibility (Figures 20.2–20.5).

20.2 RURAL HEALTHCARE

One of the most important concerns facing India's Ministry of Health is the issue of rural healthcare. Disease-related mortality is at an all-time high, with more than 70 percent of the population living in rural areas and only a minimal level of health

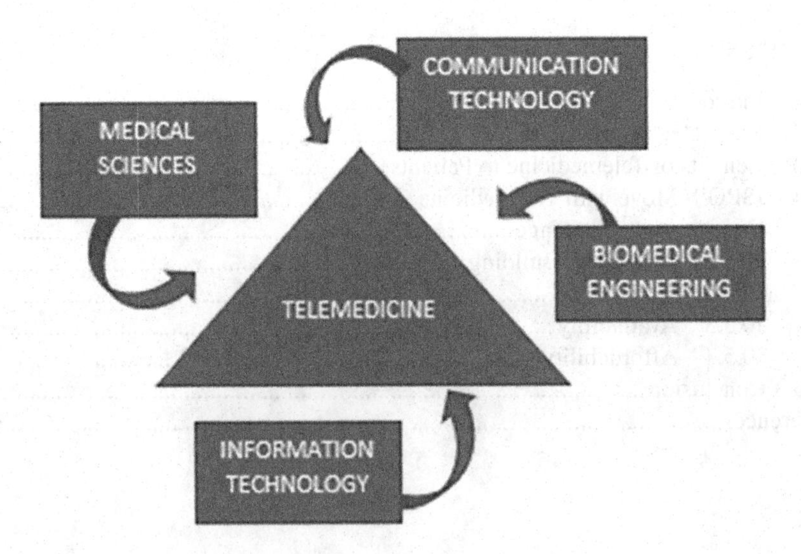

FIGURE 20.1 Telemedicine in several aspects [1].

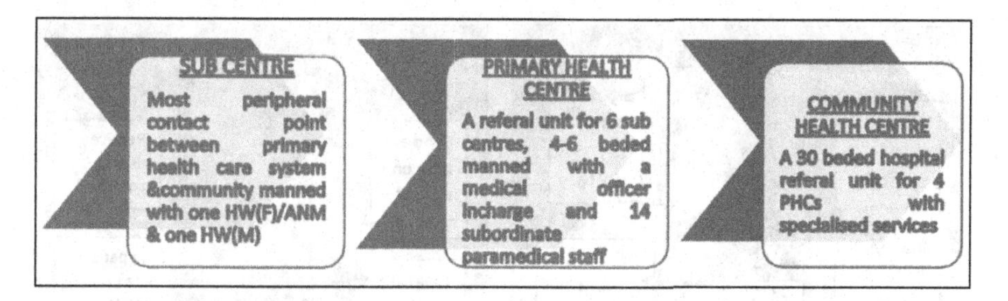

FIGURE 20.2 The rural healthcare system in India [5].

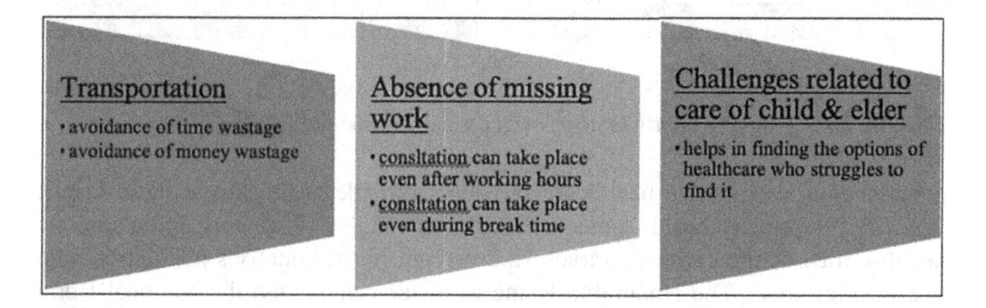

FIGURE 20.3 Advantages of telemedicine to patients [3].

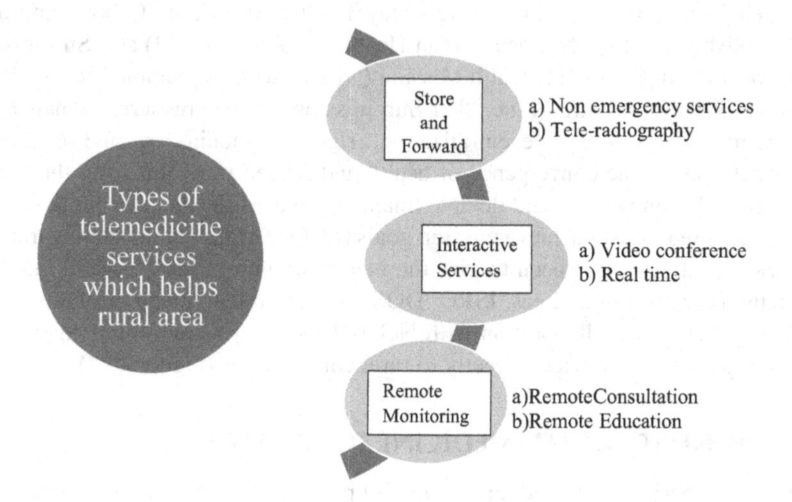

FIGURE 20.4 Types of telemedicine services [5].

services. What is the significance of rural health? Patients can focus on "getting better" rather than "going to appointments" at rural hospitals, which improves local access and allows them to focus on "getting better" rather than "getting to appointments." Rural hospitals are large enough to handle our community's health requirements, yet small enough to care, thanks to connectivity and collaboration. India's healthcare infrastructure includes primary, secondary, and tertiary care [3]. Both

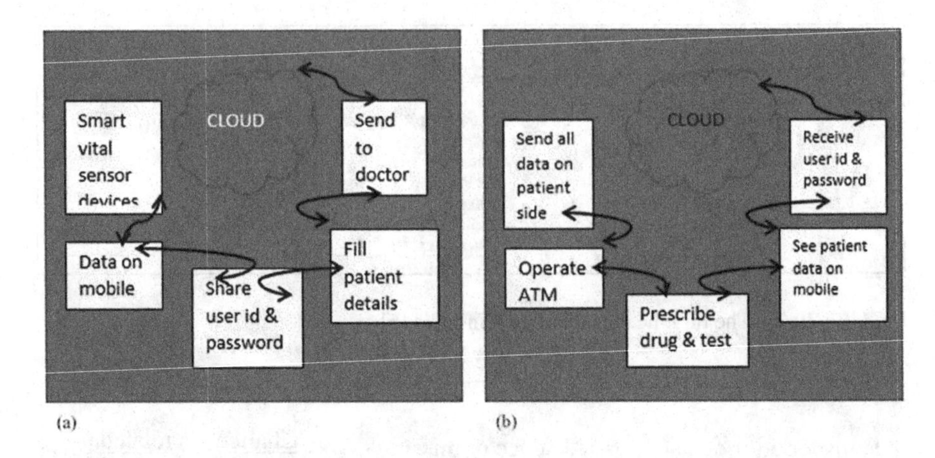

FIGURE 20.5 Sharing information between patient and doctor [8].

governmental and private healthcare providers provide healthcare at these levels. With 23,391 primary health centers (PHCs) and 145,894 subcenters, India's public health infrastructure currently serves 72.2 percent of the country's population who live in rural areas. The Honorable Prime Minister inaugurated the National Rural Health Mission (NRHM) on April 12, 2005, with the goal of providing accessible, inexpensive, and high-quality health care to the rural population, particularly disadvantaged groups. In a resolution dated May 1, 2013, the Union Cabinet approved the establishment of the National Urban Health Mission (NUHM) as a Sub-mission of the overarching National Health Mission (NHM), with the National Rural Health Mission (NRHM) serving as the other Sub-mission. Human resource management, community involvement, decentralization, rigorous monitoring and evaluation against standards, the convergence of health and related programs from the village level upwards, innovations and flexible financing, and a strategic plan are among the key features required to achieve mission goals [4]. One of the health administration's top priorities has always been to give the bare minimum of healthcare to individuals living in India's rural areas. ISRO/ DOS has started the village resource centres (VRCs) program in collaboration with NGOs/Trusts and State/Central Agencies to provide space-based services directly to rural communities (Figure 20.2).

20.3 BENEFITS OF TELEMEDICINE TO PATIENTS

- The concept of telemedicine is that 99 percent of the health problems that we encounter do not require an operation. With the ability to see a doctor remotely, patients who were previously limited to a hospital can now visit their doctor without leaving their home [7]. Further, seniors who prefer to remain in their homes can now get the medical care they need without leaving their beds. Due to the availability of telemedicine, patients no longer have to travel to a hospital or visit a doctor's office. They can now do so through the use of streaming devices. Another benefit of this technology is that it can help prevent the spread of diseases. Telemedicine, as depicted in Figure 20.3, can also benefit patients in many ways.

- A number of telemedicine services are also present which specifically shows the benefits the rural population gets in the outcome of using the various services of telemedicine. The benefits of telemedicine to the rural community are depicted in Figure 20.4.
- Information technology can be used to allow doctors and patients to communicate even though they are located in remote areas. Information can be shared between patients and doctors even though they are usually located in remote areas. This is because of the advanced skills of our providers. This can be depicted in Figure 20.5.

20.4 ISRO'S MOVE WITH TELEMEDICINE

The Indian Space Programme is motivated by the country's developmental requirements and has made an effort to reach out to the people. It has enabled the spread of tele-education and, more crucially, telemedicine/tele- health, which is of great societal significance to the country in terms of providing specialty healthcare to the country's remote, rural, and underserved populations. Telemedicine makes it easier to provide medical care from afar. It is a cost-effective approach for providing speciality healthcare to rural patients in the form of greater access and lower costs, as well as reduced professional isolation for rural doctors. Ordinary doctors may be able to execute extraordinary feats with the help of telemedicine [9]. ISRO has successfully linked hospitals and healthcare institutions in remote rural regions with specialist hospitals in cities using INSAT satellites as part of its telemedicine initiatives. As a result, there is excellent linkage between remote patients and expert doctors in urban centers [10].

ISRO has envisioned the building of "HEALTHSAT," an exclusive satellite to satisfy the country's healthcare and medical education demands, as a result of the steady growth of its telemedicine programme. When combined with wireless and terrestrial communication networks, this satellite has the potential to significantly improve the country's current healthcare delivery system. The bulk of the rural population across the country will benefit from the tireless work of several departments, such as the Department of Space and the Department of Information Technology, along with State Governments, NGOs, and private and corporate hospitals/agencies [11, 12].

20.5 DEVELOPMENT CHALLENGE

A lack of access to basic medical facilities still persists in rural and remote areas of the country. When in need of specialized care, a patient has to travel long distances simply to receive a diagnosis in exchange for a large sum of money. The things can prove to be a life changer if medical care can be obtained where they live [13, 14].

20.5.1 Awareness Building

Awareness regarding the advantages and the proper usage of telemedicine is the key to increase the adoption percentage on both sides, i.e., for healthcare workers as well as for patients. A constant effort in addressing the advantages and proper usage can increase its acceptance among both the population of patients and also health sector professionals.

20.5.2 Acceptance

"Unfamiliarity with the working of computers" is one of the main reasons given by health workers for their refusal to adopt telemedicine. Along with this, many also fear the loss of their jobs or that they will prove less useful in the workplace. In order to overcome this problem, any telemedicine program should be designed to be very simple and understandable. In one example, an Indian company, Sanjeevani, integrates older technology, such as telephones and simple document scanners, with sophisticated video conferencing technology. Thus, it helps to bridge the gap in the experience of the old and the new technologies for healthcare professionals [15][16].

20.5.3 Availability

Since there is a scarcity of doctors with license and specialists in many emerging and developing markets, low- income patients will also be able to access the highly professional care with the help of a telemedicine system. This can be a achieved through the elimination of costs of travel, for both specialists and patients. The availability of ICT infrastructure helps to determine the level of services. In those areas where the telemedicine is unreachable, mobile health clinics can help to provide access to medical care. These include, for example, mobile health clinics created by Apollo Hospitals, Philips, ISRO, and the Dhan Foundation. Where minor health queries are concerned, the patient doesn't need to travel long distances for a check-up or consultation; rather, this can be conducted from their mobile phones [17, 18].

20.5.4 Affordability

In the case of the healthcare system, telemedicine provides modern methods of dealing with patients and data which can be achieved by providing more access to the healthcare specialists at a minimum price [19–20]. It also provides the opportunity to continue the education, even on a regular basis, to the healthcare providers, even allowing for the involvement of international partners. Telemedicine thus proves a boon not only for patients but also for doctors, specialists, and hospitals. The benefits of telemedicine for doctors, patients, and hospitals are outlined in Table 20.1.

20.6 CONCLUSION

Though telemedicine has emerged as a game changer in the world of medical services, there are still a lot of obstacles to overcome. These include, for example, concerns regarding connectivity, the licensing of physicians, and security concerns. If all these concerns were taken into consideration with their solutions, the usage of telemedicine can skyrocket in all areas, ranging from rural to urban. An elevation in the field of the telemedicine thus helps in the increasing number of online centres for the medical, enlarging telemedicine across the globe, acceptability on a very large scale, and the collaboration of various health systems. Telemedicine thus can be defined as a medium which helps in delivering and managing the medical care by taking the benefits of IT and telecommunications technology. Without wastage of a

TABLE 20.1
Description of the Benefits of Telemedicine for Doctors, Patients, and the Hospitals [19, 20]

A) For Doctors:	B) For Hospitals:	C) For Patients:
• Virtually, schedule appointments over a telemedicine app • Chat over video-call or a phone call can be made with the patients. • Creation of digital prescriptions • Easily refer patients and transfer records online • Helps in Accessing the past records and decreasing the average time of consultation • Forming of a better doctor-patient relationship	• Arrange a Consultancy for the patients across the globe • Increase patient retention and follow- up • Marketing their brand • By hiring a virtual-receptionist, save on staff can be done. • Increase profit and can have availability of more doctors onboard with no physical barriers.	• Availability of a 24×7 virtual assistant over the telemedicine apps • Save money and time spent while transportation • Getting regular follow-up and medication reminders • Easy payment by online medium. • No need of fearing to lose the prescriptions and the documents. • Consulting a number of specialists from the comfort of their homes.

single minute and without any traffic, the information regarding patient, and records can transfer miles, even instantaneously. Even a live surgery under the consultation of a mentor or surgeon can be arranged online, for the health professional to provide their suggestions, guidelines, and supervision by sitting at their workstation in the city. The presence of customized medical software in telemedicine is an integration of computer hardware and medical diagnostic instruments which are connected to the commercial Vital Systems Assessment Tests (VSAT) at every location.

REFERENCES

1. Sood, S., Mbarika, V., Jugoo, S., Dookhy, R., Doarn, C. R., Prakash, N., & Merrell, R. C. (2007). What is telemedicine? A collection of 104 peer-reviewed perspectives and theoretical underpinnings. *Telemedicine and e-Health*, 13(5), 573–590.
2. Telemedicine in India, *APBN*, Vol. 10, No. 19, 2006.
3. Sood S, Mbarika V, Jugoo S, Dookhy R, Doarn CR, Prakash N, Merrell RC, Telemed JE Health, 13(5):573–590, 01 Oct 2007
4. Dasgupta A, Deb S Telemedicine: a new horizon in public health in India. Indian Journal of Community Medicine: Official Publication of Indian Association of Preventive & Social Medicine, 01 Jan 2008, 33(1):3–8 DOI: 10.4103/0970-0218.39234
5. Syed-Abdul, S., Scholl, J., Jian, W.S., Li, Y.C. (2011). Challenges and opportunities for the adoption of telemedicine in India. *Journal of Telemedicine and Telecare*, 17(6), 336–337. https://doi.org/10.1258/jtt.2011.101210
6. Mishra, S.K., Singh, I.P., Chand, R.D. (2012). Current status of telemedicine network in India and future perspective. *Proceedings of the Asia-Pacific Advanced Network*, 32, 151–163. https://doi.org/10.7125/APAN.32.19
7. Chandwani, R.K., Dwivedi, Y.K. (2015). Telemedicine in India: current state, challenges and opportunities. *Transforming Government: People, Process and Policy*, 9(4), 393–400. https://doi.org/10.1108/TG-07-2015-0029

8. Bagchi, Sanjit (2006). Telemedicine in rural India *PLoS Medicine*, 3(3):e82, DOI:10.1371/journal.pmed.0030082
9. https://sciresol.s3.us-east-2.amazonaws.com/IJST/Articles/2016/Issue-44/Article115.pdf
10. http://europepmc.org/article/med/19966987
11. Vivek, C., Vikrant, K. (2016). Tele-ECG and 24-hour physician support over telephone for rural doctors can help early treatment of acute myocardial infarction in rural areas. *Journal of Telemedicine and Telecare*, 22(3), 203–206. https://doi.org/10.1177/1357633X15592734
12. Arivanandan, M. (2016). Telemedicine programme in rural health care system of India. *Indian Journal of Applied Research*, 6(3), 458–461.
13. Mathur, P., Srivastava, S., Lalchandani, A., & Mehta, J.L. (2017). Evolving role of telemedicine in health care delivery in India. *Primary Health Care*, 7, 1–6. https://doi.org/10.4172/2167-1079.1000260
14. Ganapathy, K., Alagappan, D., Rajakumar, H., Dhanapal, B., Subbu, G.R., Nukala, L., Premanand, S., Veerla, K.M., Kumar, S. &Thaploo, V. (2018). Tele-emergency services in the Himalayas. *Telemedicine and e-Health*, 22(5), 380–390. https://doi.org/10.1089/tmj.2018.0027
15. Electronic Health Record (EHR) standards for India. Retrieved May 21, 2019 from https://mohfw.gov.in/sites/default/files/17739294021483341357_1.pdf
16. NIMHANS Telemedicine. Retrieved May 20, 2019 from http://www.nimhans.ac.in/telemedicine Patnaik, S., & Patnaik, A.N. (2015). e-Health for All – is India Ready? *National Journal of Community Medicine*, 6(4), 633–638.
17. Kustwar, Raj Kishor, Ray, Suman, eHealth and telemedicine in India: An overview on the health care need of the people. *Journal of Multidisciplinary Research in Healthcare*, 6, 25, 2020
18. Srivastava, Isha, Lal, Atil Kumar, Pandey, Mahima, Jaiswal, Ashish, Jaiswal, Ishank, Transforming healthcare in rural India by telemedicine during COVID-19 pandemic. *Journal of Evolution of Medical and Dental Sciences* / eISSN - 2278-4802, pISSN - 2278-4748 / Vol. 9(49) / Dec. 07, 2020.
19. Leite, H, Hodgkinson, IR, Gruber, T. New development: 'healing at a distance'- telemedicine and COVID -19. *Public Money and Management*, 40(6), 483–5, 2020.
20. Dash, Sambit, Aarthy, Ramasamy, Mohan, Viswanathan, Telemedicine during COVID-19 in India — a new policy and its challenges. *Journal of Public Health Policy*, 2021.

Printed in the United States
by Baker & Taylor Publisher Services

Printed in the United States
by Baker & Taylor Publisher Services